Progress in Inflammation Research

Series Editor

Prof. Dr. Michael J. Parnham
Senior Scientific Advisor
PLIVA dd
Prilaz baruna Filipovica 25
10000 Zagreb
Croatia

Advisory Board

G. Z. Feuerstein (Merck Research Laboratories, West Point, PA, USA)
M. Pairet (Boehringer Ingelheim Pharma KG, Biberach a. d. Riss, Germany)
W. van Eden (Universiteit Utrecht, Utrecht, The Netherlands)

Forthcoming titles:
Antirheumatic Therapy: Actions and Outcomes,
 R.O. Day, D.E. Furst, P.L. van Riel, B. Bresnihan (Editors), 2004
Antibiotics as Anti-inflammatory and Immunomodulatory Agents,
 B.K. Rubin, J. Tamaoki (Editors), 2004

(Already published titles see last page.)

Bone Morphogenetic Proteins: Regeneration of Bone and Beyond

Slobodan Vukicevic
Kuber T. Sampath

Editors

Birkhäuser Verlag
Basel · Boston · Berlin

Editors

Prof. Dr. Slobodan Vukicevic
Laboratory for Mineralized Tissues
School of Medicine
Salata 11
HR-10000 Zagreb
Croatia

Dr. Kuber T. Sampath
Genzyme Corporation
One Mountain Road
Framingham, MA 01701-9322
USA

A CIP catalogue record for this book is available from the Library of Congress, Washington D.C., USA

Bibliographic information published by Die Deutsche Bibliothek
Die Deutsche Bibliothek lists this publication in the Deutsche Nationalbibliografie;
detailed bibliographic data is available in the internet at http://dnb.ddb.de

ISBN 3-7643-7139-0 Birkhäuser Verlag, Basel – Boston – Berlin

© 2004 Birkhäuser Verlag, P.O. Box 133, CH-4010 Basel, Switzerland
Part of Springer Science+Business Media
Printed on acid-free paper produced from chlorine-free pulp. TCF ∞
Cover design: Markus Etterich, Basel
Cover illustration:Newly regenerated embryonic-like structures (in the middle) within a remnant kidney of aged rats in which 5/6 of their original kidney mass has been surgically removed. These animals with chronic renal failure were then systematically treated with BMP-7 for a period of 32 weeks (see p. 229).
Printed in Germany
ISBN 3-7643-7139-0

9 8 7 6 5 4 3 2 1

www.birkhauser.ch

Contents

List of contributors

Fran Borovecki, Department of Anatomy, Medical School, University of Zagreb, Salata 11, 10000 Zagreb, Croatia; e-mail: fbor@mef.hr

Susan Chubinskaya, Department of Biochemistry and Section of Rheumatology, Rush University Medical Center, Chicago, IL 60612, USA; e-mail: schubins@rush.edu

Joachim H. Clement, Department of Internal Medicine II (Oncology, Hematology, Gastroenterology, Hepatology, Infectiology), Friedrich Schiller University Jena, Erlanger Allee 101, 07740 Jena, Germany; e-mail: Joachim.Clement@med.uni-jena.de

Cosimo de Bari, UZ Gasthuisberg, Department of Rheumatology, Herestraat 49, 3000 Leuven, Belgium; present address: GKT School of Medicine, King's College London, Department of Rheumatology, 5th Floor, Thomas Guy House, SE1 9RT London, UK; e-mail: cosimo.debari@kcl.ac.uk

Dirk De Valck, UZ Gasthuisberg, Department of Rheumatology, Herestraat 49, 3000 Leuven, Belgium; e-mail: dirk.deValck@med.kuleuven.ac.be

Francesco Dell'Accio, UZ Gasthuisberg, Department of Rheumatology, Herestraat 49, 3000 Leuven, Belgium; present address: GKT School of Medicine, King's College London, Department of Rheumatology, 5th Floor, Thomas Guy House, SE1 9RT London, UK; e-mail: francesco.dellaccio@kcl.ac.uk

Gary E. Friedlaender, Department of Orthopaedics and Rehabilitation, Yale University School of Medicine, P.O. Box 208071, New Haven, CT 06520-8071, USA; e-mail: gfried@email.med.yale.edu

William A. Grasser, Pfizer Global Research and Development, Groton Laboratories, Groton, CT 06340, USA; e-mail: william_a_grasser@groton.pfizer.com

Lovorka Grgurevic, Department of Anatomy, School of Medicine, University of Zagreb, Salata 11, 10000 Zagreb, Croatia; e-mail: lgrgurev@mef.hr

Dennis Higgins, Department of Pharmacology and Toxicology, 102 Farber Hall, State University of New York, 3455 Main Street, Buffalo, NY 14214, USA; e-mail: higginsd@acsu.buffalo.edu

Søren Jepsen, Dept. of Restorative Dentistry and Periodontology, University of Bonn, Welschnonnenstr. 17, 53111 Bonn, Germany; e-mail: ...

Olexander Korchynskyi, Division of Cellular Biochemistry, The Netherlands Cancer Institute, Plesmanlaan 121, 1066 CX Amsterdam, The Netherlands; present address: Thurston Arthritis Research Center and Department of Rheumatology, School of Medicine, and Department of Endodontics, School of Dentistry, University of North Carolina at Chapel Hill, Chapel Hill, NC, USA; e-mail: olexkor@med.unc.edu, olexkor@hotmail.com

Pamela Lein, Center for Research on Occupational and Environmental Toxicology, Oregon Health & Sciences University, Portland, OR 97239, USA; e-mail: leinp@ohsu.edu

Rik Lories, UZ Gasthuisberg, Department of Rheumatology, Herestraat 49, 3000 Leuven, Belgium; e-mail: rik.lories@uz.kuleuven.ac.be

Clemens W.G.M. Löwik, Department of Endocrinology C4-R, Leiden University Medical Center, PO Box 9600, 2300 RC Leiden, The Netherlands; e-mail: C.W.G.M.Lowik@lumc.nl

Frank P. Luyten, UZ Gasthuisberg, Department of Rheumatology, Herestraat 49, 3000 Leuven, Belgium; e-mail: frank.luyten@uz.kuleuven.ac.be

Snjezana Martinovic, Department of Anatomy, Medical School, University of Zagreb, Salata 11, 10000 Zagreb, Croatia; e-mail: smart@mef.hr

Bill McKay, Biologics R & D, Medtronic Sofamor Danek, 1800 Pyramid Place, Memphis, TN 38132, USA; e-mail: bill.mckay@medtronic.com

Vishwas M. Paralkar, Pfizer Global Research and Development, Groton Laboratories, Groton, CT 06340, USA; e: mail: vishwas_m_paralkar@groton.pfizer.com

Keith A. Riccardi, Pfizer Global Research and Development, Groton Laboratories, Groton, CT 06340, USA; e: mail: Keith_A_Riccardi@groton.pfizer.com

David C. Rueger, Stryker , 35 South Street, Hopkinton, MA 01748, USA;
e-mail: david.rueger@stryker.com

Kuber T. Sampath, Genzyme Corporation, Framingham, MA 01701, USA;
e-mail: Kuber.Sampath@genzyme.com

Petra Simic, Department of Anatomy, Medical School, University of Zagreb, Salata
11, 10000 Zagreb, Croatia: e-mail: psimic@mef.hr

Peter ten Dijke, Division of Cellular Biochemistry, The Netherlands Cancer Institute, Plesmanlaan 121, 1066 CX Amsterdam, The Netherlands;
e-mail: p.t.dijke@nki.nl

Hendrik Terheyden, Dept. of Oral and Maxillofacial Surgery, University of Kiel,
Arnold Heller Str. 16, 24105 Kiel, Germany; e-mail: terheyden@mkg.uni-kiel.de

David D. Thompson, Pfizer Global Research and Development, Groton Laboratories, Groton, CT 06340, USA; e: mail: David_D_Thompson@groton.pfizer.com

Rutger L. van Bezooijen, Department of Endocrinology C4-R, Leiden University
Medical Center, PO Box 9600, 2300 RC Leiden, The Netherlands; and Division of
Cellular Biochemistry, The Netherlands Cancer Institute, Plesmanlaan 121, 1066
CX Amsterdam, The Netherlands; e-mail: R.L.van_Bezooyen@lumc.nl

Slobodan Vukicevic, Laboratory of Mineralized Tissues, Department of Anatomy,
Zagreb Medical School, Salata 11, 10000 Zagreb, Croatia; e-mail: vukicev@mef.hr

Stefan Wölfl, Department of Pharmacy and Molecular Biotechnology, University of
Heidelberg, Im Neuenheimer Feld 364, 69120 Heidelberg, Germany;
e-mail: wolfl@uni-hd.de

Preface

Since the discovery of the bone morphogenetic proteins (BMPs) more than 15 years ago, there has been an unpredicted explosion of both basic scientific discoveries and clinical reports on their use from institutions all over the world. The potent efficacy of BMPs in almost all crucial developmental events as well as during regeneration of various organs such as bone, kidney, brain, liver, heart etc., has positioned BMPs at the center of scientific interest. Many of these aspects are covered in this new PIR volume. Their role in development, biology, signal transduction, kidney regeneration, CNS functions, craniofacial skeleton reconstruction, joint and cartilage repair, long bone non-unions and acute fractures, and spinal fusion is reviewed by experts in the field. For the first time, the role of BMPs in carcinogenesis has been reviewed to provide a rationale for applying their biology in patients with bone tumors. The optimism resulting from safe and successful treatment with BMPs for various skeletal malformations of more than 10,000 patients worldwide has opened new avenues for exploring other indications for their use. The next big challenge for bringing BMPs to the benefit of mankind is in regenerating articular cartilage defects and rescuing patients with acute and chronic renal failure.

The volume editors thank all authors for the rapid preparation of their chapters in order that the book remains up-to-date for readers with specific interest in the field of regenerative medicine. The important contribution of Mrs. Morana and Mr. Branko Simat in processing all the chapters and liaising with authors is greatly acknowledged. Mr. Hans Detlef Klüber and Ms. Karin Neidhart from Birkhäuser Verlag did a great job in fast and efficient processing of the material, making it possible to finish the book in the shortest period of time.

June 2004

Slobodan Vukicevic
Kuber T. Sampath

Bone morphogenetic proteins and their role in regenerative medicine

Kuber T. Sampath

Genzyme Corporation, Framingham, MA 01701, USA

Introduction

Bone regenerates itself upon fracture by instituting a cellular process that mirrors embryonic bone development and restores function fully following its usage. Demineralized bone matrix (DBM) has been widely used as a bone graft substitute to induce new bone formation during reconstructive surgery as it reproduces the embryonic cellular events that results in new cartilage, bone and bone marrow differentiation and supports both osteoinduction and osteoconduction properties *in vivo* [1, 2]. DBM is composed of predominantly type I collagen (95%) and the remaining as non-collagenous proteins including small amounts of growth factors and naturally occurring bone morphogenetic proteins (BMPs) [3], which are responsible for the natural bone healing processes. The demonstration that proteins extracted from the demineralized bone matrix of mammals are capable of inducing new bone when reconstituted with insoluble collagen and implanted in ectopic sites [4] has allowed the discovery of BMPs, also called osteogenic proteins (OPs) [5, 6]. Subsequently, the isolation of the genes encoding these proteins from human DNA libraries has led to the identification of a large family of proteins now called bone morphogenetic proteins [7]. Since then, more than 5,000 research articles have been published on BMPs and the number is growing everyday.

BMPs' role in bone repair and regeneration

The bone morphogenetic proteins (BMPs) are growth and differentiation factors, and form a large family of proteins structurally related to TGF-βs and activins [8]. BMPs are potent bone-forming agents *in vivo*, and are capable of restoring the lost bone in post-fetal life by recapitulating the cellular events that are involved during embryonic bone development. Regulatory agencies in the US, Europe, Canada and Australia have approved BMP-7 (OP-1™) and BMP-2 (InFuse™)-containing osteogenic devices as bone graft substitutes for the treatment of long bone fractures and inter-body fusion of vertebrae in human [9]. BMPs are the first recombinant proteins approved for orthopedic use, thus demonstrating a significant promise that

Bone Morphogenetic Proteins: Regeneration of Bone and Beyond, edited by Slobodan Vukicevic and Kuber T. Sampath events responsible for tissue formation during embryogenesis can form therapeutic strategies in adult regenerative medicine [10]. The current status on the clinical application for the use of OP-1™ and InFuse™ has been provided in details by G.E. Friedlaender and B. McKay.

The carrier plays an important biological role as a component of BMP device to effect new bone formation. As of now, type I collagen is the preferred carrier for BMP as it is a natural component of bone and undergoes resorption comparable to that of extracellular matrix components of bone and considered as the "gold standard" for comparison. The BMP devices used in clinical trials have all employed type I collagen. A fibril acid soluble type I collagen prepared from bovine Achilles tendon has been used to deliver BMP-2 (also called InFuse™). Bovine diaphyseal bone-derived particulate type I collagen has been used to deliver BMP-7 (also called OP-1™). The next preferred carrier for BMP is mineral, which is the major component of bone. Studies using hydroxyapatite and BMP-2 or BMP-7/OP-1 are being evaluated currently in preclinical and clinical studies. Future studies should be directed in the development of biodegradable and biocompatible biomaterials that are suitable to support BMP induced bone repair without provoking inflammation and foreign body reaction.

BMPs' role in articular cartilage repair and regeneration

In addition to bone forming activity, BMPs are potent chondrogenic morphogens and are capable of inducing differentiation of mesenchymal stem cells (MSCs) into cell lineage of hyaline cartilage and maintenance of the expression of markers associated with chondrocyte phenotype *in vitro* and *in vivo*. Several studies have demonstrated that BMPs, when applied alone or in combination with appropriate scaffold onto chondral or osteochondral defects, are capable of inducing new articular cartilage formation *in vivo*. However, the newly formed chondrocytes failed to maintain the cellular morphology and expression of articular cartilage phenotype over time as examined in the long term preclinical studies. It is likely that providing BMPs intermittently or continuously for a sustainable time (instead of one time application in the beginning as used to repair bone fractures) may help to maintain the regenerated cartilage to attain function under mechanical loading. Furthermore, the combination of responding cells with an appropriate scaffold and providing BMP signaling as *ex vivo* gene therapy will have added advantage to enhance chondrocyte differentiation and the maintenance of phenotypic expression of articular cartilage in order to sustain its function for a long time. An update on the progress made on the role of BMPs in articular cartilage repair and regeneration using *in vitro* and *in vivo* models is provided by David C Rueger and Susan Chubinskaya. With regards to the role of BMPs in the formation of joints and their potential therapeutic use in osteoarthritis is provided by Frank F. Luyten et al.

BMPs' role in dentin and peridontium repair and regeneration

Another highly mineralized tissue in the body is tooth. Demineralized tooth matrix has been shown to induce new cartilage, bone and bone marrow differentiation when implanted in ectopic sites. This bone inducing activity of tooth matrix has been attributed to bone morphogenetic proteins that reside in the matrix. Several preclinical studies using BMP-2 and BMP-7 containing matrix have shown to induce repair and regeneration of dental tissues including periodontium (periodontal ligament, cementum and alveolar bone) and dentin [11]. It is possible to regenerate the lost dentin and periodontal tissues in the adult by applying a suitable scaffold and optimal concentration of specific BMPs at the repair site provided responding cells and vascular components are available. More details on the role of BMPs in maxillofacial and periodontal tissue repair and regeneration are presented in the Chapter by H. Terheyden and S. Jepsen.

BMPs affect bone remodeling *in vivo*

Given that BMPs induce new bone formation *in vivo*, promote the recruitment and growth of osteoblast progenitor, and maintain the expression of osteoblast phenotype in cultures, it is conceivable that providing an optimal dose of a BMP in the circulation may help to trigger the osteogenic responses, as an endocrine signal, to restore the loss of bone mass and quality in osteoporosis and related metabolic bone diseases. Recent studies have shown that systemic administration of BMP-7 was able to restore the impaired remodeling associated with the aplastic bone disorder of renal osteodystrophy in a mouse model following 5/6 cortical ablation of the kidney [12], supporting an endocrine role for BMP-7 since the highest level of its expression was found in the adult kidney. It is likely that bone morphogenetic proteins can exert their tissue morphogenesis function both locally in autocrine and paracrine manner during development and systemically in an endocrine mode during growth and in the adult organism. In the future, the efficacy of the systemic administration of BMPs to affect bone mass and bone quality need to be evaluated systematically using various dose and dosing regiments (intermittent *versus* continuous) alone and in combination with existing anti-resorptive therapies (estrogen and bisphosphonates).

BMPs act beyond bone

While BMPs are expressed at the earliest stages of skeletal development and are required for the formation of specific skeletal structure, high levels of mRNA transcripts for several BMPs are observed in several extra-skeletal organs during embryogenesis, suggesting BMPs may play morphogenic roles in tissues beyond bone. BMPs are an evolutionarily conserved family of proteins from *Drosophila* to man. The human equivalent of BMP-2, *DPP* and BMP-7, *60A* proteins have been shown to induce bone in mammals when implanted with an appropriate collagen carrier, indicating the specific biological activity of BMP is dictated by microenvironment and the responding cells available at the site of injury [13].

Studies on gain and loss of function indicate that in addition to a morphogenesis role in the musculoskeletal system, BMPs serve as inductive signals for overall tissue development during embryogenesis [14, 15]. For example, loss of BMP-2 function in mice displays a defective cardiac morphogenesis and amniotic fold closure and in contrast BMP-7 mutant mice die at birth due to defective kidney and eye. Targeted mutation of the murine genes encoding BMP-4 (a closely related member of BMP-2) results in defects in mesoderm formation and patterning during gastrulation. Mice lacking BMP-5 or BMP-6 (closely related members of BMP-7) on the other hand, are viable and fertile and show perturbance in certain selective skeletal structure, short ear in BMP-5$^{-/-}$ and sternal defects in BMP-6$^{-/-}$ mice. Double BMP mutant studies among some members gave additional uncovered defects (e.g., BMP-5/GDF-6; BMP-5/BMP-7) and in other cases, displayed exacerbation of the defect (e.g., BMP-5/BMP-6).

In accordance with these findings, BMPs are shown to induce cardiac myocyte differentiation [16] and inhibit vascular smooth muscle proliferation *in vitro* [17], and reduces intima thickening in the rat balloon angioplasty model [18]. The importance of BMPs has been observed in the ovary as a putative luteinizing factor for the action of FSH to monitor estrogen and progesterone synthesis in adult [19]. The biology of BMPs in the context of tissue and organ development and regenerative medicine has been detailed in chapters by S. Martinovic et al. and P. Simic/S. Vukicevic. The importance of BMP-7 in renal development and its potential therapeutic utility for renal diseases are described in the chapter by F. Borovecki et al. The effects of BMPs in adult neural tissue and their potential utility in the treatment of stroke and neurodegenerative diseases are discussed in the chapter by P. Lein and D. Higgins. BMPs play an important role in epithelial morphogenesis during development and epithelial growth and maintenance in post-fetal life. It is well documented that prostate and mammary cancer, when metastasizing, invariably goes to skeletal tissue. The role of BMPs in prostate and breast cancer will be discussed in the chapter by J. Clement and S. Wölfl.

BMP receptors and down stream signaling

Like TGF-β, all members of BMP elicit their cellular effects by ligand-induced association of specific heterodimeric complexes of two related type I and type II serine/threonine kinase receptors [20]. Different BMPs bind with varying affinity to the type I receptors implying that proteins in the TGF-β superfamily may have broader biological function than previously contemplated. Upon BMP-induced heterodimeric complexes formation, the constitutively active type II receptor kinase phosphorylates type I receptor and subsequently activates intracellular signaling by phosphorylating downstream components, including nuclear effector proteins known as *Smads*.

Smads can be divided into three distinct subclasses: signal transducing receptor-regulated Smads (R-Smads, i.e., Smad1, Smad5 and Smad8) and common mediator Smads (C-Smads, i.e., smad4) and inhibitory Smads (I-Smads, i.e., Smad6 and Smad7). Upon BMP receptor activation, BMPR-Smad1/5/8 forms heterodimeric complexes with C-Smad4 and then translocates to nucleus. Within the nucleus, R-Smad/C-Smad complexes act directly and/or in cooperation with other transcription factors, to regulate transcription of target genes. Inhibitory Smad6/7 specifically inhibits BMP and TGF-β signaling which BMP and TGF-β also induce thus a negative regulation is controlled to keep the activation in balance. Expression, receptor-mediated phosphorylation (activation) and translocation and accumulation in the nucleus of R-Smads and subsequently their stability determine that Smads are transcription factors. Smads upon nuclear translocation are shown to bind cooperatively to specific DNA sequences in the promoters of several BMP targeting genes by forming a complex in the nucleus to affect a specific cellular function in the cell. Smad-dependent transcriptional regulations are very diverse and cross-talk/interface with other signaling pathways that include TNF-α/NF-κB, EGF/MAPK, and Wnt/GSK [21]. Ending of BMP signaling is achieved through ubiquitination and proteosome-mediated degradation of R-Smads directly by Smurf-1. In addition, Smurf-1 forms a stable complex with Smad7 and targets the complex to plasma membrane to inhibit phosphorylation of R-Smads and ubiquitination in the process. BMPs activity is tightly regulated by gene expression, protein processing from its precursor and by binding to the naturally occurring secreted soluble BMP antagonists at the extracellular space. As TGF-β binds to α2-macroglobulin and small proteoglycan decorin, and activin binds to follistatin, BMPs bind to noggin, chordin, DAN/cereberus, gremlin, sclerostin and related USPG-1 [22]. The interplay between BMPs and their antagonists governs BMP activity locally. For more details see the chapter by O. Korchynskyi et al.

Perspectives

Local bone formation induced by the demineralized matrix of bone serves as a prototype for tissue engineering, in that the collagenous matrix acts as a substratum for migration and proliferation of mesenchymal cells and morphogenetic proteins resided within the bone matrix to signal the differentiation of mesenchymal cells into endochondral bone. Thus, this biological principle of bone formation has enabled us to identify a family of BMPs, which then led to the development of therapeutic osteogenic devices for use in the repair and regeneration of bone. The availability of human BMP in an unlimited quantity through recombinant technology has now made it possible to evaluate its clinical utility for bone repair in various indications, including normal, delayed and non-union fractures. No doubt, our future generation will benefit by off-the-shelf recombinant BMP devices instead of harvesting the iliac crest bone and thus avoiding associated donor site morbidity.

Although BMP devices offer tremendous promises as bone graft substitutes, there are still numerous challenges we are faced with. Even though the current BMP devices can induce bone readily, they lack biocompatibility for mechanical stability of the graft during the procedure. There is a need for optimal delivery systems for BMPs with varying geometry and resorptive time depending on location and mechanical loading of the defects. The handling property of the device is also important as the surgical approximation and retaining of the BMP device at the site of repair are required to affect a reproducible outcome. Since the availability of vascular components and responding cells, and the mechanical stability at the repair site determine the efficacy of BMP-induced bone formation, efforts should be taken to provide appropriate microenvironment during the procedure. Furthermore, internal and/or external fixation has to be modernized according to the rate of osteogenesis induced by BMP devices.

Recombinant BMP containing osteogenic devices are currently very expensive and faced with reimbursement issues for their wider usage. In the future, autologous bone marrow derived mesenchymal cells could be engineered in the surgical suite to deliver the BMP *via ex vivo* gene therapy with an appropriate carrier matrix to induce new bone formation [23]. There have been numerous reports describing the use of MSCs to deliver morphogenic signaling genes with appropriate extracellular matrix components as scaffolds to be implanted at sites of injury to initiate bone differentiation and speed-up the regeneration of functional bone. The local delivery of bone morphogenetic protein *via* viral and non-viral vectors transfected into MSCs and/or provided in conjunction with appropriate resorbable matrix at the repair site is a very cost-effective and readily available therapy for therapeutic tissue engineering in the future. The "cell-gene-matrix" device exploits the donor site to make the protein for a period of time sufficient enough to induce the differentiation of adjacent mesenchymal stem cells from bone marrow, periosteum and muscle into bone and restore the function.

Identification and characterization of the BMPs specific type I and II receptor complex and subsequent intracellular signaling *via* BMP specific Smads, and the identification of BMP responding elements in tissue specific target genes provides a basis for the rapid screening of orally active compounds to mimic BMP activity. It is likely that advances made in identification and characterization of the BMP promoter sequences and the regulation of BMP gene expression will allow the development of small molecules for endogenous upregulation of BMPs in individuals with osteoporosis and various metabolic bone diseases [24]. The use of small molecules in conjunction with ceramics for fracture healing is presented in the chapter by V.M. Paralkar et al. The biology of BMPs and their role in regenerative medicine indubitably holds a tremendous promise for the future generation to come.

References

1   Urist MR (1965) Bone formation by autoinduction. *Science* 150: 893–899
2   Reddi AH, Huggins CB (1972) Biochemical sequence in the transformation of fibroblasts into cartilage and bone. *Proc Natl Acad Sci USA* 69: 1601–1605
3   Urist MR, Strates BS (1970) Bone morphogenetic protein. *J Dent Res* 50: 1392–1406
4   Sampath TK, Reddi AH (1981) Dissociative extraction and reconstitution of extracellular matrix components involved in local bone differentiation. *Proc Natl Acad Sci USA* 78: 7599–7602
5   Wozney JM, Rosen V, Celeste AJ, Mitsock LM, Kriz RW, Hewick RM, Wang EA (1988) Novel regulators of bone formation: Molecular clones and activities. *Science* 242: 1528–1534
6   Ozkaynak E, Rueger DC, Drier EA, Corbett C, Ridge, RJ, Sampath TK, Oppermann H (1990) OP-1 cDNA clones, an osteogenic protein in the TGF-β family. *EMBO J* 9: 2085–2093
7   Reddi AH (1998) Role of morphogenetic proteins in skeletal tissue engineering and regeneration *Nat Biotechnol* 16: 247–252
8   Massague J (1990) The transforming growth factor-β family. *Annu Rev Cell Biol* 6: 597–641
9   Wozney JM (2002) Overview of bone morphogenetic proteins. *Spine* 73: 1020–1209
10 S Vukicevic, KT Sampath (eds): *Bone morphogenetic protein: From laboratory to clinical practice.* Birkhäuser Verlag, Basel/Switzerland 2003
11 Nakashima M, Reddi AH (2003) The application of bone morphogenetic proteins to dental tissue engineering. *Nat Biotechnol* 21: 1025–1032
12 Gonzalez EA, Lund RJ, Martin KJ, McCartney JE, Tondravi MM, Sampath TK, Hruska KA (2002). Treatment of a murine model of high-turnover renal osteodystrophy by exogenous BMP-7. *Kidney Int* 61: 1322–1331
13 Sampath TK, Rashka KE, Doctor JS, Tucker RF, Hoffmann FM (1993) *Drosophila* transforming growth factor β superfamily proteins induce endochondral bone formation in mammals. *Proc Natl Acad Sci USA* 90: 6004–6008
14 Hogan BL (1996) Bone morphogenetic proteins in development. *Curr Opin Genet Dev* 6: 432–438
15 Vukicevic S, Latin V, Chen P, Batorsky R, Reddi AH, Sampath TK (1994) Localization of osteogenic protein-1 (bone morphogenetic protein-7) during human embryonic development: high affinity to basement membrane. *Biochem Biophys Res Commun* 198: 693–700

16   Monzen K, Hiroi Y, Kudoh S, Akazawa H, Oka T, Takimoto E, Hayashi D, Hosoda T, Kawabata M, Miyazono K et al (2001) Smads, TAK1, and their common target ATF-2 play a critical role in cardiomyocyte differentiation. *J Cell Biol* 153: 687–698

17   Dorai H, Vukicevic S, Sampath TK (2000) Bone morphogenetic protein-7 (osteogenic protein-1) inhibits smooth muscle cell proliferation and stimulates the expression of markers that are characteristic of SMC phenotype *in vitro*. *J Cell Phys* 184: 37–45

18   Nakaoka T, Gonda K, Ogita T, Otawara-Hamamoto Y, Okabe F, Kira Y, Harii K, Miyazono K, Takuwa Y, Fujita T (1997) Inhibition of rat vascular smooth muscle proliferation *in vitro* and *in vivo* by bone morphogenetic protein-2. *J Clin Invest* 100: 2824–2832

19   Shimasaki S, Zahow RJ, Li D, Kim H, Iemura S-I, Ueno N, Sampath TK, Chang RJ, Erickson GF (1999) A functional bone morphogenetic system in ovary. *Proc Natl Acad Sci USA* 96: 7282–7287

20   ten Dijke P, Yamashita H, Sampath TK, Reddi AH, Estevez M, Riddle DL, Ichijo H, Heldin C-H, Miyazono K (1994) Identification of type I receptors for osteogenic protein-1 and bone morphogenetic protein-4. *J Biol Chem* 269: 16985–16988

21   Xiao C, Shim JH, Kluppel M, Zhang SS, Dong C, Flavell RA, Fu XY, Wrana JL, Hogan BL, Ghosh S (2003) Ecsit is required for Bmp signaling and mesoderm formation during mouse embryogenesis. *Genes Dev* 17: 2933–2949

22   Yanagita M, Oka M, Watabe T, Iguchi H, Niida A, Takahashi S, Akiyama T, Miyazono K, Yanagisawa M, Sakurai T (2004) USAG-1: a bone morphogenetic protein antagonist abundantly expressed in the kidney. *Biochem Biophys Res Commun* 316: 490–500

23   Scaduto AA, Lieberman JR (1999) Gene therapy for osteoinduction. *Orthop Clin North Am* 30: 625–633

24   Paralkar VM, Grasser WA, Mansolf AL, Baumann AP, Owen TA, Smock SL, Martinovic S, Borovecki F, Vukicevic S, Ke HZ et al (2002) Regulation of BMP-7 expression by retinoic acid and prostaglandin E (2). *J Cell Physiol* 190(2): 207–217

Bone morphogenetic protein receptors and their nuclear effectors in bone formation

Olexander Korchynskyi[1,2,3], Rutger L. van Bezooijen[1,4], Clemens W.G.M. Löwik[4] and Peter ten Dijke[1]

[1]Division of Cellular Biochemistry, The Netherlands Cancer Institute, Plesmanlaan 121, 1066 CX Amsterdam, The Netherlands; [2]Thurston Arthritis Research Center and Department of Rheumatology, School of Medicine, University of North Carolina, Chapel Hill, North Carolina, USA; [3]Department of Endodontics, School of Dentistry, University of North Carolina, Chapel Hill, North Carolina, USA; [4]Department of Endocrinology, Leiden University Medical Center, Leiden, The Netherlands

Introduction

Pioneering studies on the ability of extracts from decalcified bone matrix to promote ectopic bone and cartilage formation [1] led to searches for the identity of these morphogens which define skeletal patterning. With the advent of powerful methods for protein purification, capability to determine amino acid sequences on small amounts of protein and DNA cloning, bone morphogenetic proteins (BMPs) were discovered [2–4]. The amino acid sequences predicted from their cDNA sequences revealed that BMP-2, BMP-3 and BMP-4 (BMP-1 is a member of the astacin family of metalloproteases) are members of the TGF-β superfamily, which also includes the TGF-βs and activins [5]. Mainly through their sequence homology with other BMPs, approximately 20 members in the BMP subgroup have now been identified and can be divided in multiple groups of structurally related proteins, e.g., BMP-2 and BMP-4 are highly related, BMP-6, BMP-7 and BMP-8 form another subgroup, and growth and differentiation factor (GDF)-5 (also termed cartilage-derived morphogenetic protein (CDMP)-1), GDF-7 (also termed CDGF-2) and GDF-7 are similar to each other. *In vitro* BMPs were found to have potent effects on various cells implicated in cartilage and bone formation, e.g., they induce proteoglycan synthesis in chondroblasts and stimulate alkaline phosphatase activity and type I collagen synthesis in osteoblasts [4]. When injected into muscle of rats, BMPs can induce a biological cascade of cellular events leading to ectopic bone formation [3, 4]. GDF-5, GDF-6 and GDF-7 induce tendon and cartilage-like structures more efficiently [6, 7]. Preclinical studies of certain BMPs in primates and other mammals have demonstrated their effectiveness in restoring large segmental bone defects [8, 9].

Bone Morphogenetic Proteins: Regeneration of Bone and Beyond, edited by Slobodan Vukicevic and Kuber T. Sampath
© 2004 Birkhäuser Verlag Basel/Switzerland

Like other members of the TGF-β family, BMPs are multifunctional proteins with effects on cell types not related to bone formation, e.g., epithelial cells, monocytes and neuronal cells [10, 11]. In addition, BMPs were found to be expressed not only in skeletal tissues, but also in many soft tissues. Consistent with these results, phenotypes of mice with mutated BMP genes revealed that they are multifunctional proteins that posses distinct roles in bone formation and many other morphogenic processes (Tab. 1) [12]. Interestingly, several different mouse and human skeletal disorders have been linked to genetic alterations in BMP genes. Two mouse skeletal disorders – short ear and brachipodism – are caused by a null mutations in BMP-5 [13] and GDF-5 [14], respectively. Double muscle cattle were found to have mutations in GDF-8 (also called myostatin) [15]. Hunter-Thompson type chondrodysplasia has been linked to mutations in human cartilage-derived morphogenetic protein [16].

In this chapter we review the BMP signal transduction pathways leading to bone formation. In particular, we will discuss the latest advances towards our understanding of the function of BMP receptors and their nuclear effector proteins, termed Smads, in controlling target gene expression.

Identification and structure of BMP receptors

TGF-β family members, which include BMPs, elicit their cellular effects by inducing specific heteromeric complexes of two related serine/threonine kinase receptors, i.e., type I receptor and type II receptors [17, 18]. Among the TGF-β family of receptors, the cDNAs encoding mouse activin and human TGF-β type II receptor were isolated first by an expression cloning strategy [19, 20]. Subsequently, other mammalian type II and type I receptors, including those for BMPs, were isolated based upon their sequence similarity with other serine/threonine kinase receptors [21–32]. Both receptor types contain glycosylated cysteine-rich extracellular ligand-binding domains, short transmembrane domains and intracellular serine-threonine kinase domains (Fig. 1) [17, 18]. A shared feature for type I receptors is that they have a glycine/serine residue-rich stretch in the juxtamembrane region, which is essential for type I receptor activation [17, 18, 32]. Three mammalian BMPR-Is have been described to date [24, 29], i.e., activin receptor-like kinase (ALK)2, BMPR-IA (also termed ALK3) and BMPR-IB (also termed ALK6). Initially, ALK2 has been referred to as a type I receptor for TGF-β [33] or activin (ActR-I) [22], but recent studies suggest that ALK2 is most important in BMP signaling [24, 29, 34, 35]. Different BMPs bind with different affinity to the type I receptors. For example, BMP-4 binds preferentially to BMPR-A and –IB [24], BMP-7 binds with higher affinity to ALK2 and BMPR-IB than to BMPR-IA [24] and GDF-5 binds preferentially to BMPR-IB, when compared with other type I receptors [36]. Functional importance of BMPR-Is in bone formation was shown by the induction of chondroblast and osteoblast differ-

Table 1 - Phenotypes of organisms with disruption of genes for BMPs, their receptors or their downstream Smads*

Mutated gene	Phenotype	Refs.
BMP		
BMP-2	Embryonic death (E7.5-E10.5). Defects in amnion/ chorion formation and cardiac development.	[215]
BMP-3	Viable. Increased bone mass.	[41]
BMP-4	Embryonic death (E7.5-E9.5). Block of mesoderm formation.	[57]
BMP-5	Viable. Skeletal abnormalities, short ear, brachypodism.	[13]
BMP-6	Viable. Delay in developing sternum ossification.	[216]
BMP-7/OP-1	Perinatal lethality. Severe defects in kidney and eyes. Abnormalities of rib cage, skull and hindlimbs.	[217] [218]
BMP5/7	Embryonic death (E10.5). Retarded heart development.	[219]
BMP-8B	Viable. Defects in spermatogenesis.	[220]
BMP-15 (GDF9B)	Viable. Increased ovulation rate leading to twins and triple births in heterozygotes and infertility in homozygotes.	[221]
GDF-5	Viable. Skeletal abnormalities, short ear, brachypodism.	[14]
GDF-8	Viable. Increased skeletal muscle mass and body size	[15]
Receptors		
ActR-IA/ALK-2	Embryonic death (E9.5). Block of mesoderm formation.	[42]
BMPR-IA/ALK3	Embryonic death (E7.5-E9.5). Block of mesoderm formation.	[56]
BMPR-IB/ALK6	Viable. Defects in limb development.	[46]
BMPR-II	Embryonic death (E9.5). Block of mesoderm formation.	[49]
ActR-IIA and ActR-IIB	Embryonic death (E9.5). Arrest at the egg cylinder stage and block of mesoderm formation.	[59]
Smads		
Smad1	Embryonic death (E9.5). Defects in allantois formation.	[222, 223]
Smad4	Embryonic death (E6.5-E8.5). Block of mesoderm formation.	[95]
Smad5	Embryonic death (E9.5-E10.5). Defects in angiogenesis.	[224]

*All gene mutations are in mice except for sheep BMP-15 and bovine GDF-8.

entiation upon ectopic expression of mutant constitutively active BMPR-Is in mesenchymal precursor cells, and by observations that overexpression of dominant negative BMPR-Is interfered with BMP-induced osteoblast differentiation [37–40]. Surprisingly, BMP-3, which is one of the most abundant BMPs in adult bone, func-

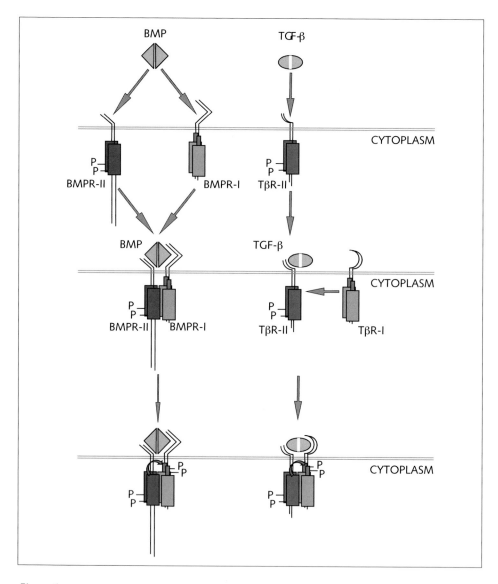

Figure 1

Activation of BMP and TGF-β receptors

BMPs binds with weak affinity to type I or in particular type II receptors alone, but with high affinity to type I/type II heteromeric complex. Upon BMP-induced heteromeric complex formation, the constitutively active type II serine/threonine kinase of type II receptor phosphorylates type I receptor in its GS-domain. TGF-β binds first to TGF-β type II receptor, and subsequently recruits TGF-β type I receptor and initiates signaling in a similar fashion as described for BMP receptor activation.

tions as an antagonist of BMP signaling, and is claimed to signal *via* the activin type IB receptor (ActR-IB)/ALK4 [41].

Three distinct type II receptors have been implicated in BMP signaling: BMPR-II, activin type II receptor (ActR-II) and ActR-IIB [26, 27, 30]. However, binding affinities of ActR-II and ActR-IIB for BMPs are lower than those for activins [30]. Type II receptors, but not type I receptors, have extensions rich in serine and threonine residues distal from the kinase domains.

Expression of BMP type I and type II receptors

During mouse embryogenesis, ALK2 is expressed primarily in the extraembryonic visceral endoderm before gastrulation and it is widely expressed in midgestation embryos [42, 43]. BMPR-IA is also found broadly expressed, but it is absent in the liver during embryogenesis [44]. Among the three BMPR-Is, BMPR-IB expression is the one that is most tissue or developmental stage restricted in its expression pattern [44, 45]; BMPR-IB is predominantly expressed in mesenchymal cells representing the primordia of long bones and later in development, it is widely expressed in skeleton components [46]. During chicken limb development, BMPR-IB is strongly expressed in undifferentiated mesechymal cells, condensations prefiguring the future cartilage primordium. Expression of chicken BMPR-IA, however, is restricted to the prehypertrophic chondrocytes [45].

All three type II receptors (BMPR-II, ActR-II and ActR-IIB) are differentially expressed during mouse embryogenesis [47–49]. BMPR-II mRNA is detected in one-cell, two-cell and blastocyst stage embryos [50] and it is present in both embryonic and extraembryonic regions [49]. ActR-II and ActR-IIB, however, are mainly expressed in extraembryonic ectoderm [47]. All three BMP type II receptors are expressed in hypertrophic cartilage and ossified tissue [51, 52]. Interestingly, BMP receptor expression is enhanced at sites of fracture repair [53]. Furthermore, during pathological ossification in the spinal ligaments, hypertrophic chondrocytes were found to express high levels of BMP receptors, and these sites colocalized with high levels of BMP expression [51, 54, 55]. Aberrant expression of BMPs and their receptors, possibly induced by mechanical stress, may be involved in the pathogenesis of orthotopic ossification [55].

Determination of *in vivo* function of BMP type I and type II receptors through gene targeting approaches

BMP type I and type II receptors were found to be critically important for embryogenesis (Tab. 1) [12]. Mice lacking ALK3 and BMPR-II are lethal due to absence of mesodermal development [49, 56] and have a phenotype similar to BMP-4 knockout mice [57]. ALK2-deficient embryos are much smaller than their normal litter-mates, and lack a morphologically discernible primitive streak and die prior to or during early gastrulation [42]. ALK6-deficient mice are viable and exhibit mainly appendicular skeleton defects [46]. Mice lacking ActR-II or ActR-IIB are viable and were found to have a milder phenotype compared to a deficiency of one of their lig-ands. Some of ActR-II-deficient animals had mandibular hypoplasia and other skeletal and facial abnormalities [48]. ActR-IIB knock-out mice showed cardiac defects, abnormal anteroposterior and left-right body axis patterning [58]. Howev-er, ActR-II and ActR-IIB double-knockout homozygous showed strong lethal embryonic abnormalities; these mice were growth arrested at the egg cylinder stage and did not form mesoderm [59]. The stronger phenotype in the double knock-out *versus* the single knock-outs suggests a functional redundancy for ActR-II and ActR-IIB in the mouse.

Mechanism of BMP receptor activation

Like other TGF-β family members, both type I and type II receptors are required for BMP signaling [17, 18]. BMPs bind with weak affinity to type II or type I receptors alone and with high affinity to a heteromeric complex of the two receptor types [24, 26–30] (Fig. 1). The affinity of BMPR-I for ligand binding is higher that of BMPR-II and it is thus plausible that BMPR-I binds ligand initially and then recruits BMPR-II into the ligand-receptor complex [60]. This is in contrast to TGF-β and activin, which first bind to type II receptors and subsequently recruit type I recep-tors [21–23, 61] (Fig. 1). The mechanism of receptor activation has been best char-acterized for TGF-β [32], but it is likely to occur in an analogous fashion for BMPs [17, 18]. Upon BMP-induced heteromeric complex formation, the constitutively active type II receptor kinase phosphorylates type I receptor predominantly in its GS domain. The type I receptor thus acts downstream of type II, and consistent with this notion it has been shown to confer signaling specificity to the type I/type II het-eromeric complex [62] (Fig. 1). The activated type I receptor initiates intracellular signaling by phosphorylating downstream components, including the nuclear effec-tor proteins known as Smads. The L45 loop regions in the kinase domain of type I receptors were found to be important determinants for signaling specificity [63–65].

BMPR-II is distinct from the other type II receptors in that it has a long carboxy-terminal (C-)tail extension [26, 27]. This tail is not important in the transactivating of BMPR-I, since BMPR-II lacking the C-tail is fully functional in this respect [28]. However, patients with familial primary pulmonary hypertension syndrome have been genetically linked to mutations in BMPR-II, and certain of these mutations result in a partial truncation of the C-tail [66–69]. Recently, the tail region was found to bind LIM kinase 1 (LIMK1), a key regulator of actin dynamics as it phos-phorylates and inactivates cofilin, an actin depolymerizing factor [70]. A BMPR-II mutant containing the smallest C-terminal truncation described in these patients failed to bind or inhibit LIMK1.

Identification and structure of Smad proteins

Our understanding of BMP intracellular signaling has dramatically increased through genetic studies in *Drosophila* and *Caenorhabditis* (*C.*) *elegans*, in which Mothers against DPP (Mad) [71] and *Sma* (Small body size) genes [72], respectively, were identified. MAD and SMA proteins were found to possess a critical role downstream of BMP-like proteins in these organisms. Thus far nine mammalian Mad and Sma related (Smad) proteins have been identified, which perform a pivotal function in TGF-β family intracellular signaling [17, 18]. Based upon their functional properties, Smads can be divided into three distinct subclasses: signal transducing receptor-regulated Smads (R-Smads) and common-mediator Smads (Co-Smads, i.e., Smad4) and inhibitory Smads (I-Smads, i.e., Smad6 and Smad7) which inhibit the activation of R- and Co-Smads [54, 73, 74] (Fig. 2). R- and Co-Smads have conserved amino and carboxy regions, known as Mad homology (MH1) domain and MH2 domains, respectively. Both domains are separated by a variable proline-rich linker region. Whereas the I-Smads have an MH2 domain, their amino-terminal regions show only weak sequence similarity to the MH1 domains (Fig. 2) [17, 18].

Activation and function of Smad proteins

R-Smads interact transiently with and become phosphorylated by the activated type I receptor (Fig. 3); whereas Smad1, Smad5 and Smad8 act in the BMP pathway and Smad2 and Smad3 are activated by TGF-β and activin type I receptors [17, 18]. The L3 loop of R-Smad was shown to interact with the L45 loop in TGF-β and BMP type I receptors, a region which determines signaling specificity among different type I receptors [75]. Smad2 and Smad3 have been shown to be presented to TGF-β receptor complex through phospholipid binding FYVE-domain containing proteins, termed Smad anchor for receptor activation (SARA) [76] and Hrs [77]. However, SARA/Hrs-like proteins that facilitate BMP type I receptor-mediated activation of R-Smads remain to be identified. R-Smad phosphorylation by the activated type I receptor occurs at the two most carboxy-terminal serine residues in a SSXS motif [34, 78–80]. In osteoblasts, BMP was found to induce the C-terminal phosphorylation of Smad5, and to a lesser extent Smad1 [81–83].

Upon BMP receptor activation BMP R-Smads form heteromeric complexes with Co-Smad4, i.e., Smad4 [84]. Preferentially trimeric Smad complexes are formed [85–87] (of which the exact stoichiometry needs further investigation) that effi-

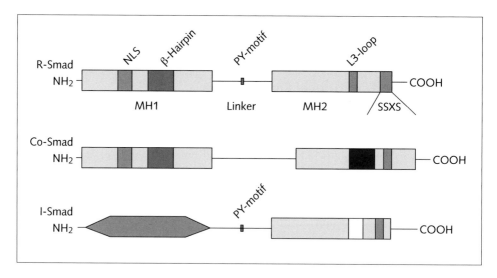

Figure 2
Structure of Smad proteins
Receptor-regulated Smads (R-Smads) and common partner Smads (Co-Smads) consist of two highly conserved MH1 and MH2 domains that are separated by a proline-rich linker region. The amino-terminal region of inhibitory Smads (I-Smads) has only weak similarity to MH1 domains of R- and Co-Smads. The L3-loop in R-Smads interacts with activated type I receptors. Activated BMPR-I phosphorylates R-Smads in their C-terminal SXS motif, which is not present in Co-Smads and I-Smads. Nuclear localization signal (NLS) and DNA binding motif (β-hairpin) are conserved in the MH1 domains of R-Smads and Co-Smad. The PY motif is important for interaction with WW-containing HECT E3 ligases.

ciently translocate to the nucleus (Fig. 3) [17, 18]. Nuclear accumulation of BMP R-Smads and Smad4 was observed in osteoblasts after stimulation with BMP [81, 82]. The osteoblast-induced differentiation of mesenchymal precursor cell lines by ectopic expression of Smad1 or Smad5 became more pronounced when co-expressed with Smad4 and greatly enhanced by addition of BMP, which strongly promotes R-Smad/Co-Smad nuclear accumulation [40].

A nuclear localization signal (NLS)-like sequence in the MH1 domain of Smad3, that is conserved among all R- and Co-Smads, was shown to be required for TGF-β-induced nuclear import [88, 89]. In Smad4 a functional leucine-rich nuclear export sequence (NES) was identified, ensuring cytoplasmic location of Smad4 in unstimulated cells. TGF-β-induced complex formation of Smad4 with R-Smads was found to inactivate the NES [90, 91]. Nuclear entry of the Smad4/R-Smad complex may be stimulated upon unmasking of the NLS on the R-Smad and/or Co-Smad

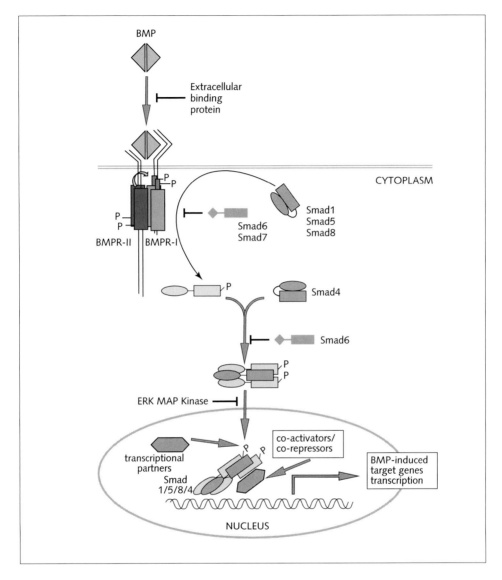

Figure 3

Signaling from activated BMP receptors to nucleus by Smad proteins

Upon BMP receptor activation R-Smads are phosphorylated by the activated BMP type I receptor. Activated R-Smads can form a heteromeric complex with Co-Smad and translocate into the nucleus where they can directly or through their transcriptional partners bind to specific sequences in the promoters of BMP target genes and activate transcription of those genes. I-Smads block BMP signaling. Phosphorylation of Smads by Erk/MAPK into linker region inhibits nuclear translocation of Smads.

upon heteromeric complex formation. Within the nucleus, R-Smad/Co-Smad complexes act directly and/or in cooperation with other transcription factors, to regulate the transcription of target genes (see below) [7, 92, 93].

Gene disruption of Smad genes in mice has begun to reveal specific and developmental functions of Smads that are implicated in BMP signaling. Whereas mice lacking Smad1, Smad4 or Smad5 are developmentally arrested, Smad6 mice make it to term (Tab. 1) [12]. To study the role of Smads in cartilage and bone formation, conditional knock-outs in e.g. mesenchymal precursor cells and osteoblasts are eagerly awaited.

Expression and stability of Smad proteins

In a recent study, the expression of Smad1 to Smad6 was examined in the 15th day of gestation of the mouse embryo. All tissues were found to express Smad4 and at least one on the R-Smads. Among the Smads, Smad6 expression was found most restricted [94]. At sites of endochondral ossification, expression patterns of BMPs and their receptors were found to overlap with Smad1, Smad5 and Smad4 expression in proliferating chondrocytes and in the maturing chondrocytes [52, 94, 95]. Highest expression of inhibitory Smads was shown in zones of mature chondrocytes. These findings suggest that Smad expression is an important determinant in regulating BMP signaling during the different phases of the bone forming process [52, 94, 95].

The stability of Smad proteins appears also to be carefully regulated. Smad ubiquitination regulatory factor 1 (Smurf-1) was identified as a homology to E6-AP carboxyl terminal (HECT) domain containing E3 ubiquitin ligase for BMP R-Smads [96]. The WW motifs in Smurf1 interact with the PY motif (PPXY) in the linker regions of Smad1 and Smad5. Increased expression of Smurf1 leads to a selective decrease in BMP R-Smads, thereby decreasing the cellular competence to BMP-mediated responses [96] and inhibiting osteoblast differentiation and bone formation *in vitro* and *in vivo* [97]. The proteasome-mediated degradation of Smad1/5 by Smurf1 is independent from their activation by ligands. Whether Smads can also be modified in order to make them more stable is an interesting area for future research; e.g. the conjugation of ubiquitin to lysine residues in Smads may be blocked by acetylation of those same residues, and ligation of small ubiquitin-related and modifier (SUMO) to Smads may inhibit their ubiquitin-mediated degradation.

Smads are transcription factors

R-Smads (except for Smad2) and Smad4 were found to recognize specific sequences *via* their MH1 domains in the promoters of Smad target genes [98–101]. The affinity of Smad3 and Smad4 to DNA is much higher than BMP R-Smads. An *in vitro* screen of random DNA oligonucleotides that specifically bound to MH1-linker domain subdomains of Smad3 and Smad4 revealed that these Smads bind with highest affinity to sequences containing GTCT sequence (called also Smad-binding element, SBE) [102]. Multimers of SBE when placed in front of a minimal promoter reporter construct provide a strong enhancer function for TGF-β family members [98, 99, 102]. SBE-like sequences have been shown to be critically important for both TGF-β-inducibility of multiple TGF-β responsive genes [92, 93] and for activation of multiple BMP target genes [101, 103, 104]. TGF-β induced activation of several TGF-β induced genes, including Smad7 [105–109], plasminogen activator inhibitor-1 [98, 110] α2(I) collagen [111] and type VII collagen [112] is critically dependent on SBE sequences, which have been found in multiple copies in promoters of these genes. The Smad1 MH1 domain was shown to bind SBE [113] and a reporter construct containing a multimerized SBE present in JunB promoter is activated by BMP [99]. BMP R-Smads (and also Smad3 and Smad4) have also been shown to bind to GCAT motifs [114] or to GCCG-rich sequences present in promoters of different BMP target genes [115, 116]. Mutation of these sequences significantly decreased BMP-induced response [116, 117]. However, BMP-inducibility of reporter constructs containing multimerized GCCG-rich sequences is very low [116–118] and requires high levels of Smad overexpression [116, 117]. The true physiological significance of the low affinity interaction of BMP R-Smads with GCCG-rich sequences or GCAT motifs remains to be shown. Most likely, inducibility of GCCG-rich motifs can be explained by the partial overlapping of these sequences with GGCG/CGCC motifs. Robust BMP-induced activation of *Id1* promoter (and other sequences) requires cooperation between three distinct sequence motifs, i.e., SBE or other transcription factors binding sites and GGCG/GGCGCC and CAGC [101, 103, 119, 120]. Interestingly, multimerization of GGCGCC-containing motifs together with SBE motifs creates a potent, highly sensitive BMP-specific enhancer, that does not require overexpression of Smads [101].

The DNA affinity of Smads, and of BMP R-Smads in particular, is weak. Smads thus need to cooperate with other DNA binding factors in order to bind efficiently to the promoters of target genes [92, 93, 121]. The 30-zinc finger nuclear protein OAZ was the first identified DNA-binding factor that associates with BMP R-Smads in response to BMP [121, 122]. OAZ interacts with the MH2 domains of Smad1 and also Smad4. Expression of OAZ is tissue and cell type-specific and OAZ cannot be detected in different cells, including mesenchymal precursors [122]. Interestingly, a member of core binding factor (CBF) family of transcriptional factors Cbfa1 (also called osteoblast-specific factor (Osf) 2, *Runt*-related gene 2 (RUNX2) acute myeloid leukemia (AML) protein 3 (AML3) or polyomavirus enhancer core-binding protein-2αA (PEBPα2A) and its homologues Cbfa2 and Cbfa3 were shown to interact directly with Smad1/5 (as well as Smad2 and Smad3) [123, 124]. Cbfa1 precedes the appearance of osteoblasts and mice deficient in Cbfa1 lack osteoblasts and the bone ossification is completely blocked [125]. Cbfa1 is also critically important for already differentiated osteoblasts and acts as a maintenance factor for mature osteoblasts by regulating the rate of bone matrix deposition [126]. The Cbfa1 genetic locus has been linked to one of the most frequent human skeletal disorders termed *cleidocranial dysplasia* (CCD) syndrome [127, 128]. CCD patients express truncated mutant Cbfa1 proteins that retain the ability to bind DNA by their *runt* domains, but fail to interact with Smads. These data suggest that Cbfa1 and Smad cooperate in BMP-induced osteoblast differentiation [129].

Initially, the MH2 domains of R- and Co-Smads were found to have transactivation properties when fused to a GAL4-DNA binding domain [130, 131]. Subsequent studies have provided a mechanistic explanation for this; Smad1 as well as Smad2 and Smad4 were found to interact with transcriptional co-activators CBP/p300 which possess intrinsic acetyltransferase activity [132]. P300 and CBP facilitate transcription by both decreasing the chromosome condensation through histone acetylation and increasing the accessibility of Smad with components of the basal transcriptional machinery. CBP/p300 interacts with many different transcription factors. MSG1 non-DNA-binding transactivator binds to CBP/p300 co-activators and enhances their functional link to Smads [133]. Two-handed zinc finger transcription factor Zeb1/Zfhx1a/δEF1 was also shown as a transcriptional co-activator of BMP-Smad and TGF-β-Smad signaling due to functioning as an adaptor of Smad-CBP/p300 interaction [134, 135]. Similarly, SMIF was found to be a Smad4-interacting co-activator of both TGF-β and BMP signaling and to require CBP/p300 for its function [136]. The synergy between BMP and leukemia inhibitory factor (LIF) in the induction of differentiation of neuronal progenitors into astrocytes was shown to be mediated by cooperative binding of Smad1 and STAT3 to CBP/p300 [137].

Negative regulation of BMP/Smad pathway

Negative regulation occurs at nearly every step in the BMP/Smad pathway. Extracellularly, BMP activity is inhibited by antagonists and pseudoreceptors. The extracellular antagonists include noggin, chordin, follistatin, follistatin-related gene (FLRG), and the DAN family of proteins that include DAN, cerberus, gremlin, protein related to Dan and cerberus (PRDC), and Dante. Chordin and noggin have been shown to antagonize BMP signaling by blocking the binding of BMPs to their receptors [138, 139]. A similar mechanism has been proposed for members of the DAN family that bind BMPs [140–142]. In constrast, follistatin is noncompetitive with BMP/BMP receptor interaction and forms a trimeric complex [143].

Noggin and gremlin have been shown to regulate osteoblastic differentiation and function [144]. A new member of the cysteine-knot family of growth factors, the SOST gene product sclerostin, was predicted to be a BMP antagonist based on its amino acid sequence [145, 146]. It specificly binds BMPs, although with low affinity (10^{-7}–10^{-8} M), and inhibits osteoblastic differentiation and BMP-stimulated alkaline phosphatase activity with no clear preferences for any of the BMPs [147–149]. Interestingly, sclerostin did not prevent all BMP responses, suggesting that it only acts as a direct BMP antagonist in a certain cellular context [149]. Similarly, chordin has been shown to act in combination with the co-factor twisted gastrulation [150]. Alternatively, another bone formation stimulating ligand may be the actual ligand of sclerostin. Wnt family of growth factors may be such BMP-induced factors that are antagonized by sclerostin, since they have been reported to mediate BMP-stimulated osteoblastic differentiation [151]. This would make sclerostin a Wnt antagonist similar to the DAN family member Cerberus [141]. Interestingly again, gain-of-function mutations in low-density lipoprotein receptor-related protein 5 (LRP5), a co-receptor for Wnt proteins, result in high bone mass due to increased bone formation [152, 153], a clinical feature very similar to sclerosteosis. Furthermore, the recently cloned Wnt antagonist Wise showed the highest amino acid identity with sclerostin [154].

The transmembrane factor BMP and activin membrane-bound inhibitor (Bambi) is related to TGF-β-family type I receptors, but lacks the intracellular kinase domain and has high sequence similarity to human NMA and zebrafish *nma* [155–157]. It was found to act as a pseudo type I receptor that inhibits signaling possibly by preventing type I receptor homomeric complex formation *via* its intracellular domain.

At the intracellular level, BMP signaling is inhibited at several levels. I-Smads, i.e. Smad6 and Smad7, interfere with TGF-β family intracellular signaling [54, 73, 74]. Smad7 functions as a general inhibitor of TGF-β, activin and BMP pathways, whereas Smad6 specifically inhibits the BMP signaling [158]. Overexpression of I-Smads in mesenchymal precursor cells potently interfered with BMP-induced osteoblast differentiation [40]. I-Smads interact efficiently with activated type I receptors, and the initial mechanism described for I-Smad antagonism was by competing with R-Smads for type I receptor interaction [73, 74, 159]. However, other mechanisms by which I-Smads antagonize TGF-β family/Smad pathways have now been described. Smad7 has been found to constitutively interact with HECT-domain ubiquitin ligase, Smurf2 [160] and more recently with Smurf1 as well [161]. Binding of Smad7 to Smurf induces the export of Smad7/Smurf complex from the nucleus. Upon recruitment of the complex to the activated TGF-β receptor, Smurf1 or Smurf2 induces TGF-β receptor degradation through proteosomal and lysosomal pathways. Smad7 may thus function as an adapter protein to mediate degradation of TGF-β receptor complex [160, 161]. Smurf2 has also been reported to bind Smad6 and target the BMP receptor for degradation [160]. Other mechanisms for Smad6-inhibition of BMP signaling have been proposed: (i) by competing with Smad4 for heteromeric complex formation with activated R-Smads [162]; (ii) by acting as a direct transcriptional corepressor [163], and; (iii) by inhibiting the action of TAK1, a MAPKKK implicated downstream of BMP receptor signaling to apoptosis [164]. Further studies are needed to determine the physiological importance of these inhibitory mechanisms for I-Smads. The activity of I-Smads themselves is under negative regulation of two cytoplasmic proteins, associated molecule with the SH3 of domain STAM (AMSH) and AMSH2, that inhibit binding of I-Smads to activated type I receptors or R-Smads [165, 166].

The group of Smad binding proteins that inhibit BMP/Smad signaling is continuously growing. Tob, a member of a family of antiproliferative proteins, was shown to bind R-Smads and to negatively regulate osteoblast proliferation and differentiation by suppressing the BMP R-Smads transcriptional activity [118]. Mice deficient in Tob showed increased bone mass due to increased numbers of osteoblasts. Another negative regulator is the transcriptional corepressor Ski, which can interact with Smad4 [167] and Smad1 or Smad5 through their MH2 domains [168]. Ectopic expression of Ski was found to inhibit BMP-2-induced osteoblast differentiation of murine W-20-17 cells [168]. SIP1, a member of the deltaEF1/Zfh-1 family of two-handed zinc finger/homeodomain proteins, also interacts with the MH2 domain of R-Smads and binds to 5'-CACCT sequences in promoters preventing transcription [169]. XMAN1, a family member of the inner nuclear membrane proteins implicated in gene regulation by interacting with chromatin, nuclear lamina, and intranuclear proteins, binds *via* its C-terminal region to Smad1, Smad5, and Smad8 and antagonizes BMP signaling [170]. Furthermore, the U-box-dependent E3 ubiquitin ligase CHIP mediates degradation of Smad1/Smad4 complexes [171]. Activation of extracellular signal-regulated kinase (ERK) can lead to inhibition of BMP signaling, since ERK MAPkinase mediated phosphorylation of Smad1 in its linker region was found to inhibit BMP-induced Smad1 nuclear accumulation [172]. Lastly, the Cas-interacting zinc finger protein (CIZ) that acts as a transcription factor or modulator *via* binding to consensus DNA elements, (G/C)AAAAA, has been shown to inhibit BMP/Smad signaling and to suppress BMP-stimulated osteoblastic differentiation [173].

BMP receptor-initiated signaling distinct from Smad activation

Ectopic expression of BMP R-Smads can recapitulate osteoblast differentiation, but not chondrogenic differentiation [40]. Thus, BMP-induced osteoblast differentiation appears to occur mainly *via* the Smad pathway, whereas BMP-induced chondrogenic differentiation is mediated *via* Smad-dependent and Smad-independent pathways [40]. Other pathways distinct from Smad pathway that are initiated downstream of ligand-induced activation of BMP receptor complex have been identified (Fig. 4). TGF-β-activated kinase 1 (TAK1), a MAP kinase kinase kinase (MKKK), can be activated by TAK1 binding protein (TAB1) in response to BMP and activate both SAPK and p38 pathways [174, 175]. X-chromosome-linked inhibitor of apoptosis (XIAP) may provide the direct link between TAB1 and type I receptor as it was shown to interact with both proteins [175]. p38 MAP kinase activation induces the

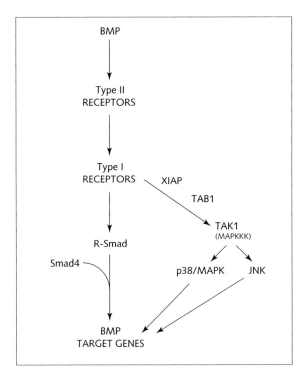

Figure 4
BMP signaling through Smad-dependent and Smad-independent pathways
BMP activates the Smad pathway as well as other signaling pathways. Abbreviations: JNK,
c-Jun N-terminal kinase; TAB, TAK-1 binding protein; TAK, TGF-β activated kinase; XIAP, X-
linked inhibitor of apoptosis protein.

phosphorylation of transcriptional factor ATF-2, and both ATF-2 and Smads were shown to act synergistically in transcriptional regulation [176]. BMP-induced apoptosis was shown to be mediated by TAK1-p38 kinase pathway [164]. In ATDC5 cells activation of p38 kinase by GDF-5 contributes to chondrogenesis [177]. Further studies are needed to demonstrate the physiological and general importance of Smad-independent pathways in BMP signal transduction.

BMP target genes

A number of extracellular matrix proteins, including osteocalcin, collagen type α, bone sialoprotein and decorin are potently induced by BMP [178–180]. Some of them may be direct targets for BMPs (such as collagen), whereas others (such as osteocalcin) are indirect and are only induced after prolonged exposure to BMPs (Tab. 2). The BMP-induced expression of alkaline phosphatase, a late BMP target gene, is often used as a read-out for BMP-induced osteoblast differentiation [40, 178–181]. BMPs were shown to activate osteopontin gene expression by preventing the binding of transcriptional repressor Hoxc-8 to the osteopontin promoter. Activated Smads can bind to Hoxc-8 and disloge the inhibitory Hoxc-8 from the DNA [182, 183]. In addition, a Smad binding region was identified in osteopontin promoter, and shown to be involved in BMP-mediated activation of this promoter [184]. BMP-induces expression of osteoprotegrin (OPG), an osteoblast-secreted decoy receptor, which specifically binds to the osteoclast differentiation factor and inhibits osteoclast maturation [185]. Interestingly, characterization of the OPG promoter revealed two homeodomain transcriptional factor Hoxc-8 binding sites that are essential for OPG promoter activation by BMP [185].

Connective tissue growth factor (CTGF), an important regulator of extracellular matrix formation, is also an early target gene induced by BMPs, as well as by TGF-β [186]. In the rat long bone growth plate the CTGF expression in chondroblasts is restricted to hypertrophyc region [186], which overlaps with the expression of BMP signaling components [52]. Recombinant CTGF promotes the proliferation and differentiation of chondrocytes and induces the expression of osteoblast-specific genes and bone mineralization [187]. While being viable, mice with targeted inactivation of CTGF suffers from skeletal dysmorphism due to impared chondrocyte proliferation, extracellular matrix deposition and angiogenesis in the hypertrophic chartilage zone [188]. Interestingly, CTGF directly binds both TGF-β1 and BMP4 and enhances binding of TGF-β1 to its receptors, while antagonizing binding of BMP-4 to its specific receptors, thus inhibiting BMP signaling [189].

Targeting inactivation of Indian hedgehog (Ihh) causes severe retardation of skeletal growth due to impaired chondrocyte proliferation and differentiation [190]. Recently Ihh was shown as a direct BMP-Smad target gene [191].

In many cell types (including osteoblasts), the expression of I-Smads (Smad6 and Smad7) is potently induced by BMPs [74, 192, 193]. In the Smad6 promoter a BMP responsive GC-rich elements has been identified [116]. All three BMP-responsive elements in Smad7 promoter [119] contain GGCGCC/CGCC and two of them CAGC sequences similar to such ones in Id1 promoter [101]. Interestingly, main BMP-responsive element in Smad7 promoter was found in the first intron and maximal activation of Smad7 expression requires cooperation between Smad1 and GATA transcription factors [119]. The BMP-induced I-Smads may serve a role in a negative feedback loop in Smad signaling to control the intensity and duration of BMP signaling response [74, 192, 193]. Another inhibitor of BMP function, pseudoreceptor Bambi, is also a direct BMP target gene and its expression during development overlaps with the BMP-4 expression pattern [104]. GGCGCC motifs in Bambi and Ihh promoters were shown to be crucially important for BMP-driven transcription of these genes [104, 191].

Table 2 - Genes induced by BMPs in osteoblasts or their precursors

BMP target gene	Defects resulting from gene inactivation
Components of ECM	
Ostecalcin [179]	Viable. Osteopetrosis [225].
Ostepontin [178]	Viable. Altered collagen fibrillogenesis and wound healing [226]. Resistant to to ovariectomy-induced osteoporosis [227].
Collagen Iα1 and α2 [180]	Viable. Osteogenesis imperfecta [228].
Bone sialoprotein [39]	Not determined.
Decoy receptor	
Osteoprotegrin [182, 184]	Not determined.
Enzymes	
Alkaline phosphatase [179]	Metabolic and skeletal defects. Infantile hypophosphatasia [229].
Growth factors	
CTGF* [186]	Skeletal dysmorphism due to impaired chondrocyte proliferation, extracellular proliferation and extracellular matrix composition and angiogenesis in hypertrophic zone [188].
Indian hedgehog* [191]	Viable. Severe reduction in skeletal growth due to reduced chondrocyte proliferation and differentiation [190].
Inhibitors of BMP function	
Bambi* [104]	Not determined.
Smad6* [73, 116, 193]	Cardiovascular abnormalities. Defects in endocardial cushio transformation [230].
Smad7* [74, 192]	Not determined.
Transcriptional regulators	
Msx2* [196, 203]	Viable. Defects in craniofacial bone ossification and endochondral bone formation. Tooth, mammary gland, cerebellum defects. Mutated in craniosynostosis patients. Haploinsufficiency causes parietal foramina [204].
Dlx5* [200]	Viable. Delayed membraneous ossification [201].
Id1*, Id2*, Id3* [101, 103, 195, 196]	Id1$^{-/-}$ Id3$^{-/-}$ and Id2$^{-/-}$ Id3$^{-/-}$ are not viable [231]. Haematopoietic and neural abnormalities.
JunB* [99, 194]	Embryonic death (E8.5-E10). Multiple defects in placental neovascularisation [232].
Cbfa1 [39, 181, 233]	Death after birth. No ossification. Skeleton made from chondrocytes only [125]. Mutated in CCD patients [127, 128].

*for this gene it has been demonstrated that it is a direct BMP target.

BMPs have also been found to induce many transcription factors (Tab. 2). JunB was shown in osteoblast precursor cells as a direct early BMP-2 target gene involved in the inhibition of myogenic differentiation [194]. Investigation of JunB promoter revealed the importance of multiple Smad binding elements through which this gene can be activated by ectopic BMP R-Smad expression [99]. BMPs induce the expression of helix-loop-helix proteins inhibitors of differentiation (Id) in osteoblasts and their precursors in part *via* transcriptional and post-transcriptional events [195, 196]. The induction of Id proteins by BMPs may indirectly support osteoblast differentiation of mesenchymal precursor cells by blocking their adipocyte [197] and myoblast differentiation [198, 199].

Mammalian homologs of the *Drosophila distalles* (dll) *Dlx5* and *Dlx6* are direct gene targets for BMP [200]. Overexpression of Dlx5 in cells induces their osteoblast differentiation while disruption of Dlx5 exhibits defects in the ossification of the membranous bones [201]. Dlx1 protein, that is also a BMP target gene, was shown as a Smad4 interaction partner that inhibits TGF-β, activin and BMP-4 signaling [202]. BMPs can directly induce the Msx-1 and Msx-2 homeobox genes. Mice deficient in Msx-2 [203] or Msx-1 have defects in the skull bones and show an overall decrease in bone mass [204]. Albeit not a direct BMP target, Cbfa1 induction by BMP is critical for BMP-induced osteogenesis [181]. Cbfa1 can induce extracellular matrix proteins, but Cbfa1 is not sufficient to induce the whole onset of osteoblastic differentiation without cooperation with Smad5 [129, 181]. Many new both direct and indirect target genes for BMP are reported as a result of cDNA micro array studies [205–211]. Functional importance of these targets in BMP-induced responses is an important aim for future studies.

Perspectives

Recent studies have demonstrated the pivotal role of BMP type I and type II receptors and their downstream Smad effectors in BMP-induced osteoblast differentiation. However, the molecular mechanisms that govern BMP-induced osteogenic differentiation need further study. In particular, physiological interactions between BMP family members with their receptors and Smads, and downstream gene targets in osteoblasts remain to be validated by comparing the phenotypes of mice deficient in a particular BMP ligand, receptor, Smad or target gene. In many cases a null mutation of a particular BMP signaling component leads to an embryonal lethal phenotype. Conditional knockout approaches will therefore be required to study the role of these molecules in bone formation. The repertoire of BMP Smad interacting proteins in different osteoblast (precursor) cell types or at different states of their differentiation need to be elucidated. In addition, the genetic programs that are initiated in mesenchymal precursor cells, chondroblasts and osteoblasts upon BMP stimulation *via* the various BMP intracellular pathways need to be determined. To analyze this efficiently functional genomics technologies will be useful. This approach should provide an answer to the question why different BMP type I receptors, although activating the same set of Smad proteins, can induce distinct biological responses. Stimulating mesenchymal precursor cells with a constitutively active (ca) BMPR-IA induces adipocyte differentiation, whereas (ca) BMPR-IB induces osteoblast differentiation and apoptosis [39]. During limb bud morphogenesis in the chick, BMPR-IA was found to mediate osteogenesis whereas BMPR-IB induced preferentially chondrocyte differentiation [45].

BMPs in animal models have shown to be very effective in bone repair [8, 9, 212]. Adenoviral BMP-7 gene transfer [213] and BMP-4 plasmid implantation into bone [214] have been succesfully used in mouse models of osteogenic induction. However, clinical use of BMPs as regenerative agents in humans has thus far been limited; there is a need of using high doses of BMPs to get specific effects, if any. With the elucidation of the BMP/Smad pathway numerous inhibitors of BMP signaling have been identified. An interesting possibility, which remains to be explored, is that by inhibiting the action of antagonists, like extracellular sclerostin or noggin and the intracellular I-Smad, BMP signaling can be potentiated, thereby making BMPs more effective therapeutic agents.

Acknowledgements
Our studies on BMP receptor and Smad signal transduction are supported by the Netherlands Organization for Scientific Research (NWO 916.046.017 to RvB and MW902-16-295 to PtD).

References

1   Urist MR (1965) Bone: formation by autoinduction. *Science* 150: 893–899

2   Wozney JM, Rosen V, Celeste AJ, Mitsock LM, Whitters MJ, Kriz RW, Hewick RM, Wang EA (1988) Novel regulators of bone formation: molecular clones and activities. *Science* 242: 1528–1534

3   Sampath TK, Maliakal JC, Hauschka PV, Jones WK, Sasak H, Tucker RF, White KH, Coughlin JE, Tucker MM, Pang RH et al (1992) Recombinant human osteogenic protein-1 (hOP-1) induces new bone formation *in vivo* with a specific activity comparable with natural bovine osteogenic protein and stimulates osteoblast proliferation and differentiation *in vitro*. *J Biol Chem* 267: 20352–20362

4   Vukicevic S, Luyten FP, Reddi AH (1989) Stimulation of the expression of osteogenic and chondrogenic phenotypes *in vitro* by osteogenin. *Proc Natl Acad Sci USA* 86: 8793–8797

5   Massagué J (1990) The transforming growth factor-β family. *Annu Rev Cell Biol* 6: 597–641

6   Hotten GC, Matsumoto T, Kimura M, Bechtold RF, Kron R, Ohara T, Tanaka H, Satoh Y, Okazaki M, Shirai T, Pan H, Kawai S, Pohl JS, Kudo A (1996) Recombinant human growth/differentiation factor 5 stimulates mesenchyme aggregation and chondrogenesis responsible for the skeletal development of limbs. *Growth Factors* 13: 65–74

7   Wolfman NM, Hattersley G, Cox K, Celeste AJ, Nelson R, Yamaji N, Dube JL, DiBlasio-Smith E, Nove J, Song JJ et al (1997) Ectopic induction of tendon and ligament in rats by growth and differentiation factors 5, 6, and 7, members of the TGF-β gene family. *J Clin Invest* 100: 321–330

8   Reddi AH (1994) Symbiosis of biotechnology and biomaterials: applications in tissue engineering of bone and cartilage. *J Cell Biochem* 56: 192–195

9   Reddi AH (1998) Role of morphogenetic proteins in skeletal tissue engineering and regeneration. *Nat Biotechnol* 16: 247–252

10   Cunningham NS, Paralkar V, Reddi AH (1992) Osteogenin and recombinant bone morphogenetic protein 2B are chemotactic for human monocytes and stimulate transforming growth factor β1 mRNA expression. *Proc Natl Acad Sci USA* 89: 11740–11744

11   Hogan BL (1996) Bone morphogenetic proteins in development. *Curr Opin Genet Dev* 6: 432–438

12   Goumans MJ and Mummery C (2000) Functional analysis of the TGFβ receptor/Smad pathway through gene ablation in mice. *Int J Dev Biol* 44: 253–265

13   Kingsley DM, Bland AE, Grubber JM, Marker PC, Russell LB, Copeland NG, Jenkins NA (1992) The mouse short ear skeletal morphogenesis locus is associated with defects in a bone morphogenetic member of the TGF β superfamily. *Cell* 71: 399–410

14   Storm EE, Huynh TV, Copeland NG, Jenkins NA, Kingsley DM, Lee SJ (1994) Limb alterations in brachypodism mice due to mutations in a new member of the TGF β-superfamily. *Nature* 368: 639–643

15   McPherron AC and Lee SJ (1997) Double muscling in cattle due to mutations in the myostatin gene. *Proc Natl Acad Sci USA* 94: 12457–12461

16   Thomas JT, Lin K, Nandedkar M, Camargo M, Cervenka J, Luyten FP (1996) A human chondrodysplasia due to a mutation in a TGF-β superfamily member. *Nat Genet* 12: 315–317

17   Massagué J (1998) TGF-β signal transduction. *Annu Rev Biochem* 67: 753–791

18   Heldin CH, Miyazono K, ten Dijke P (1997) TGF-β signalling from cell membrane to nucleus through SMAD proteins. *Nature* 390: 465–471

19   Mathews LS and Vale WW (1991) Expression cloning of an activin receptor, a predicted transmembrane serine kinase. *Cell* 65: 973–982

20   Lin HY, Wang XF, Ng-Eaton E, Weinberg RA, Lodish HF (1992) Expression cloning of the TGF-β type II receptor, a functional transmembrane serine/threonine kinase. *Cell* 68: 775–785

21   Ebner R, Chen RH, Shum L, Lawler S, Zioncheck TF, Lee A, Lopez AR, Derynck R (1993) Cloning of a type I TGF-β receptor and its effect on TGF-β binding to the type II receptor. *Science* 260: 1344–1348

22   Attisano L, Cárcamo J, Ventura F, Weis FM, Massagué J, Wrana JL (1993) Identifica-

tion of human activin and TGF β type I receptors that form heteromeric kinase complexes with type II receptors. *Cell* 75: 671–680

23 Franzén P, ten Dijke P, Ichijo H, Yamashita H, Schulz P, Heldin CH, Miyazono K (1993) Cloning of a TGF β type I receptor that forms a heteromeric complex with the TGF β type II receptor. *Cell* 75: 681–692

24 ten Dijke P, Yamashita H, Sampath TK, Reddi AH, Estevez M, Riddle DL, Ichijo H, Heldin CH, Miyazono K (1994) Identification of type I receptors for osteogenic protein-1 and bone morphogenetic protein-4. *J Biol Chem* 269: 16985–16988

25 ten Dijke P, Yamashita H, Ichijo H, Franzén P, Laiho M, Miyazono K, Heldin CH (1994) Characterization of type I receptors for transforming growth factor-β and activin. *Science* 264: 101–104

26 Nohno T, Ishikawa T, Saito T, Hosokawa K, Noji S, Wolsing DH, Rosenbaum JS (1995) Identification of a human type II receptor for bone morphogenetic protein-4 that forms differential heteromeric complexes with bone morphogenetic protein type I receptors. *J Biol Chem* 270: 22522–22526

27 Rosenzweig BL, Imamura T, Okadome T, Cox GN, Yamashita H, ten Dijke P, Heldin CH, Miyazono K (1995) Cloning and characterization of a human type II receptor for bone morphogenetic proteins. *Proc Natl Acad Sci USA* 92: 7632–7636

28 Liu F, Ventura F, Doody J, Massagué J (1995) Human type II receptor for bone morphogenic proteins (BMPs): extension of the two-kinase receptor model to the BMPs. *Mol Cell Biol* 15: 3479–3486

29 Koenig BB, Cook JS, Wolsing DH, Ting J, Tiesman JP, Correa PE, Olson CA, Pecquet AL, Ventura F, Grant RA (1994) Characterization and cloning of a receptor for BMP-2 and BMP-4 from NIH 3T3 cells. *Mol Cell Biol* 14: 5961–5974

30 Yamashita H, ten Dijke P, Huylebroeck D, Sampath TK, Andries M, Smith JC, Heldin CH, Miyazono K (1995) Osteogenic protein-1 binds to activin type II receptors and induces certain activin-like effects. *J Cell Biol* 130: 217–226

31 ten Dijke P, Ichijo H, Franzén P, Schulz P, Saras J, Toyoshima H, Heldin CH, Miyazono K (1993) Activin receptor-like kinases: a novel subclass of cell-surface receptors with predicted serine/threonine kinase activity. *Oncogene* 8: 2879–2887

32 Wrana JL, Attisano L, Wieser R, Ventura F, Massagué J (1994) Mechanism of activation of the TGF-β receptor. *Nature* 370: 341–347

33 Miettinen PJ, Ebner R, Lopez AR, Derynck R (1994) TGF-β induced transdifferentiation of mammary epithelial cells to mesenchymal cells: involvement of type I receptors. *J Cell Biol* 127: 2021–2036

34 Macías-Silva M, Hoodless PA, Tang SJ, Buchwald M, Wrana JL (1998) Specific activation of Smad1 signaling pathways by the BMP7 type I receptor, ALK2. *J Biol Chem* 273: 25628–25636

35 Armes NA and Smith JC (1997) The ALK-2 and ALK-4 activin receptors transduce distinct mesoderm-inducing signals during early Xenopus development but do not cooperate to establish thresholds. *Development* 124: 3797–3804

36 Nishitoh H, Ichijo H, Kimura M, Matsumoto T, Makishima F, Yamaguchi A, Yamashita

H, Enomoto S, Miyazono K (1996) Identification of type I and type II serine/threonine kinase receptors for growth/differentiation factor-5. *J Biol Chem* 271: 21345–21352

37 Akiyama S, Katagiri T, Namiki M, Yamaji N, Yamamoto N, Miyama K, Shibuya H, Ueno N, Wozney JM, Suda T (1997) Constitutively active BMP type I receptors transduce BMP-2 signals without the ligand in C2C12 myoblasts. *Exp Cell Res* 235: 362–369

38 Namiki M, Akiyama S, Katagiri T, Suzuki A, Ueno N, Yamaji N, Rosen V, Wozney JM, Suda T (1997) A kinase domain-truncated type I receptor blocks bone morphogenetic protein-2-induced signal transduction in C2C12 myoblasts. *J Biol Chem* 272: 22046–22052

39 Chen D, Ji X, Harris MA, Feng JQ, Karsenty G, Celeste AJ, Rosen V, Mundy GR, Harris SE (1998) Differential roles for bone morphogenetic protein (BMP) receptor type IB and IA in differentiation and specification of mesenchymal precursor cells to osteoblast and adipocyte lineages. *J Cell Biol* 142: 295–305

40 Fujii M, Takeda K, Imamura T, Aoki H, Sampath TK, Enomoto S, Kawabata M, Kato M, Ichijo H, Miyazono K (1999) Roles of bone morphogenetic protein type I receptors and Smad proteins in osteoblast and chondroblast differentiation. *Mol Biol Cell* 10: 3801–3813

41 Daluiski A, Engstrand T, Bahamonde ME, Gamer LW, Agius E, Stevenson SL, Cox K, Rosen V, Lyons KM (2001) Bone morphogenetic protein-3 is a negative regulator of bone density. *Nat Genet* 27: 84–88

42 Gu Z, Reynolds EM, Song J, Lei H, Feijen A, Yu L, He W, MacLaughlin DT, van den Eijnden-van Raaij, Donahoe PK et al (1999) The type I serine/threonine kinase receptor ActRIA (ALK2) is required for gastrulation of the mouse embryo. *Development* 126: 2551–2561

43 Verschueren K, Dewulf N, Goumans MJ, Lonnoy O, Feijen A, Grimsby S, Vandi Spiegle K, ten Dijke P, Morén A, Vanscheeuwijck P et al (1995) Expression of type I and type IB receptors for activin in midgestation mouse embryos suggests distinct functions in organogenesis. *Mech Dev* 52: 109–123

44 Dewulf N, Verschueren K, Lonnoy O, Morén A, Grimsby S, Vande Spiegle K, Miyazono K, Huylebroeck D, ten Dijke P (1995) Distinct spatial and temporal expression patterns of two type I receptors for bone morphogenetic proteins during mouse embryogenesis. *Endocrinology* 136: 2652–2663

45 Zou H, Wieser R, Massagué J, Niswander L (1997) Distinct roles of type I bone morphogenetic protein receptors in the formation and differentiation of cartilage. *Genes Dev* 11: 2191–2203

46 Yi SE, Daluiski A, Pederson R, Rosen V, Lyons KM (2000) The type I BMP receptor BMPRIB is required for chondrogenesis in the mouse limb. *Development* 127: 621–630

47 Manova K, De Leon V, Angeles M, Kalantry S, Giarre M, Attisano L, Wrana J, Bachvarova RF (1995) mRNAs for activin receptors II and IIB are expressed in mouse oocytes and in the epiblast of pregastrula and gastrula stage mouse embryos. *Mech Dev* 49: 3–11

48   Matzuk MM, Kumar TR, Bradley A (1995) Different phenotypes for mice deficient in either activins or activin receptor type II. *Nature* 374: 356–360

49   Beppu H, Kawabata M, Hamamoto T, Chytil A, Minowa O, Noda T, Miyazono K (2000) BMP type II receptor is required for gastrulation and early development of mouse embryos. *Dev Biol* 221: 249–258

50   Roelen BA, Goumans MJ, van Rooijen MA, Mummery CL (1997) Differential expression of BMP receptors in early mouse development. *Int J Dev Biol* 41: 541–549

51   Yonemori K, Imamura T, Ishidou Y, Okano T, Matsunaga S, Yoshida H, Kato M, Sampath TK, Miyazono K, ten Dijke P et al (1997) Bone morphogenetic protein receptors and activin receptors are highly expressed in ossified ligament tissues of patients with ossification of the posterior longitudinal ligament. *Am J Pathol* 150: 1335–1347

52   Sakou T, Onishi T, Yamamoto T, Nagamine T, Sampath T, ten Dijke P (1999) Localization of Smads, the TGF-β family intracellular signaling components during endochondral ossification. *J Bone Miner Res* 14: 1145–1152

53   Ishidou Y, Kitajima I, Obama H, Maruyama I, Murata F, Imamura T, Yamada N, ten Dijke P, Miyazono K, Sakou T (1995) Enhanced expression of type I receptors for bone morphogenetic proteins during bone formation. *J Bone Miner Res* 10: 1651–1659

54   Hayashi K, Ishidou Y, Yonemori K, Nagamine T, Origuchi N, Maeda S, Imamura T, Kato M, Yoshida H, Sampath TK et al (1997) Expression and localization of bone morphogenetic proteins (BMPs) and BMP receptors in ossification of the ligamentum flavum. *Bone* 21: 23–30

55   Okano T, Ishidou Y, Kato M, Imamura T, Yonemori K, Origuchi N, Matsunaga S, Yoshida H, ten Dijke P, Sakou T (1997) Orthotopic ossification of the spinal ligaments of Zucker fatty rats: a possible animal model for ossification of the human posterior longitudinal ligament. *J Orthop Res* 15: 820–829

56   Mishina Y, Suzuki A, Ueno N, Behringer RR (1995) Bmpr encodes a type I bone morphogenetic protein receptor that is essential for gastrulation during mouse embryogenesis. *Genes Dev* 9: 3027–3037

57   Winnier G, Blessing M, Labosky PA, Hogan BL (1995) Bone morphogenetic protein-4 is required for mesoderm formation and patterning in the mouse. *Genes Dev* 9: 2105–2116

58   Oh SP, Li E (1997) The signaling pathway mediated by the type IIB activin receptor controls axial patterning and lateral asymmetry in the mouse. *Genes Dev* 11: 1812–1826

59   Song J, Oh SP, Schrewe H, Nomura M, Lei H, Okano M, Gridley T, Li E (1999) The type II activin receptors are essential for egg cylinder growth, gastrulation, and rostral head development in mice. *Dev Biol* 213: 157–169

60   Gilboa L, Nohe A, Geissendorfer T, Sebald W, Henis YI, Knaus P (2000) Bone morphogenetic protein receptor complexes on the surface of live cells: a new oligomerization mode for serine/threonine kinase receptors. *Mol Biol Cell* 11: 1023–1035

61   Wrana JL, Attisano L, Cárcamo J, Zentella A, Doody J, Laiho M, Wang XF, Massagué J (1992) TGF β signals through a heteromeric protein kinase receptor complex. *Cell* 71: 1003–1014

62   Cárcamo J, Weis FM, Ventura F, Wieser R, Wrana JL, Attisano L, Massagué J (1994) Type I receptors specify growth-inhibitory and transcriptional responses to transforming growth factor β and activin. *Mol Cell Biol* 14: 3810–3821

63   Chen YG, Hata A, Lo RS, Wotton D, Shi Y, Pavletich N, Massagué J (1998) Determinants of specificity in TGF-β signal transduction. *Genes Dev* 12: 2144–2152

64   Feng XH, Derynck R (1997) A kinase subdomain of transforming growth factor-β (TGF-β) type I receptor determines the TGF-β intracellular signaling specificity. *EMBO J* 16: 3912–3923

65   Persson U, Izumi H, Souchelnytskyi S, Itoh S, Grimsby S, Engstrom U, Heldin CH, Funa K, ten Dijke P (1998) The L45 loop in type I receptors for TGF-β family members is a critical determinant in specifying Smad isoform activation. *FEBS Lett* 434: 83–87

66   Lane KB, Machado RD, Pauciulo MW, Thomson JR, Phillips JA, Loyd JE, Nichols WC, Trembath RC (2000) Heterozygous germline mutations in BMPR2, encoding a TGF-β receptor, cause familial primary pulmonary hypertension. The International PPH Consortium. *Nat Genet* 26: 81–84

67   Machado RD, Pauciulo MW, Thomson JR, Lane KB, Morgan NV, Wheeler L, Phillips JA, Newman J, Williams D, Galie N et al (2001) BMPR2 haploinsufficiency as the inherited molecular mechanism for primary pulmonary hypertension. *Am J Hum Genet* 68: 92–102

68   Thomson JR, Machado RD, Pauciulo MW, Morgan NV, Humbert M, Elliott GC, Ward K, Yacoub M, Mikhail G, Rogers P et al (2000) Sporadic primary pulmonary hypertension is associated with germline mutations of the gene encoding BMPR-II, a receptor member of the TGF-β family. *J Med Genet* 37: 741–745

69   Wilkins MR, Gibbs JS, Shovlin CL (2000) A gene for primary pulmonary hypertension. *Lancet* 356: 1207–1208

70   Foletta VC, Lim MA, Soosairajah J, Kelly AP, Stanley EG, Shannon M, He W, Das S, Massagué J, Bernard O, Soosairaiah J (2003) Direct signaling by the BMP type II receptor *via* the cytoskeletal regulator LIMK1. *J Cell Biol* 162: 1089–1098

71   Sekelsky JJ, Newfeld SJ, Raftery LA, Chartoff EH, Gelbart WM (1995) Genetic characterization and cloning of mothers against dpp, a gene required for decapentaplegic function in *Drosophila melanogaster*. *Genetics* 139: 1347–1358

72   Savage C, Das P, Finelli AL, Townsend SR, Sun CY, Baird SE, Padgett RW (1996) Caenorhabditis elegans genes sma-2, sma-3, and sma-4 define a conserved family of transforming growth factor β pathway components. *Proc Natl Acad Sci USA* 93: 790–794

73   Imamura T, Takase M, Nishihara A, Oeda E, Hanai J, Kawabata M, Miyazono K (1997) Smad6 inhibits signalling by the TGF-β superfamily. *Nature* 389: 622–626

74   Nakao A, Afrakhte M, Morén A, Nakayama T, Christian JL, Heuchel R, Itoh S, Kawabata M, Heldin NE, Heldin CH et al (1997) Identification of Smad7, a TGFβ-inducible antagonist of TGF-β signalling. *Nature* 389: 631–635

75   Lo RS, Chen YG, Shi Y, Pavletich NP, Massagué J (1998) The L3 loop: a structural motif

determining specific interactions between SMAD proteins and TGF-β receptors. *EMBO J* 17: 996–1005

76 Tsukazaki T, Chiang TA, Davison AF, Attisano L, Wrana JL (1998) SARA, a FYVE domain protein that recruits Smad2 to the TGFβ receptor. *Cell* 95: 779–791

77 Miura S, Takeshita T, Asao H, Kimura Y, Murata K, Sasaki Y, Hanai JI, Beppu H, Tsukazaki T, Wrana JL et al (2000) Hgs (Hrs), a FYVE domain protein, is involved in Smad signaling through cooperation with SARA. *Mol Cell Biol* 20: 9346–9355

78 Abdollah S, Macías-Silva M, Tsukazaki T, Hayashi H, Attisano L, Wrana JL (1997) TβRI phosphorylation of Smad2 on Ser465 and Ser467 is required for Smad2-Smad4 complex formation and signaling. *J Biol Chem* 272: 27678–27685

79 Kretzschmar M, Liu F, Hata A, Doody J, Massagué J (1997) The TGF-β family mediator Smad1 is phosphorylated directly and activated functionally by the BMP receptor kinase. *Genes Dev* 11: 984–995

80 Souchelnytskyi S, Tamaki K, Engstrom U, Wernstedt C, ten Dijke P, Heldin CH (1997) Phosphorylation of Ser465 and Ser467 in the C terminus of Smad2 mediates interaction with Smad4 and is required for transforming growth factor-β signaling. *J Biol Chem* 272: 28107–28115

81 Ebisawa T, Tada K, Kitajima I, Tojo K, Sampath TK, Kawabata M, Miyazono K, Imamura T (1999) Characterization of bone morphogenetic protein-6 signaling pathways in osteoblast differentiation. *J Cell Sci* 112: 3519–3527

82 Nishimura R, Kato Y, Chen D, Harris SE, Mundy GR, Yoneda T (1998) Smad5 and DPC4 are key molecules in mediating BMP-2-induced osteoblastic differentiation of the pluripotent mesenchymal precursor cell line C2C12. *J Biol Chem* 273: 1872–1879

83 Tamaki K, Souchelnytskyi S, Itoh S, Nakao A, Sampath K, Heldin CH, ten Dijke P (1998) Intracellular signaling of osteogenic protein-1 through Smad5 activation. *J Cell Physiol* 177: 355–363

84 Lagna G, Hata A, Hemmati-Brivanlou A, Massagué J (1996) Partnership between DPC4 and SMAD proteins in TGF-β signalling pathways. Nature 383: 832–836

85 Correia JJ, Chacko BM, Lam SS, Lin K (2001) Sedimentation studies reveal a direct role of phosphorylation in Smad3:Smad4 homo- and hetero-trimerization. *Biochemistry* 40: 1473–1482

86 Kawabata M, Inoue H, Hanyu A, Imamura T, Miyazono K (1998) Smad proteins exist as monomers *in vivo* and undergo homo- and hetero-oligomerization upon activation by serine/threonine kinase receptors. *EMBO J* 17: 4056–4065

87 Shi Y, Hata A, Lo RS, Massagué J, Pavletich NP (1997) A structural basis for mutational inactivation of the tumour suppressor Smad4. *Nature* 388: 87–93

88 Xiao Z, Liu X, Henis YI, Lodish HF (2000) A distinct nuclear localization signal in the N terminus of Smad 3 determines its ligand-induced nuclear translocation. *Proc Natl Acad Sci USA* 97: 7853–7858

89 Xiao Z, Liu X, Lodish HF (2000) Importin β mediates nuclear translocation of Smad 3. *J Biol Chem* 275: 23425–23428

90 Pierreux CE, Nicolas FJ, Hill CS (2000) Transforming growth factor β-independent shuttling of Smad4 between the cytoplasm and nucleus. *Mol Cell Biol* 20: 9041–9054

91 Watanabe M, Masuyama N, Fukuda M, Nishida E (2000) Regulation of intracellular dynamics of Smad4 by its leucine-rich nuclear export signal. *EMBO Rep* 1: 176–182

92 Massagué J, Wotton D (2000) Transcriptional control by the TGF-β/Smad signaling system. *EMBO J* 19: 1745–1754

93 ten Dijke P, Miyazono K, Heldin CH (2000) Signaling inputs converge on nuclear effectors in TGF-β signaling. *Trends Biochem Sci* 25: 64–70

94 Flanders KC, Kim ES, Roberts AB (2001) Immunohistochemical expression of Smads 1-6 in the 15-day gestation mouse embryo: signaling by BMPs and TGF-βs. *Dev Dyn* 220: 141–154

95 Sirard C, de la Pompa JL, Elia A, Itie A, Mirtsos C, Cheung A, Hahn S, Wakeham A, Schwartz L, Kern SE et al (1998) The tumor suppressor gene Smad4/Dpc4 is required for gastrulation and later for anterior development of the mouse embryo. *Genes Dev* 12: 107–119

96 Zhu H, Kavsak P, Abdollah S, Wrana JL, Thomsen GH (1999) A SMAD ubiquitin ligase targets the BMP pathway and affects embryonic pattern formation. *Nature* 400: 687–693

97 Zhao M, Qiao M, Harris SE, Oyajobi BO, Mundy GR, Chen D (2003) Smurf1 inhibits osteoblast differentiation and bone formation *in vitro* and *in vivo*. *J Biol Chem; in press*

98 Dennler S, Itoh S, Vivien D, ten Dijke P, Huet S, Gauthier JM (1998) Direct binding of Smad3 and Smad4 to critical TGF β-inducible elements in the promoter of human plasminogen activator inhibitor-type 1 gene. *EMBO J* 17: 3091–3100

99 Jonk LJ, Itoh S, Heldin CH, ten Dijke P, Kruijer W (1998) Identification and functional characterization of a Smad binding element (SBE) in the JunB promoter that acts as a transforming growth factor-β, activin, and bone morphogenetic protein-inducible enhancer. *J Biol Chem* 273: 21145–21152

100 Yingling JM, Datto MB, Wong C, Frederick JP, Liberati NT, Wang XF (1997) Tumor suppressor Smad4 is a transforming growth factor β-inducible DNA binding protein. *Mol Cell Biol* 17: 7019–7028

101 Korchynskyi O, ten Dijke P (2002) Identification and functional characterization of distinct critically important bone morphogenetic protein-specific response elements in the Id1 promoter. *J Biol Chem* 277: 4883–4891

102 Zawel L, Dai JL, Buckhaults P, Zhou S, Kinzler KW, Vogelstein B, Kern SE (1998) Human Smad3 and Smad4 are sequence-specific transcription activators. *Mol Cell* 1: 611–617

103 Lopez-Rovira T, Chalaux E, Massagué J, Rosa JL, Ventura F (2002) Direct binding of Smad1 and Smad4 to two distinct motifs mediates bone morphogenetic protein-specific transcriptional activation of Id1 gene. *J Biol Chem* 277: 3176–3185

104 Karaulanov E, Knochel W, Niehrs C (2004) Transcriptional regulation of BMP4 synexpression in transgenic Xenopus. *EMBO J* 23: 344–356

105 Brodin G, Ahgren A, ten Dijke P, Heldin CH, Heuchel R (2000) Efficient TGF-β induc-

tion of the Smad7 gene requires cooperation between AP-1, Sp1, and Smad proteins on the mouse Smad7 promoter. *J Biol Chem* 275: 29023–29030

106 Denissova NG, Pouponnot C, Long J, He D, Liu F (2000) Transforming growth factor β-inducible independent binding of SMAD to the Smad7 promoter. *Proc Natl Acad Sci USA* 97: 6397–6402

107 Nagarajan RP, Zhang J, Li W, Chen Y (1999) Regulation of Smad7 promoter by direct association with Smad3 and Smad4. *J Biol Chem* 274: 33412–33418

108 Stopa M, Anhuf D, Terstegen L, Gatsios P, Gressner AM, Dooley S (2000) Participation of Smad2, Smad3, and Smad4 in transforming growth factor β (TGF-β)-induced activation of Smad7. THE TGF-β response element of the promoter requires functional Smad binding element and E-box sequences for transcriptional regulation. *J Biol Chem* 275: 29308–29317

109 von Gersdorff G, Susztak K, Rezvani F, Bitzer M, Liang D, Bottinger EP (2000) Smad3 and Smad4 mediate transcriptional activation of the human Smad7 promoter by transforming growth factor β. *J Biol Chem* 275: 11320–11326

110 Stroschein SL, Wang W, Luo K (1999) Cooperative binding of Smad proteins to two adjacent DNA elements in the plasminogen activator inhibitor-1 promoter mediates transforming growth factor β-induced smad-dependent transcriptional activation. *J Biol Chem* 274: 9431–9441

111 Chen SJ, Yuan W, Lo S, Trojanowska M, Varga J (2000) Interaction of smad3 with a proximal smad-binding element of the human α2(I) procollagen gene promoter required for transcriptional activation by TGF-β. *J Cell Physiol* 183: 381–392

112 Vindevoghel L, Lechleider RJ, Kon A, de Caestecker MP, Uitto J, Roberts AB, Mauviel A (1998) SMAD3/4-dependent transcriptional activation of the human type VII collagen gene (COL7A1) promoter by transforming growth factor β. *Proc Natl Acad Sci USA* 95: 14769–14774

113 Shi Y, Wang YF, Jayaraman L, Yang H, Massagué J, Pavletich NP (1998) Crystal structure of a Smad MH1 domain bound to DNA: insights on DNA binding in TGF-β signaling. *Cell* 94: 585–594

114 Henningfeld KA, Rastegar S, Adler G, Knochel W (2000) Smad1 and Smad4 are components of the bone morphogenetic protein-4 (BMP-4)-induced transcription complex of the Xvent-2B promoter. *J Biol Chem* 275: 21827–21835

115 Kim J, Johnson K, Chen HJ, Carroll S, Laughon A (1997) Drosophila Mad binds to DNA and directly mediates activation of vestigial by Decapentaplegic. *Nature* 388: 304–308

116 Ishida W, Hamamoto T, Kusanagi K, Yagi K, Kawabata M, Takehara K, Sampath TK, Kato M, Miyazono K (2000) Smad6 is a Smad1/5-induced smad inhibitor. Characterization of bone morphogenetic protein-responsive element in the mouse Smad6 promoter. *J Biol Chem* 275: 6075–6079

117 Kusanagi K, Inoue H, Ishidou Y, Mishima HK, Kawabata M, Miyazono K (2000) Characterization of a bone morphogenetic protein-responsive Smad-binding element. *Mol Biol Cell* 11: 555–565

118 Yoshida Y, Tanaka S, Umemori H, Minowa O, Usui M, Ikematsu N, Hosoda E, Imamura T, Kuno J, Yamashita T et al (2000) Negative Regulation of BMP/Smad Signaling by Tob in Osteoblasts. *Cell* 103: 1085–1097

119 Benchabane H, Wrana JL (2003) GATA- and Smad1-dependent enhancers in the Smad7 gene differentially interpret bone morphogenetic protein concentrations. *Mol Cell Biol* 23: 6646–6661

120 Itoh F, Itoh S, Goumans MJ, Valdimarsdottir G, Iso T, Dotto GP, Hamamori Y, Kedes L, Kato M, Dijke Pt P (2004) Synergy and antagonism between Notch and BMP receptor signaling pathways in endothelial cells. *EMBO J* 23: 541–551

121 Derynck R, Zhang Y, Feng XH (1998) Smads: transcriptional activators of TGF-β responses. *Cell* 95: 737–740

122 Hata A, Seoane J, Lagna G, Montalvo E, Hemmati-Brivanlou A, Massagué J (2000) OAZ uses distinct DNA- and protein-binding zinc fingers in separate BMP-Smad and Olf signaling pathways. *Cell* 100: 229–240

123 Hanai J, Chen LF, Kanno T, Ohtani-Fujita N, Kim WY, Guo WH, Imamura T, Ishidou Y, Fukuchi M, Shi MJ et al (1999) Interaction and functional cooperation of PEBP2/CBF with Smads. Synergistic induction of the immunoglobulin germline Cα promoter. *J Biol Chem* 274: 31577–31582

124 Pardali E, Xie XQ, Tsapogas P, Itoh S, Arvanitidis K, Heldin CH, ten Dijke P, Grundstrom T, Sideras P (2000) Smad and AML proteins synergistically confer transforming growth factor β1 responsiveness to human germ-line IgA genes. *J Biol Chem* 275: 3552–3560

125 Komori T, Yagi H, Nomura S, Yamaguchi A, Sasaki K, Deguchi K, Shimizu Y, Bronson RT, Gao YH, Inada M et al (1997) Targeted disruption of Cbfa1 results in a complete lack of bone formation owing to maturational arrest of osteoblasts. *Cell* 89: 755–764

126 Ducy P, Starbuck M, Priemel M, Shen J, Pinero G, Geoffroy V, Amling M, Karsenty G (1999) A Cbfa1-dependent genetic pathway controls bone formation beyond embryonic development. *Genes Dev* 13: 1025–1036

127 Mundlos S, Mulliken JB, Abramson DL, Warman ML, Knoll JH, Olsen BR (1995) Genetic mapping of cleidocranial dysplasia and evidence of a microdeletion in one family. *Hum Mol Genet* 4: 71–75

128 Mundlos S, Otto F, Mundlos C, Mulliken JB, Aylsworth AS, Albright S, Lindhout D, Cole WG, Henn W, Knoll JH et al (1997) Mutations involving the transcription factor CBFA1 cause cleidocranial dysplasia. *Cell* 89: 773–779

129 Zhang YW, Yasui N, Ito K, Huang G, Fujii M, Hanai J, Nogami H, Ochi T, Miyazono K, Ito Y (2000) A RUNX2/PEBP2α A/CBFA1 mutation displaying impaired transactivation and Smad interaction in cleidocranial dysplasia. *Proc Natl Acad Sci USA* 97: 10549–10554

130 Liu F, Hata A, Baker JC, Doody J, Cárcamo J, Harland RM, Massagué J (1996) A human Mad protein acting as a BMP-regulated transcriptional activator. *Nature* 381: 620–623

131 Meersseman G, Verschueren K, Nelles L, Blumenstock C, Kraft H, Wuytens G, Remacle

J, Kozak CA, Tylzanowski P, Niehrs C et al (1997) The C-terminal domain of Mad-like signal transducers is sufficient for biological activity in the Xenopus embryo and transcriptional activation. *Mech Dev* 61: 127–140

132 Pouponnot C, Jayaraman L, Massagué J (1998) Physical and functional interaction of SMADs and p300/CBP. *J Biol Chem* 273: 22865–22868

133 Yahata T, de Caestecker MP, Lechleider RJ, Andriole S, Roberts AB, Isselbacher KJ, Shioda T (2000) The MSG1 non-DNA-binding transactivator binds to the p300/CBP coactivators, enhancing their functional link to the Smad transcription factors. *J Biol Chem* 275: 8825–8834

134 Postigo AA, Depp JL, Taylor JJ, Kroll KL (2003) Regulation of Smad signaling through a differential recruitment of coactivators and corepressors by ZEB proteins. *EMBO J* 22: 2453–2462

135 Postigo AA (2003) Opposing functions of ZEB proteins in the regulation of the TGFβ/BMP signaling pathway. *EMBO J* 22: 2443–2452

136 Bai RY, Koester C, Ouyang T, Hahn SA, Hammerschmidt M, Peschel C, Duyster J (2002) SMIF, a Smad4-interacting protein that functions as a co-activator in TGFβ signalling. *Nat Cell Biol* 4: 181–190

137 Nakashima K, Yanagisawa M, Arakawa H, Kimura N, Hisatsune T, Kawabata M, Miyazono K, Taga T (1999) Synergistic signaling in fetal brain by STAT3-Smad1 complex bridged by p300. *Science* 284: 479–482

138 Piccolo S, Sasai Y, Lu B, De Robertis EM (1996) Dorsoventral patterning in Xenopus: inhibition of ventral signals by direct binding of chordin to BMP-4. *Cell* 86: 589–598

139 Zimmerman LB, Jesus-Escobar JM, Harland RM (1996) The Spemann organizer signal noggin binds and inactivates bone morphogenetic protein 4. *Cell* 86: 599–606

140 Hsu DR, Economides AN, Wang X, Eimon PM, Harland RM (1998) The Xenopus dorsalizing factor Gremlin identifies a novel family of secreted proteins that antagonize BMP activities. *Mol Cell* 1: 673–683

141 Piccolo S, Agius E, Leyns L, Bhattacharyya S, Grunz H, Bouwmeester T, De Robertis EM (1999) The head inducer Cerberus is a multifunctional antagonist of Nodal, BMP and Wnt signals. *Nature* 397: 707–710

142 Yokouchi Y, Vogan KJ, Pearse RV, Tabin CJ (1999) Antagonistic signaling by Caronte, a novel Cerberus-related gene, establishes left-right asymmetric gene expression. *Cell* 98: 573–583

143 Iemura S, Yamamoto TS, Takagi C, Uchiyama H, Natsume T, Shimasaki S, Sugino H, Ueno N (1998) Direct binding of follistatin to a complex of bone-morphogenetic protein and its receptor inhibits ventral and epidermal cell fates in early *Xenopus* embryo. *Proc Natl Acad Sci USA* 95: 9337–9342

144 Canalis E, Economides AN, Gazzerro E (2003) Bone morphogenetic proteins, their antagonists, and the skeleton. *Endocr Rev* 24: 218–235

145 Balemans W, Ebeling M, Patel N, Van Hul E, Olson P, Dioszegi M, Lacza C, Wuyts W, Van Den Ende J, Willems P et al (2001) Increased bone density in sclerosteosis is due to the deficiency of a novel secreted protein (SOST). *Hum Mol Genet* 10: 537–543

146 Brunkow ME, Gardner JC, Van Ness J, Paeper BW, Kovacevich BR, Proll S, Skonier JE, Zhao L, Sabo PJ, Fu Y et al (2001) Bone dysplasia sclerosteosis results from loss of the SOST gene product, a novel cystine knot-containing protein. *Am J Hum Genet* 68: 577–589

147 Kusu N, Laurikkala J, Imanishi M, Usui H, Konishi M, Miyake A, Thesleff I, Itoh N (2003) Sclerostin is a novel secreted osteoclast-derived bone morphogenetic protein antagonist with unique ligand specificity. *J Biol Chem* 278: 24113–24117

148 Winkler DG, Sutherland MK, Geoghegan JC, Yu C, Hayes T, Skonier JE, Shpektor D, Jonas M, Kovacevich BR, Staehling-Hampton K et al (2003) Osteocyte control of bone formation *via* sclerostin, a novel BMP antagonist. *EMBO J* 22: 6267–6276

149 van Bezooijen RL, Roelen BAJ, Visser A, Wee-Pals L, de Wilt E, Karperien M, Hamersma H, Papapoulos SE, ten Dijke P, Lowik CWGM (2004) Sclerostin is an osteocyte-expressed negative regulator of bone formation, but not a classical BMP antagonist. *J Exp Med* 199: 805–814

150 Oelgeschlager M, Larrain J, Geissert D, De Robertis EM (2000) The evolutionarily conserved BMP-binding protein Twisted gastrulation promotes BMP signalling. *Nature* 405: 757–763

151 Rawadi G, Vayssiere B, Dunn F, Baron R, Roman-Roman S (2003) BMP-2 controls alkaline phosphatase expression and osteoblast mineralization by a Wnt autocrine loop. *J Bone Miner Res* 18: 1842–1853

152 Little RD, Carulli JP, Del Mastro RG, Dupuis J, Osborne M, Folz C, Manning SP, Swain PM, Zhao SC, Eustace B et al (2002) A mutation in the LDL receptor-related protein 5 gene results in the autosomal dominant high-bone-mass trait. *Am J Hum Genet* 70: 11–19

153 Boyden LM, Mao J, Belsky J, Mitzner L, Farhi A, Mitnick MA, Wu D, Insogna K, Lifton RP (2002) High bone density due to a mutation in LDL-receptor-related protein 5. *N Engl J Med* 346: 1513–1521

154 Itasaki N, Jones CM, Mercurio S, Rowe A, Domingos PM, Smith JC, Krumlauf R (2003) Wise, a context-dependent activator and inhibitor of Wnt signalling. *Development* 130: 4295–4305

155 Degen WG, Weterman MA, van Groningen JJ, Cornelissen IM, Lemmers JP, Agterbos MA, Geurts van Kessel A, Swart GW, Bloemers HP (1996) Expression of nma, a novel gene, inversely correlates with the metastatic potential of human melanoma cell lines and xenografts. *Int J Cancer* 65: 460–465

156 Onichtchouk D, Chen YG, Dosch R, Gawantka V, Delius H, Massagué J, Niehrs C (1999) Silencing of TGF-β signalling by the pseudoreceptor BAMBI. *Nature* 401: 480–485

157 Tsang M, Kim R, de Caestecker MP, Kudoh T, Roberts AB, Dawid IB (2000) Zebrafish nma is involved in TGFβ family signaling. *Genesis* 28: 47–57

158 Ishisaki A, Yamato K, Hashimoto S, Nakao A, Tamaki K, Nonaka K, ten Dijke P, Sugino H, Nishihara T (1999) Differential inhibition of Smad6 and Smad7 on bone mor-

phogenetic protein- and activin-mediated growth arrest and apoptosis in B cells. *J Biol Chem* 274: 13637–13642

159 Hayashi H, Abdollah S, Qiu Y, Cai J, Xu YY, Grinnell BW, Richardson MA, Topper JN, Gimbrone MA Jr, Wrana JL et al (1997) The MAD-related protein Smad7 associates with the TGFβ receptor and functions as an antagonist of TGFβ signaling. *Cell* 89: 1165–1173

160 Kavsak P, Rasmussen RK, Causing CG, Bonni S, Zhu H, Thomsen GH, Wrana JL (2000) Smad7 binds to Smurf2 to form an E3 ubiquitin ligase that targets the TGFβ receptor for degradation. *Mol Cell* 6: 1365–1375

161 Ebisawa T, Fukuchi M, Murakami G, Chiba T, Tanaka K, Imamura T, Miyazono K (2001) Smurf1 interacts with transforming growth factor-β type I receptor through Smad7 and induces receptor degradation. *J Biol Chem* 276: 12477–12480

162 Hata A, Lagna G, Massagué J, Hemmati-Brivanlou A (1998) Smad6 inhibits BMP/Smad1 signaling by specifically competing with the Smad4 tumor suppressor. *Genes Dev* 12: 186–197

163 Bai S, Shi X, Yang X, Cao X (2000) Smad6 as a transcriptional corepressor. *J Biol Chem* 275: 8267–8270

164 Kimura N, Matsuo R, Shibuya H, Nakashima K, Taga T (2000) BMP2-induced apoptosis is mediated by activation of the TAK1-p38 kinase pathway that is negatively regulated by Smad6. *J Biol Chem* 275: 17647–17652

165 Itoh F, Asao H, Sugamura K, Heldin CH, ten Dijke P, Itoh S (2001) Promoting bone morphogenetic protein signaling through negative regulation of inhibitory Smads. *EMBO J* 20: 4132–4142

166 Ibarrola N, Kratchmarova I, Nakajima D, Schiemann WP, Moustakas A, Pandey A, Mann M (2004) Cloning of a novel signaling molecule, AMSH-2, that potentiates transforming growth factor β signaling. *BMC Cell Biol* 5: 2

167 Luo K, Stroschein SL, Wang W, Chen D, Martens E, Zhou S, Zhou Q (1999) The Ski oncoprotein interacts with the Smad proteins to repress TGFβ signaling. *Genes Dev* 13: 2196–2206

168 Wang W, Mariani FV, Harland RM, Luo K (2000) Ski represses bone morphogenic protein signaling in *Xenopus* and mammalian cells. *Proc Natl Acad Sci USA* 97: 14394–14399

169 Verschueren K, Remacle JE, Collart C, Kraft H, Baker BS, Tylzanowski P, Nelles L, Wuytens G, Su MT, Bodmer R et al (1999) SIP1, a novel zinc finger/homeodomain repressor, interacts with Smad proteins and binds to 5'-CACCT sequences in candidate target genes. *J Biol Chem* 274: 20489–20498

170 Osada S, Ohmori SY, Taira M (2003) XMAN1, an inner nuclear membrane protein, antagonizes BMP signaling by interacting with Smad1 in *Xenopus* embryos. *Development* 130: 1783–1794

171 Li L, Xin H, Xu X, Huang M, Zhang X, Chen Y, Zhang S, Fu XY, Chang Z (2004) CHIP mediates degradation of Smad proteins and potentially regulates Smad-induced transcription. *Mol Cell Biol* 24: 856–864

172 Kretzschmar M, Doody J, Massagué J (1997) Opposing BMP and EGF signalling pathways converge on the TGF-β family mediator Smad1. *Nature* 389: 618–622

173 Shen ZJ, Nakamoto T, Tsuji K, Nifuji A, Miyazono K, Komori T, Hirai H, Noda M (2002) Negative regulation of bone morphogenetic protein/Smad signaling by Cas-interacting zinc finger protein in osteoblasts. *J Biol Chem* 277: 29840–29846

174 Yamaguchi K, Shirakabe K, Shibuya H, Irie K, Oishi I, Ueno N, Taniguchi T, Nishida E, Matsumoto K (1995) Identification of a member of the MAPKKK family as a potential mediator of TGF-β signal transduction. *Science* 270: 2008–2011

175 Yamaguchi K, Nagai S, Ninomiya-Tsuji J, Nishita M, Tamai K, Irie K, Ueno N, Nishida E, Shibuya H, Matsumoto K (1999) XIAP, a cellular member of the inhibitor of apoptosis protein family, links the receptors to TAB1-TAK1 in the BMP signaling pathway. *EMBO J* 18: 179–187

176 Sano Y, Harada J, Tashiro S, Gotoh-Mandeville R, Maekawa T, Ishii S (1999) ATF-2 is a common nuclear target of Smad and TAK1 pathways in transforming growth factor-β signaling. *J Biol Chem* 274: 8949–8957

177 Nakamura K, Shirai T, Morishita S, Uchida S, Saeki-Miura K, Makishima F (1999) p38 mitogen-activated protein kinase functionally contributes to chondrogenesis induced by growth/differentiation factor-5 in ATDC5 cells. *Exp Cell Res* 250: 351–363

178 Ahrens M, Ankenbauer T, Schroder D, Hollnagel A, Mayer H, Gross G (1993) Expression of human bone morphogenetic proteins-2 or -4 in murine mesenchymal progenitor C3H10T1/2 cells induces differentiation into distinct mesenchymal cell lineages. *DNA Cell Biol* 12: 871–880

179 Katagiri T, Yamaguchi A, Komaki M, Abe E, Takahashi N, Ikeda T, Rosen V, Wozney JM, Fujisawa-Sehara A, Suda T (1994) Bone morphogenetic protein-2 converts the differentiation pathway of C2C12 myoblasts into the osteoblast lineage. *J Cell Biol* 127: 1755–1766

180 Maliakal JC, Asahina I, Hauschka PV, Sampath TK (1994) Osteogenic protein-1 (BMP-7) inhibits cell proliferation and stimulates the expression of markers characteristic of osteoblast phenotype in rat osteosarcoma (17/2.8) cells. *Growth Factors* 11: 227–234

181 Lee KS, Kim HJ, Li QL, Chi XZ, Ueta C, Komori T, Wozney JM, Kim EG, Choi JY, Ryoo HM et al (2000) Runx2 is a common target of transforming growth factor β1 and bone morphogenetic protein 2, and cooperation between Runx2 and Smad5 induces osteoblast-specific gene expression in the pluripotent mesenchymal precursor cell line C2C12. *Mol Cell Biol* 20: 8783–8792

182 Shi X, Yang X, Chen D, Chang Z, Cao X (1999) Smad1 interacts with homeobox DNA-binding proteins in bone morphogenetic protein signaling. *J Biol Chem* 274: 13711–13717

183 Yang X, Ji X, Shi X, Cao X (2000) Smad1 domains interacting with Hoxc-8 induce osteoblast differentiation. *J Biol Chem* 275: 1065–1072

184 Hullinger TG, Pan Q, Viswanathan HL, Somerman MJ (2001) TGFβ and BMP-2 activation of the OPN promoter: roles of smad- and hox-binding elements. *Exp Cell Res* 262: 69–74

185 Wan M, Shi X, Feng X, Cao X (2001) Transcriptional mechanisms of BMP-induced osteoprotegrin gene expression. *J Biol Chem* 276: 10119–10125

186 Nakanishi T, Kimura Y, Tamura T, Ichikawa H, Yamaai Y, Sugimoto T, Takigawa M (1997) Cloning of a mRNA preferentially expressed in chondrocytes by differential display-PCR from a human chondrocytic cell line that is identical with connective tissue growth factor (CTGF) mRNA. *Biochem Biophys Res Commun* 234: 206–210

187 Nishida T, Nakanishi T, Asano M, Shimo T, Takigawa M (2000) Effects of CTGF/Hcs24, a hypertrophic chondrocyte-specific gene product, on the proliferation and differentiation of osteoblastic cells *in vitro*. *J Cell Physiol* 184: 197–206

188 Ivkovic S, Yoon BS, Popoff SN, Safadi FF, Libuda DE, Stephenson RC, Daluiski A, Lyons KM (2003) Connective tissue growth factor coordinates chondrogenesis and angiogenesis during skeletal development. *Development* 130: 2779–2791

189 Abreu JG, Ketpura NI, Reversade B, De Robertis EM (2002) Connective-tissue growth factor (CTGF) modulates cell signalling by BMP and TGF-β. *Nat Cell Biol* 4: 599–604

190 St Jacques B, Hammerschmidt M, McMahon AP (1999) Indian hedgehog signaling regulates proliferation and differentiation of chondrocytes and is essential for bone formation. *Genes Dev* 13: 2072–2086

191 Seki K, Hata A (2004) Indian hedgehog gene is a target of the bone morphogenetic protein signaling pathway. *J Biol Chem* 279: 18544–18549

192 Afrakhte M, Morén A, Jossan S, Itoh S, Sampath K, Westermark B, Heldin CH, Heldin NE, ten Dijke P (1998) Induction of inhibitory Smad6 and Smad7 mRNA by TGF-β family members. *Biochem Biophys Res Commun* 249: 505–511

193 Takase M, Imamura T, Sampath TK, Takeda K, Ichijo H, Miyazono K, Kawabata M (1998) Induction of Smad6 mRNA by bone morphogenetic proteins. *Biochem Biophys Res Commun* 244: 26–29

194 Chalaux E, Lopez-Rovira T, Rosa JL, Bartrons R, Ventura F (1998) JunB is involved in the inhibition of myogenic differentiation by bone morphogenetic protein-2. *J Biol Chem* 273: 537–543

195 Ogata T, Wozney JM, Benezra R, Noda M (1993) Bone morphogenetic protein 2 transiently enhances expression of a gene, Id (inhibitor of differentiation), encoding a helix-loop-helix molecule in osteoblast-like cells. *Proc Natl Acad Sci USA* 90: 9219–9222

196 Hollnagel A, Oehlmann V, Heymer J, Ruther U, Nordheim A (1999) Id genes are direct targets of bone morphogenetic protein induction in embryonic stem cells. *J Biol Chem* 274: 19838–19845

197 Moldes M, Lasnier F, Feve B, Pairault J, Djian P (1997) Id3 prevents differentiation of preadipose cells. *Mol Cell Biol* 17: 1796–1804

198 Jen Y, Weintraub H, Benezra R (1992) Overexpression of Id protein inhibits the muscle differentiation program: *in vivo* association of Id with E2A proteins. *Genes Dev* 6: 1466–1479

199 Melnikova IN and Christy BA (1996) Muscle cell differentiation is inhibited by the helix-loop-helix protein Id3. *Cell Growth Differ* 7: 1067–1079

200 Miyama K, Yamada G, Yamamoto TS, Takagi C, Miyado K, Sakai M, Ueno N, Shibuya

H (1999) A BMP-inducible gene, dlx5, regulates osteoblast differentiation and mesoderm induction. *Dev Biol* 208: 123–133

201 Acampora D, Merlo GR, Paleari L, Zerega B, Postiglione MP, Mantero S, Bober E, Bar-
bieri O, Simeone A, Levi G (1999) Craniofacial, vestibular and bone defects in mice
lacking the Distal-less-related gene Dlx5. *Development* 126: 3795–3809

202 Chiba S, Takeshita K, Imai Y, Kumano K, Kurokawa M, Masuda S, Shimizu K, Nakamura S, Ruddle FH, Hirai H (2003) Homeoprotein DLX-1 interacts with Smad4 and
blocks a signaling pathway from activin A in hematopoietic cells. *Proc Natl Acad Sci
USA* 100: 15577–15582

203 Sirard C, Kim S, Mirtsos C, Tadich P, Hoodless PA, Itie A, Maxson R, Wrana JL, Mak
TW (2000) Targeted disruption in murine cells reveals variable requirement for Smad4
in transforming growth factor β-related signaling. *J Biol Chem* 275: 2063–2070

204 Satokata I, Ma L, Ohshima H, Bei M, Woo I, Nishizawa K, Maeda T, Takano Y, Uchiyama M, Heaney S et al (2000) Msx2 deficiency in mice causes pleiotropic defects in bone
growth and ectodermal organ formation. *Nat Genet* 24: 391–395

205 Wahl M, Shukunami C, Heinzmann U, Hamajima K, Hiraki Y, Imai K (2004) Tran-
scriptome analysis of early chondrogenesis in ATDC5 cells induced by bone morphogenetic protein 4. *Genomics* 83: 45–58

206 Peng Y, Kang Q, Cheng H, Li X, Sun MH, Jiang W, Luu HH, Park JY, Haydon RC, He
TC (2003) Transcriptional characterization of bone morphogenetic proteins (BMPs)-
mediated osteogenic signaling. *J Cell Biochem* 90: 1149–1165

207 Korchynskyi O, Dechering KJ, Sijbers AM, Olijve W, ten Dijke P (2003) Gene array
analysis of bone morphogenetic protein type I receptor-induced osteoblast differentiation. *J Bone Miner Res* 18: 1177–1185

208 Balint E, Lapointe D, Drissi H, van der Meijden C, Young DW, van Wijnen AJ, Stein JL,
Stein GS, Lian JB (2003) Phenotype discovery by gene expression profiling: mapping of
biological processes linked to BMP-2-mediated osteoblast differentiation. *J Cell
Biochem* 89: 401–426

209 Vaes BL, Dechering KJ, Feijen A, Hendriks JM, Lefevre C, Mummery CL, Olijve W, van
Zoelen EJ, Steegenga WT (2002) Comprehensive microarray analysis of bone morphogenetic protein 2-induced osteoblast differentiation resulting in the identification of
novel markers for bone development. *J Bone Miner Res* 17: 2106–2118

210 de Jong DS, van Zoelen EJ, Bauerschmidt S, Olijve W, Steegenga WT (2002) Microarray analysis of bone morphogenetic protein, transforming growth factor β, and activin
early response genes during osteoblastic cell differentiation. *J Bone Miner Res* 17:
2119–2129

211 Gu K, Zhang L, Jin T, Rutherford RB (2004) Identification of potential modifiers of
Runx2/Cbfa1 activity in C2C12 cells in response to bone morphogenetic protein-7.
Cells Tissues Organs 176: 28–40

212 Service RF (2000) Tissue engineers build new bone. *Science* 289: 1498–1500

213 Franceschi RT, Wang D, Krebsbach PH, Rutherford RB (2000) Gene therapy for bone

formation: *in vitro* and *in vivo* osteogenic activity of an adenovirus expressing BMP7. *J Cell Biochem* 78: 476–486

214 Fang J, Zhu YY, Smiley E, Bonadio J, Rouleau JP, Goldstein SA, McCauley LK, Davidson BL, Roessler BJ (1996) Stimulation of new bone formation by direct transfer of osteogenic plasmid genes. *Proc Natl Acad Sci USA* 93: 5753–5758

215 Zhang H, Bradley A (1996) Mice deficient for BMP2 are nonviable and have defects in amnion/chorion and cardiac development. *Development* 122: 2977–2986

216 Solloway MJ, Dudley AT, Bikoff EK, Lyons KM, Hogan BL, Robertson EJ (1998) Mice lacking Bmp6 function. *Dev Genet* 22: 321–339

217 Luo G, Hofmann C, Bronckers AL, Sohocki M, Bradley A, Karsenty G (1995) BMP-7 is an inducer of nephrogenesis, and is also required for eye development and skeletal patterning. *Genes Dev* 9: 2808–2820

218 Dudley AT, Lyons KM, Robertson EJ (1995) A requirement for bone morphogenetic protein-7 during development of the mammalian kidney and eye. *Genes Dev* 9: 2795–2807

219 Solloway MJ, Robertson EJ (1999) Early embryonic lethality in Bmp5;Bmp7 double mutant mice suggests functional redundancy within the 60A subgroup. *Development* 126: 1753–1768

220 Zhao GQ, Deng K, Labosky PA, Liaw L, Hogan BL (1996) The gene encoding bone morphogenetic protein 8B is required for the initiation and maintenance of spermatogenesis in the mouse. *Genes Dev* 10: 1657–1669

221 Galloway SM, McNatty KP, Cambridge LM, Laitinen MP, Juengel JL, Jokiranta TS, McLaren RJ, Luiro K, Dodds KG, Montgomery GW et al (2000) Mutations in an oocyte-derived growth factor gene (BMP15) cause increased ovulation rate and infertility in a dosage-sensitive manner. *Nat Genet* 25: 279–283

222 Lechleider RJ, Ryan JL, Garrett L, Eng C, Deng C, Wynshaw-Boris A, Roberts AB (2001) Targeted mutagenesis of Smad1 reveals an essential role in chorioallantoic fusion. *Dev Biol* 240: 157–167

223 Tremblay KD, Dunn NR, Robertson EJ (2001) Mouse embryos lacking Smad1 signals display defects in extra-embryonic tissues and germ cell formation. *Development* 128: 3609–3621

224 Chang H, Huylebroeck D, Verschueren K, Guo Q, Matzuk MM, Zwijsen A (1999) Smad5 knock-out mice die at mid-gestation due to multiple embryonic and extraembryonic defects. *Development* 126: 1631–1642

225 Ducy P, Desbois C, Boyce B, Pinero G, Story B, Dunstan C, Smith E, Bonadio J, Goldstein S, Gundberg C et al (1996) Increased bone formation in osteocalcin-deficient mice. *Nature* 382: 448–452

226 Liaw L, Birk DE, Ballas CB, Whitsitt JS, Davidson JM, Hogan BL (1998) Altered wound healing in mice lacking a functional osteopontin gene (spp1). *J Clin Invest* 101: 1468–1478

227 Yoshitake H, Rittling SR, Denhardt DT, Noda M (1999) Osteopontin-deficient mice are

resistant to ovariectomy-induced bone resorption. *Proc Natl Acad Sci USA* 96: 8156–8160

228 Willing MC, Pruchno CJ, Atkinson M, Byers PH (1992) Osteogenesis imperfecta type I is commonly due to a COL1A1 null allele of type I collagen. *Am J Hum Genet* 51: 508–515

229 Fedde KN, Blair L, Silverstein J, Coburn SP, Ryan LM, Weinstein RS, Waymire K, Narisawa S, Millan JL, MacGregor GR et al (1999) Alkaline phosphatase knock-out mice recapitulate the metabolic and skeletal defects of infantile hypophosphatasia. *J Bone Miner Res* 14: 2015–2026

230 Galvin KM, Donovan MJ, Lynch CA, Meyer RI, Paul RJ, Lorenz JN, Fairchild-Huntress V, Dixon KL, Dunmore JH, Gimbrone MA et al (2000) A role for smad6 in development and homeostasis of the cardiovascular system. *Nat Genet* 24: 171–174

231 Lyden D, Young AZ, Zagzag D, Yan W, Gerald W, O'Reilly R, Bader BL, Hynes RO, Zhuang Y, Manova K et al (1999) Id1 and Id3 are required for neurogenesis, angiogenesis and vascularization of tumour xenografts. *Nature* 401: 670–677

232 Schorpp-Kistner M, Wang ZQ, Angel P, Wagner EF (1999) JunB is essential for mammalian placentation. *EMBO J* 18: 934–948

233 Ducy P, Zhang R, Geoffroy V, Ridall AL, Karsenty G (1997) Osf2/Cbfa1: a transcriptional activator of osteoblast differentiation. *Cell* 89: 747–754

Biology of bone morphogenetic proteins

Snjezana Martinovic, Petra Simic, Fran Borovecki and Slobodan Vukicevic

Department of Anatomy, Medical School, University of Zagreb, Salata 11, 10000 Zagreb, Croatia

Introduction

TGF-β superfamily of proteins consists of conserved families of signaling molecules. One of the largest of these multifunctional families is that of the bone morphogenetic proteins (BMPs), with more than 20 members identified in organisms ranging from sea urchin to mammals. BMPs were first named by the ability to induce ectopic cartilage and endochondral bone when implanted in experimental animals [1]. It is now clear that the name is misleading because there is strong evidence that these molecules regulate biological processes as diverse as cell proliferation, apoptosis, differentiation, cell fate determination and morphogenesis [2]. Besides skeleton, BMPs play a role in the development of other organ and tissue systems that form *via* mesenchymal-epithelial interactions and possibly function to deliver or interpret positional information in a wide variety of organisms [3, 4] (see the chapter by Simic/Vukicevic).

BMPs are synthetized and folded as large dimeric proteins in the cytoplasm and cleaved by proteases during secretion. The functional carboxyl region is then released into the extracellular compartment to bind membrane receptors on target cells [5]. There is also a number of BMP inhibitors present extracellulary to restrict the movement and functions of mature BMPs, such as Noggin, Chordin, and Follistatin [6].

BMPs bind to heteromeric receptor complexes that contain type I and type II serine/threonine protein-kinase receptor subunits. Both receptor types are essential for signal transduction. Receptors transduce the signal to downstream target proteins: SMAD proteins (see the chapter by Korchynsky et al.).

TGF-β superfamily members

BMPs are grouped into several classes based on the degree of sequence identity or homology in the mature carboxyl domain [2] (Fig. 1). The DPP and 60 classes are the best characterized. The *Drosophila* BMP homologue, *decapentaplegic (dpp)*

Bone Morphogenetic Proteins: Regeneration of Bone and Beyond, edited by Slobodan Vukicevic and Kuber T. Sampath
© 2004 Birkhäuser Verlag Basel/Switzerland

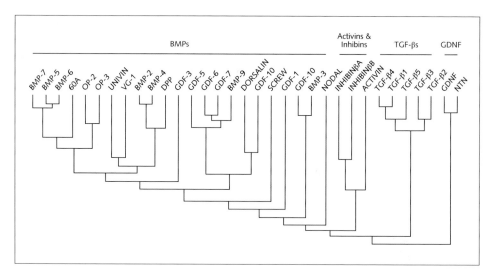

Figure 1
TGF-β superfamily

protein [7], responsible for proper dorsal-ventral patterning of the early embryo is closely related to mammalian BMP-2 and BMP-4 (75% homogeneity). Indeed, *Drosophila dpp* protein can induce bone and cartilage when implanted in 700 million years distant mammals [8] and mammalian BMP-4 can rescue defects caused by *dpp* mutations [9]. Furthermore, Vg1 is a *Xenopus* factor whose messenger RNA (mRNA) is localized to the vegetal hemisphere of the oocyte; although its exact function has remained elusive, it too is believed to be involved in embryonic development [10]. Another subgroup represented by BMP-5, BMP-6 and BMP-7 and last discovered member of this subset, BMP-8 [11], is closely related to 60A, a *Drosophila* protein of unknown function expressed in the early embryo [12, 13]. Although BMP-3 lies outside these two groups of proteins, it is the next most closely related TGF-β superfamily member. Other TGF-β superfamily members clearly lie outside the BMP family. Five TGF-βs have been identified: TGF-β1, -β2 and -β3 in humans, TGF-β4 in chicken, and TGF-β5 in *Xenopus* and show an average of only about 37% amino acid identity in the seven-cysteine region to the BMP molecules. Mullerian inhibiting substance and inhibin α are also quite distantly related (Fig. 1).

BMP function using gene disruption and overexpression

Little is known about the structure of BMP genes. BMP-2 and BMP-4 genes, for example, show high similarity to the *Drosophila dpp* gene, with conserved position of a single intron within the coding region [14], and the BMP-7 gene is structurally related to murine *Vgr-1* gene [15]. BMP genes have been linked to specific chromosomes in mouse as well as in the human genome (Tab. 1). It is likely that they are candidate genes for some known developmental anomalies [16, 17]. The roles of individual BMPs have been studied through identification of mutated genes in classic mouse mutants or through conventional gene-targeting approaches, gene disruption and overexpression of genes encoding members of the BMP family, BMP receptors and SMAD proteins (see also the chapter by Korchynsky et al.). Collectively, these studies confirmed that BMPs have significant roles in the development of the skeleton, nervous system, eye, kidney and heart [18–23].

However, gene disruption experiments did not always deduct the total extent of the BMP function. Namely, some homozygous knockout animals were embryonic lethal, which prevented the disclosure of the true impact of the disruption. In TGF-β knockout mice, the function was also masked by the fact that the maternal protein in heterozygous mothers crossed the placental barrier at early stages of development, resulting in the maternal rescue of offspring [24]. Whether BMPs circulating in biological fluids [25] of heterozygous mice can also cross the placental barrier and mask the true developmental role will be discussed in the chapter by Paralkar. This merely indicates that gene disruption will not necessarily result in a protein deficiency.

Disruption of the gene encoding BMP-2 expresses the most highly malformed phenotype. Homozygous mice are embryonic lethal between E7.0 and E10.5 [26] (Fig. 2). This is caused by the persistence of the proamniotic canal, a transient embryonic structure, the preservation of which leads to malformation of the amnion and the chorion. In mutant embryos, the heart develops in the exocoelomic cavity or does not develop at all. Delay in allantois development, open neural tubes and overall slower growth of these embryos is also observed. The defects are consistent with previously detected patterns of expression of BMP-2 in the extraembryonic mesoderm and promyocardium [27]. The mutation of this gene localized on chromosome 2 showed that it is a candidate gene for the tight skin (*tsk*) mutant (Tab. 1). These animals show increased bone, cartilage and tendon growth with excessive collagen deposition in the subcutaneous connective tissue. Overexpression of BMP-2 in the developing embryo of *Xenopus laevis* leads to ventralization, through inhibition of dorsalizing factors, such as β-tubulin and α-actin [28]. In chick embryos, BMP-2 is expressed in mesenchyme surrounding early cartilage condensations in the developing limb.

In humans, BMP-2 gene is assigned to chromosome 20 (with sublocalization to p12) and it has a positive linkage to Holt-Oram syndrome [29, 30], characterized by defects in cardiac and skeletal development resulting in septal and upper limb deformities.

The mouse BMP-3 gene is localized on chromosome 5, but the human homologue has been assigned to chromosome 4 (between p14 and q21). Interestingly,

Table 1 - Chromosomal localizations of the BMPs

	Human chromosome/ disease	Mouse chromosome/ mutation	Refs.
BMP-2	20p12 / Holt-Oram syndrome	2 / tight skin syndrome (*tsk*), delayed primitive streak, small alantois, lack of amnion, heart defects, decreased number of PGC	[13, 14, 25–29, 36, 41]
BMP-3	4p14.8-q21 / Dentino-genesis imperfecta II	5 / increased bone density	[15, 28, 32, 33]
BMP-3B	10		
BMP-4	14q22-23 / Holt-Oram syndrome	14 / pugnose (*pn*), no mesoderm formation, lack of allantois and PGCs, posterior truncation, heart defects, and lack of optic vesicle; heterozygotes-cystic kidney, craniofacial malformations, microphthalmia, and preaxial poly-dactyly of the right hindlimb.	[13, 34–37, 39, 40]
BMP-5	6	9 / Short-ear (*se*) including defects in skeleton, lung, and kidney	[41–44, 54]
BMP-6	6	13 / congenital hydrocephalus, delayed sternum ossification	[44–48]
BMP-7	20 / Holt-Oram syndrome	2 / impaired kidney and eye development, skeletal defects	[14, 18, 21, 22, 41, 44]
BMP-8A	1	4 / defects in spermatogenesis and epididymis	[49]
BMP-8B	1	4 / defects in PGC formation, testis cord formation, and spermatogenesis	[50]
GDF-1	19	16/ defects in left/right asymetry	[52]
CDMP-1 (GDF-5, BMP-14)	20 / Grebe syndrome, Hunter-Thompson disease	2 / brachypodism	[53–60]
CDMP-2 (GDF-6, BMP-13)	2	4	[60]
CDMP-3 (GDF-7, BMP-15)	2	12 / improper development of dorsal spinal cord	[60, 61]
GDF-8	2	1 / increased skeletal muscle mass	62
GDF-9	5	11 / infertility, impaired folliculogenesis	[63, 64]
GDF-10	10	14 / none	[65]
GDF-11	12	10 / defects in A-P patterning of axial skeleton	[66]

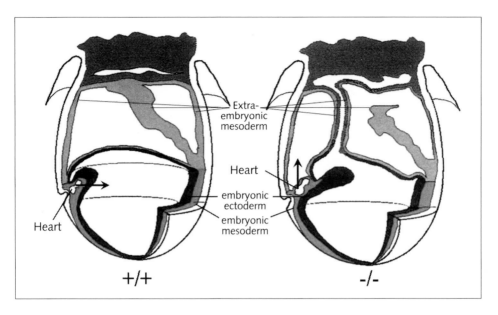

Figure 2
Bmp-2$^{-/-}$ embryos show persistence of the proamniotic canal, malformation of the amnion and the chorion, and development of heart in the exocoelomic cavity.

dentinogenesis imperfecta type II, a disease of tooth development has been associated with human chromosome 4 (Tab. 1). BMP-3 (osteogenin) is the most ample member of the BMP family in demineralized bone, accounting for more than 60% of the total amount of BMPs [31], suggesting an important role in the skeletal homeostasis [32]. *In vitro* studies have shown that BMP-3 antagonizes the osteoinductive capacity of BMP-2 [1, 33]. A recent study on homozygous BMP-3 deficient mice showed that mutants (Fig. 3B), although possessing normal skeletal phenotype, have increased bone density, and total trabecular bone volume twice that of the wild-type animals [33]. The increased bone density is not a consequence of the reduced osteoclast number or increased number of osteoblasts. BMP-3, as a negative regulator of bone density *in vivo*, probably produces its effect through regulation of osteoclast function and subtle effects on osteoblast proliferation and/or differentiation [33]. Experiments using bone chambers in rats have shown that mechanical stimuli decrease expression of BMP-3 [34], which is in line with its role as a negative regulator of osteogenesis in BMP-3-deficient animals (Fig. 4).

The BMP-4 gene is localized to chromosome 14 both in mouse and human genome. It may be a candidate for the pugnose mutation (*pn*) in mice, characterized by abnormalities in skull bone development, and has a possible association

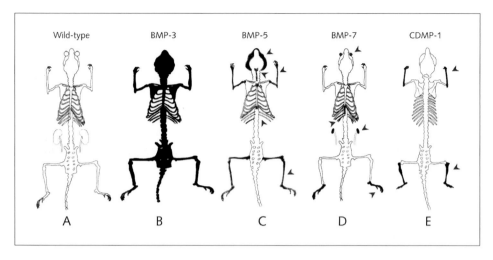

Figure 3
Schematic depiction of a normal mouse and BMP-3, -5, -7 and GDF-5 deficient mice.
(B) BMP-3 deficient mice show increased bone mineral density throughout the whole skeleton, while BMP-5 deficient mice exhibit deformities of the appendicular skeleton, thorax and auricular cartilage (arrows) (C). (D) Mice lacking a functional BMP-7 gene have impaired development of the eye, malformed ribs and feet and undeveloped kidneys (arrows). (E) GDF-5 deficient mice posses shortened appendicular skeleton.

with Holt-Oram syndrome described in humans. Another BMP-4-related gene has been assigned to mouse X chromosome, but as human homologue of this gene has not been found, the mouse sequence might be a pseudogene. Inactivation by homologous recombination of the BMP-4 gene leads to anomalies in extraskeletal tissues and embryonic lethality between E6.5 and E9.5, and a variable phenotype in homozygous animals. A majority of mutant embryos show highly impaired mesodermal differentiation [35]. Some homozygous mutants develop to the head fold or beating heart/early somite stage or beyond, and are developmentally retarded with disorganized posterior structures and a reduction in the extraembryonic mesoderm, including blood islands. Heterozygous BMP-4 mutant mice exhibit craniofacial malformation, microphtalmia and preaxial polydactyly. The plethora and diversity of abnormalities observed indicate that BMP-4 is crucial for normal gastrulation and mesoderm formation. This is also corroborated by previous findings that BMP-4 is needed for differentiation and proliferation of the posterior mesoderm, from which the extraembryonic mesoderm of the amnion, allantois and yolk-sack develops, as well as the ventral-lateral mesoderm [36]. BMP-4 is normally expressed in the perichondrium of the developing cartilage elements. Over-expression of BMP-2 and BMP-4 produced by using retroviral vectors caused

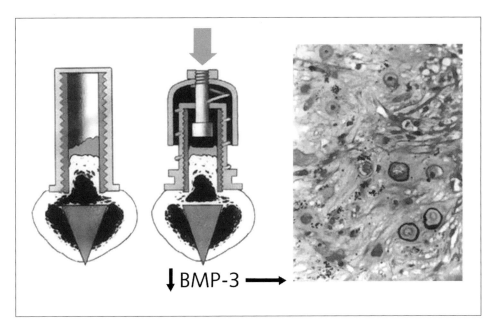

Figure 4
Mechanical loading leads to reduced expression of BMP-3 and skeletal differentiation

enlarged and malformed cartilage elements and joint fusions by increasing the matrix production and number of chondrocytes [37]. Formation of the periosteum was considerably delayed. An overexpression of BMP-4 has been found in lymphocytes and fibroblast-like cells derived from fibroproliferative lesions in patients with fibrodysplasia ossificans progressiva (FOP), a rare human autosomal-dominant disorder characterized by progressive heterotopic ossification and congenital malformation of the big toes [38]. Given the osteogenic capability of BMPs, any of BMP genes could be a candidate for FOP. But, overexpression of the BMP-4 gene has been found in lymphocytes of patients with FOP, suggesting the disease could result from an error in the regulation of this gene [38]. Normal lymphocytes do not produce BMPs, but express ALK-3, a BMP specific receptor [39]. Therefore, in patients with FOP, lymphocytes capable of expressing BMP-4 are presumably recruited to the connective tissue from the bloodstream after soft tissue injury. Increased doses of BMP-4 in the connective tissue may lead to fibroproliferative lesions. Gastric cancer cells also show increased expression of BMP-4 mRNA. These cells can be classified as poorly and well differentiated. The poorly differentiated type shows greater tendency towards bone metastasis and patients with this type of cancer have a decreased life expectancy. Expression of BMP mRNA has been examined in seven different gastric cancer cell lines and results have shown increased expression of BMP-4 [40].

Salivary pleomorphic adenomas, which are often associated with ectopic cartilaginous tissue formation, have also been examined in regards to expression of different members of the BMP family. A marked increase in expression of BMP-2, BMP-4 and BMP-7 mRNA was found. However, chondroid formation and expression of the type II collagen was most frequently observed in pleomorphic adenomas overexpressing BMP-2 mRNA. BMP-2 was also detected in modified myoepithelia cells around the chondroid tissue and basement membranes [41].

The BMP-5 gene is localized on chromosome 6, and the phenotype resulted from its mutation has been studied for over 40 years [42, 43]. The mutation of the BMP-5 gene alters size, shape and number of many different skeletal elements with greatly reduced size of the external ear, named a short ear mutation. The short ear mouse displays numerous skeletal abnormalities (Fig. 3C), such as reduction in body size, absence of the xyphoid process, reduction of ventral processes of the cervical vertebrae, deletion of one pair of ribs and, the most prominent change of all, a reduced size of the auricle [44]. Mutant adult animals also have a reduced capacity to repair rib fractures. Short ear mice also develop a number of other extraskeletal abnormalities, like hydronephrosis, as well as misplacement of gonads, lung cysts, liver granulomas and neuromuscular tail kinks. BMP-5 is expressed in the mesenchyme of the affected skeleton elements and in the periosteum. It is also expressed in liver, lung, bladder and intestine [42]. The expression pattern corresponds to the localization of the affected tissues and organs (Fig. 3).

The BMP-6 gene is present on human chromosome 6 with no reported disease association, and on mouse chromosome 13, possibly near the congenital hydrocephalus (*ch*) locus, which is associated with abnormalities in the growth and differentiation of the skeletal system and kidney [45]. However, mice with targeted null mutation at the BMP-6 locus are viable and fertile, and show no obvious difference in the skeleton to the wild-type animals. Upon closer examination of skeletogenesis in late pregnancy, delayed ossification of the developing sternum is observed [46]. As other members of the BMP family overlap with the BMP-6 expression, especially BMP-2, this apparent lack of defects in mutant mice could be the result of the functional redundancy. BMP-6 is expressed during the development of the epidermis, coinciding with the commencement of stratification. It declines one week after the birth. To study the effects of increased expression of BMP-6 in the epidermis, transgenic mice with inherent overexpression of BMP-6 in suprabasal layers of the intrafollicular epidermis were created [47]. The pattern of transgene expression influences the effects on proliferation and differentiation to a large extent. Consistent and strong expression of BMP-6 leads to lessened cell proliferation in the embryonic and perinatal epidermis, but has hardly any effect on differentiation. Weaker and irregular expression induces hyperproliferation and parakeratosis in the adult epidermis and disturbed differentiation. Histologically, the later findings show high similarity to psoriasis.

Misexpression of BMP-6 results in complete agenesis of the pancreas and reduction in the size of the stomach and spleen causing fusion of the liver and duodenum [48]. BMP-6 is also expressed in the smooth muscle cells of the intestine [49]. It is, therefore, likely that the expression of BMP-6 increases smooth muscle formation, possibly by means of recruitment of mesenchyme to smooth muscle and thereby disrupt the development of the pancreas [48].

The gene for BMP-7 is localized to chromosome 20, both in mouse and human genome. In humans, both chromosomes 2 and 20 have been implicated in Holt-Oram syndrome, so that BMP-2, BMP-4 or BMP-7 might be involved. Deletion in the mature domain of the BMP-7 coding gene produces no apparent malformations in heterozygous animals. However, crosses between these heterozygotes produce a very distinctive phenotype in a quarter of neonates. Mice are smaller in size, have polydactyly in the hindlimbs, exhibit abnormally formed thoracic skeleton and have either anophtalmia or microphtalmia. Most importantly, these animals die of uremia within 48 hours of birth due to small dysgenic kidneys with hydroureters (Fig. 3D). The kidneys have no identifiable metanephric mesenchyme without evidence of glomeruli formation in the cortical region [22, 23] (Fig. 5).

Mice lacking BMP-8a exert normal phenotype throughout embryonic and postnatal development. However, in 47% of homozygous mutants, germ-cell degeneration occurs. A small proportion of homozygous mutants also show degeneration of the epididymal epithelium. BMP-8a thus plays a pivotal role in spermatogenesis and regulation of epididymal function [50]. Targeted mutation of the BMP-8b gene also leads to germ-cell deficiency and sterility. This occurs because of impaired proliferation and differentiation of germ cells as well as premature apoptosis of spermatocytes [51].

BMP-9 shows highly restricted hepatic expression profile, with the mature protein expressed in the liver endothelial, stellate and Kupffer cells. BMP-9 inhibits gluconeogenesis and modulates the expression of regulators of lipid metabolism. It is the first liver-specific factor shown to regulate blood glucose concentrations [52].

Malformations described, both skeletal and extraskeletal, are numerous, but studies of localization of different BMPs imply that deficient phenotypes should be more severe. This apparent discrepancy is, most likely, caused by mechanisms, which are still not fully known. Firstly, BMPs show overlapping, both in localization and function. Only at localizations in which one BMP is predominant, like BMP-7 in the kidney mesenchyme, will the deficiency of that morphogen lead to impaired development and function. Secondly, maternal morphogens might also play an important role in early embryonic development, disguising or totally eliminating deficiencies, which might lead to irregular or impaired development. This has been shown to be the case in TGF-β-deficient mice, and is probably in the root of variations of phenotypes in BMP-4-deficient animals. Early mesenchyme induction in BMP-7-deficient animals could also be linked to the maternal BMP-7 circulating

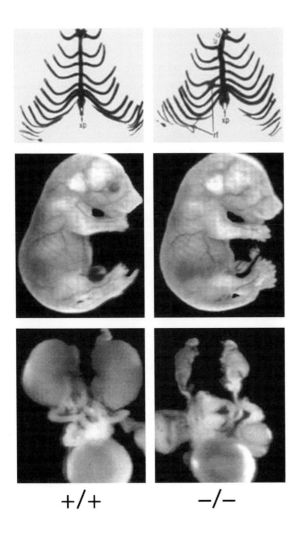

+/+   −/−

Figure 5
Bmp-7$^{-/-}$ mice have malformed ribs, impaired eye development and undeveloped kidneys.

in the bloodstream of BMP-7 deficient embryos. This indicates that genetic and functional evidence, when determining the role of a certain morphogen, often differs greatly. The genetic findings, which mainly derive from studies in cell cultures and on gene-deficient animals, although valuable, do not always hold ground when put to a test in a physiological surrounding. It is only through the combination of genetic and functional data that one can reveal the complex web of interactions, which weave the delicate balance of a gene function.

GDFs function using gene disruption and overexpression

GDF-1 was originally isolated from an early mouse embryo cDNA library and is expressed specifically in the nervous system in late-stage embryos and adult mice. *Gdf-1* is expressed initially throughout the embryo proper and then most prominently in the primitive node, ventral neural tube, and intermediate and lateral plate mesoderm. *Gdf-1*$^{-/-}$ mice exhibit a spectrum of defects related to left-right axis formation, including visceral *situs inversus*, right pulmonary isomerism and a range of cardiac anomalies [53].

Genes encoding cartilage derived morphogenetic proteins, CDMP-1 (GDF-5), CDMP-2 (GDF-6) and CDMP-3 (GDF-7) are localized on human chromosomes 22, 8 and 2, respectively. However, the *brachypodism* mouse phenotype has been studied long before the discovery of BMPs/CDMPs (GDFs). The most prominent feature of these animals is reduction in length of the appendicular skeleton. The axial skeleton is largely unaffected. The defects in the limbs affect metacarpals and metatarsals, along with altered patterning segments in the digits of the limbs. *Brachypodism* is a direct result of three independent mutations in the GDF-5 gene [54, 55]. GDF-5 is expressed during joint formation *in vivo* [56, 57] and malformations in *brachypodism* mouse could be due to impaired chondrogenesis. However, ear, sternum, rib or vertebral morphology is not affected (Fig. 4E). The only known human mutation in a gene encoding a member of the TGF-β superfamily described is the mutation of CDMP-1 gene (*cdmp-1*), a human homologue of GDF-5 [58]. *Cdmp-1* mutations have been implicated in two recessive chondrodysplasias: the Hunter-Thompson chondrodysplasia [59] and the chondrodysplasia Grebe type [60] (Figs. 6, 7). The Hunter-Thompson chondrodysplasia is caused by insertion of 22 bp in the mature region of the *cdmp-1* gene, while the cause of the chondrodysplasia Grebe type seems to be a single replacement of cysteine by tyrosine in a mature TGF-β domain of the cdmp-1 gene. In both cases, the appendicular skeleton is severely shortened, while the axial skeleton remains largely intact (see the chapter by Luyten et al.). It has been shown that a recombinant GDF-5 protein implanted subcutaneously in a bone collagen carrier induces tendon and ligament structures in the subcutaneous bone induction assay in rats [61].

GDF-7 is selectively expressed in the cells of the roof plate in the developing central nervous system [62]. GDF-7 null mutant embryos lack a specific class of neurons, which are important for dorsal spinal cord development. GDF-7 could play a crucial role in the assignment of neuronal identity within the mammalian central nervous system (CNS) – see the chapter by P. Lein and D. Higgins.

GDF-8 deficient animals with induced mutation in mice and spontaneous mutation in double-muscled Belgian blue and piedmontese cattle exert an extensive increase in skeletal muscle mass. The weight of individual muscles in mutants is increased two- or three-fold when compared to wild-type animals. This suggests a role of GDF-8 as negative regulator of the skeletal muscle growth [63].

TTGCGCTCCCACCTGGAGCCCA

Figure 6
The Hunter-Thompson chondrodysplasia is caused by insertion of 22 bp in the mature region of the cdmp-1 *gene. The appendicular skeleton is severely shortened, while the axial skeleton is intact.*

GDF-9 is a member of the BMP family important in the development and maintenance of the reproductive system in mice. It is expressed at high levels in the mammalian oocyte and mice lacking GDF-9 are infertile. This occurs because of impaired folliculogenesis [64, 65].

GDF-10 is expressed during development in the craniofacial region and the vertebral column of the skeleton. During adult life it is most highly expressed in the brain and in the uterus [66]. Mice carrying null mutation for the GDF-10 gene, however, do not show any obvious abnormality in the development, confirming that gene knockout experiments do not necessarily have functional consequences.

GDF-11 has an important role in establishing skeletal pattern. During early mouse embryogenesis, *gdf-11* is expressed in the primitive streak and tail bud regions, which are sites where new mesodermal cells are generated. Homozygous mutant mice carrying a targeted deletion of *gdf-11* exhibit anteriorly directed homeotic transformations throughout the axial skeleton and posterior displacement

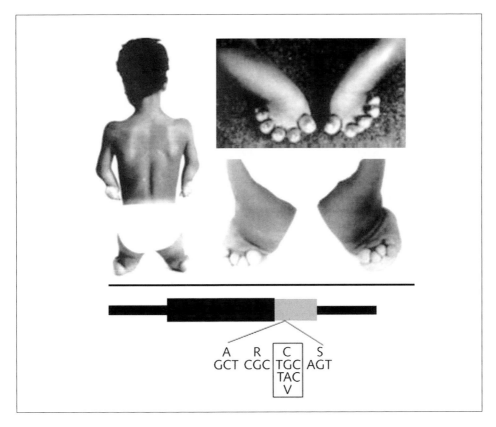

Figure 7
Chondrodysplasia Grebe type is a single replacement of cysteine by tyrosine in a mature TGF-β domain of the cdmp-1 *gene. The appendicular skeleton is severely shortened, while the axial skeleton is intact.*

of the hindlimbs. *Gdf-11* is a secreted signal that acts globally to specify positional identity along the anterior/posterior axis [67].

Double deficiencies in genes encoding BMPs

Absence of malformations observed in some mutants lacking functional BMP encoding genes, and simultaneous expression of several BMP members in different tissues have pointed to a possible functional redundancy in the role of these proteins. Therefore, several phenotypes have been investigated in which the function of two genes encoding different members of the BMP family has been disrupted.

Functionally, BMP-2 exhibits many of the same activities as BMP-4. Both BMP-2 and BMP-4 play a role in primordial germ cells (PGC) generation. BMP-2$^{+/-}$; BMP-4$^{+/-}$ double mutants have fewer PGC generation [68].

Doubly heterozygous BMP-4 and BMP-7 mice develop defects in the rib cage and distal part of the limbs [69]. These two morphogens seem to act in cooperation in the mesenchymal condensations of the affected skeletal regions, possibly through regulation of apoptosis.

BMP-5/-6 double mutants show sternal defects similar to those found in BMP-6 single mutants. However, these defects tend to be slightly exacerbated in the double mutant.

Mice with simultaneous deficiency in BMP-5 and BMP-7 show the most severe phenotype. Coexpression of both morphogens seems to be pivotal for development of allantois, heart, branchial arches, somites and the forebrain since mutant embryos die at E10.5 with extensive defects of the aforementioned tissues [70].

Null mutants with simultaneous deficiency in BMP-5 and GDF-5/CDMP-1 exert defects, which cannot be observed in either of the single mutants. Disruption of the sternebrae within the sternum and abnormal formation of fibrocartilagi-nous joints between the sternebrae and the ribs are the most prominent of those defects [55].

BMP-6 and BMP-7 are expressed in overlapping and adjacent sites, including the cardiac cushions during mouse embryonic development. Previous analyses demon-strated that neither of these BMPs was required during cardiogenesis, but analysis of BMP-6:BMP-7 double mutants uncovers a marked delay in the formation of the outflow tract endocardial cushions. A proportion of BMP-6:BMP-7 mutants also display defects in valve morphogenesis and chamber septation, and the embryos die between 10.5 and 15.5 dpc due to cardiac insufficiency [71].

BMP-8a and BMP-8b are expressed in similar patterns in male germ cells. BMP-8a$^{+/-}$:BMP-8b$^{+/-}$ double mutants show germ cell defects during the initiation of sper-matogenesis and defects in epididymis [72]. In BMP-7/BMP-8a double mutants the removal of one allele of BMP-7 exacerbates the phenotype of BMP-8a null mutants in spermatogenesis and epididymis of the adult mice. These indicate that, similar to BMP-8a, BMP-7 plays a role in both the maintenance of spermatogenesis and epi-didymal function. It further suggests that BMP-8 and BMP-7 signal through the same or similar receptors in these two systems [73].

BMP antagonists

Noggin was isolated from *Xenopus* based on rescue of dorsal development in ultra-violet-induced ventralized embryos [74]. Noggin is produced by the Spemann orga-nizer and antagonizes the action of BMPs, induces neural tissue and dorsalizes ven-tral mesoderm [75]. Noggin binds to BMP-2 and BMP-4 with high affinity and

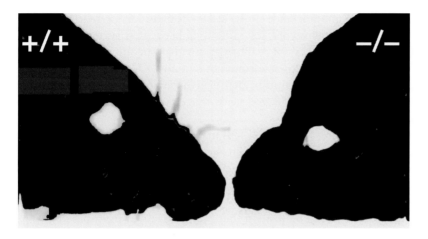

Figure 8
Noggin^{-/-} mutant mice show blunt cranial skeleton.

blocks interaction with BMP receptor. BMP-2 in return induces the expression of noggin in osteoblasts [76] and bone marrow cells [77].

Heterozygous noggin-deficient mice possess normal phenotypes. Skeletal structures in homozygous animals however exhibit abnormalities (Fig. 8). The defects are especially striking in the vertebrae, ribs and limbs, with the severity of axial defects increasing caudally [78]. The cranial skeleton is blunt (Fig. 8), cervical vertebrae are basically normal, but thoracic vertebrae are fused. They also fail to close dorsally. Ribs are reduced in number and have abnormal morphology. The appendicular skeleton in mutant animals is also shortened. All these processes seem to arise from a lack of noggin leading to increased BMP activity after the chondrogenesis has started. A majority of the joints is also fused. Elbows and digits are fused and have cartilaginous spurs as a result of a failure to specify the joint. Unregulated expression of GDF-5/CDMP-1 in the joint regions seems to play a pivotal role in those processes. Absence of local regulation of BMP members, especially BMP-6, which is expressed in the hypertrophic zone of cartilage in the joints, most probably also plays a role in impaired articular development.

During the course of investigation on *Xenopus* pattern formation, chordin was identified as BMP-4-binding protein [79]. Chordin has a homolog in *Drosophila*, called short gastrulation (sog), known to bind to decapentaplegic, a BMP 2/4 homolog, showing that BMP antagonists have been conserved for several million years. Chordin binds to BMP-2 and BMP-4, and is further regulated by a metalloprotease BMP-1 and tolloid, and Xenopus xolloid [80]. Antagonists such as chordin govern the pattern formation by BMPs, and in turn are proteolytically inactivated by a metalloprotease BMP-1.

Chordin homozygous mutant mice show, at low penetrance, early lethality and a ventralized gastrulation phenotype. The mutant embryos that survive die perinatally, displaying an extensive array of malformations that encompass most features of DiGeorge and Velo-Cardio-facial syndromes in humans [81]. Chordin is required for the correct expression of transcription factors involved in the development of the pharyngeal region and leads to head and neck congenital malformations that frequently occur in humans [81].

Ventroptin is an antagonist of BMP-4. It is mainly expressed in the ventral retina of the eye. Misexpression of ventroptin alters expression patterns of several topographic genes suggesting its importance in specifying the topographic retinotectal projection [82]. Ventroptin also precedes noggin in the modulation of BMP activity in the developing cartilage [83].

Follistatin is an activin-binding protein that prevents activin from binding to its receptors and neutralizes its activity. Follistatin also binds BMPs. Follistatin-deficient mice have numerous embryonic defects, including shiny, taut skin, growth retardation, and cleft palate leading to death within hours of birth. Gain-of-function mutant mice in which mouse follistatin is overexpressed had defects in the testis, ovary, and hair [84].

Follistatin-related gene (FLRG) was isolated in a case of B-cell chronic lymphocytic leukaemia [85]. The recombinant mouse FLRG proteins were found to have binding activity for both activin and BMP-2. Like follistatin, FLRG has higher affinity for activin than for BMP-2. The FLRG protein inhibited activin-induced and BMP-2-induced transcriptional responses in a dose-dependent manner [86].

DAN (differential screening-selected gene aberrative in neuroblastoma, also known as N03) is a member of a class of glycoproteins shown to be secreted inhibitors of TGF-β and BMP pathways [87]. The DAN gene was isolated as a candidate tumor suppressor gene in a differential hybridization screen [88]. Secreted DAN suppressed DNA synthesis in transformed cells. DAN helps to regulate BMP activity spatially and temporally and patterns medial otic tissue between the endolymphatic duct/sac and medially derived inner ear structures [89].

The head inducer gene Cerberus codes for a secreted protein and can induce heads in *Xenopus* embryos, and it is related to DAN in the cysteine rich domain [90]. "Knockdown" of Cerberus function by antisense morpholino oligonucleotides does not impair head formation in the embryo. In contrast, targeted increase of BMP, nodal and Wnt signaling in the anterior dorsal-endoderm (ADE) results in synergistic loss of anterior head structures, without affecting more posterior axial ones. Remarkably, those head phenotypes are aggravated by simultaneous depletion of Cerberus, demonstrating that endogenous Cerberus protein can inhibit BMP, Nodal and Wnt factors *in vivo*. Conjugates of dorsal ectoderm (DE) and ADE explants in which Cerberus function is "knocked down" reveal the requirement of Cerberus in the ADE for the proper induction and patterning of the neuroectoderm [91].

Cerberus-like, member of Cerberus/Dan family, is a secreted BMP and nodal antagonist not essential for mouse development. Homozygous null mutants of the mouse Cer-l gene show no anterior patterning defects, are born alive, and are fertile [92].

Coco is a member of Cerberus/Dan family of secreted BMP inhibitors, which was identified in a screen for Smad7-induced genes. This gene is expressed maternally in an animal to vegetal gradient, and its expression levels decline rapidly following gastrulation. Coco is broadly expressed in the ectoderm until the end of gastrulation. Function of Coco is to block BMP and TGF-β signals in the ectoderm in order to regulate cell fate specification and competence prior to the onset of neural induction. In addition, Coco can act as a neural inducer and induce ectopic head-like structures in neurula staged embryos [93].

Gremlin is a *Xenopus* homolog related to DAN that inhibits BMP-2 action and was identified by screening an ovarian cDNA library for activities inducing the secondary axis [94].

During limb outgrowth, signaling by BMPs must be moderated to maintain the signaling loop between the zone of polarizing activity (ZPA) and the apical ectodermal ridge (AER). Gremlin's ability to fulfill this function was tested by mutating the mouse gene encoding gremlin. In the mutant limb, the feedback loop between the ZPA and the AER is interrupted, resulting in abnormal skeletal pattern showing that gremlin is the principal BMP antagonist required for early limb outgrowth and patterning. Gremlin mutation is allelic to the limb deformity mutation (ld) [95].

Gremlin negatively modulates BMP-4 induction of embryonic lung branching morphogenesis. Reduction of endogenous gremlin expression with antisense oligonucleotides enhances peripheral lung epithelial branching morphogenesis, the same gain-of-function phenotype as exogenous BMP-4. On the other hand, adenoviral overexpression of gremlin blocks the stimulatory effects of exogenous BMP-4 [96]. Increased expression of gremlin has been demonstrated in several models of diabetic nephropathy. Gremlin arrests the cell cycle in mesangial cells and has also been shown to be upregulated in transdifferentiated renal proximal tubular cells [97].

Protein related to DAN and cerberus (PRDC) is a secreted protein with a cystine knot structure identified by gene trapping in embryonic stem cells. PRDC transcripts are widely expressed showing higher levels in ovary, brain, and spleen. PRDC is expressed in granulosa cells of the ovaries and could be involved in follicular development by antagonizing the actions of BMPs [98].

Sclerostin is an osteocyte-expressed negative regulator of bone formation. Loss of the SOST gene product, sclerostin, leads to sclerosteosis, a skeletal disorder characterized by high bone mass due to increased osteoblast activity [99]. Sclerostin competes with the type I and type II BMP receptors for binding to BMPs, decreases BMP signaling and suppresses mineralization of osteoblastic cells. Transgenic mice overexpressing SOST exhibit low bone mass and decreased bone strength as a result of a significant reduction in osteoblast activity and subsequently bone formation [100].

Disruptions in the genes encoding BMP receptors

BMPs bind to heteromeric receptors that contain type I and type II serine/threonine receptor subunits. Type II receptors have kinase activity that phosphorylates type I receptors upon ligand-receptor formation (see the chapter by Korchynsky et al.). Phosphorylated type I receptors transduce the signal to downstream target proteins [101]. Several receptors for BMPs have been identified including the type I receptors ALK1, ALK2, ALK3 (type IA or BMPR), and ALK6 (or type IB), and type II receptors BMPRII (TALK), ActRIIA and ActRIIB [5] (see the chapter by Korchynsky et al.).

Alk3, encoding receptor type I, is expressed in the entire embryo throughout the development. Therefore, inactivation of Alk3 results in early embryonic lethality, due to impaired growth of the epiblast and a lack of mesoderm or gastrulation [102]. The inactivation of Alk3 represents the most severe phenotype among all genes in BMP signaling pathways. Conditional mutants of Alk3 receptor have impaired cardiac and limb development [103, 104].

Inactivation of Alk1 receptor leads to defects in embryonic angiogenesis [105]. Alk2 is primarily expressed in visceral endoderm. Knocking-out Alk2 gene results in defective visceral endoderm which is unable to induce the epiblast and therefore consequently leads to defects in mesoderm formation [106]. Alk4 inactivation impairs epiblast differentiation and leads to lack of mesoderm formation [107]. Alk6 null mutants have defects in seminal vesicle development, female reproduction and limb skeletal formation [108, 109].

BmprII encodes a type II receptor subunit for BMP-2 and BMP-4 [110]. BmprII null mutants have defects in gastrulation because of the lack of mesoderm [110], a phenotype similar to inactivation of Alk3. Both ActrIIa and ActrIIb are expressed in the epiblast. ActrIIa null mutants have deficient reproduction due to suppressed FSH and mild defects in skeletal development [111]. ActrIIb null mutants show defects in axial patterning and left-right asymmetry [112]. ActRIIb[-/-] mice die after birth with complicated cardiac defects, including randomized heart position, malposition of the great arteries, and ventricular and atrial septal defects; right pulmonary isomerism and splenic abnormalities, recapitulating the clinical symptoms of the human asplenia syndrome [112]. Single ActrIIa or ActrIIb null mutation doesn't affect mesoderm formation, while their double mutants do not form mesoderm and show a phenotype resembling Alk3 or BmprII null mutants [113].

Phenotypes of single BMP mutants are milder than those of Alk3 or BmprII null mutants, since BMPs can compensate for one another.

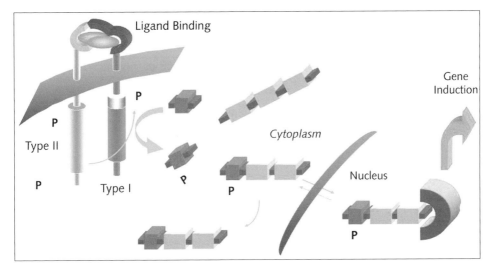

Figure 9
BMPs bind to heteromeric receptor complexes that contain type I and type II serine/threo-
nine protein-kinase receptor subunits. Receptors transduce the signal to downstream target
proteins: SMAD proteins. The phosphorylated SMAD-1 oligomerize with SMAD-4 in the
cytoplasm and translocates into the nucleus to serve as transcriptional regulator.

Disruptions in the genes encoding SMAD

A major family of downstream targets of TGF-β superfamily is the SMAD protein family (Fig. 9). SMAD-1, -5 and -8 transduce signals of BMPs, while SMAD-2 and -3 transduce signals of TGF-βs and Activins. SMAD-6 and -7 have inhibitory functions. As the ligand binds, receptor type II phosphorylates type I subunit, which then phosphorylates SMAD proteins. The phosphorylated SMAD-1, -5 or -8 as well as SMAD-2 or -3 oligomerize with SMAD-4 in the cytoplasm and translocate into the nucleus to serve as transcriptional regulators (see the chapter by Korchynsky et al.).

Smad1 null mutants show defects in visceral endoderm, extraembryonic meso-derm and lack or reduced number of primordial germ cells [114, 115]. Surprisingly, despite the key involvement of BMP signaling in tissues of the embryo proper, Smad1-deficient embryos develop remarkably normally. An examination of the expression domains of Smad1, Smad5 and Smad8 in early mouse embryos show that, while Smad1 is uniquely expressed in the visceral endoderm at 6.5 dpc, in other tissues Smad1 is co-expressed with Smad5 and/or Smad8 [24].

Inactivation of Smad2 leads to defects in visceral endoderm causing abnormal mesoderm formation, anterior-posterior patterning and left-right patterning [116]. The rescued postgastrulation embryos show malformation of head structures, abnormal embryo turning, and cyclopia [117]. Homozygous mutant embryonic stem cells fail to contribute to gut endoderm [117].

Smad3 null mutants comprise metastatic colorectal cancer, impaired local inflammatory response, impaired mucosal immunity and diminished T cell responsiveness to TGF-β, accelerated wound healing and abnormal hypertrophic chondrocyte differentiation [118–121].

SMAD4 proteins are shared by multiple pathways, therefore Smad4$^{-/-}$ mutants display a severe phenotype of no mesoderm, caused by gastrulation defect [122]. Heterozygotes have gastric polyposis [123].

Inactivation of Smad5 leads to impaired angiogenesis in yolk sac, reduced mesoderm, defects in progenitor germ cells and left-right asymmetry [124–126]. Smad6 null mutants show hyperplasia of the cardiac valves and defects in outflow tract septation [127]. Smad7 null mutants have scleroderma-like phenotype, an autoimmune disease that leads to fibrosis of affected organs [128].

References

1   Wozney JM, Rosen V, Celeste AJ, Mitsock LM, Whitters MJ, Kriz RW, Hewick RM, Wang EA (1988) Novel regulators of bone formation: molecular clones and activities. *Science* 242: 1528–1534

2   Hogan BLM (1996) Bone morphogenetic proteins-multifunctional regulators of vertebrate development. *Gen Develop* 10: 1580–1594

3   Hogan BLM (1996) Bone morphogenetic proteins in development. *Curr Opin Gen Dev* 6: 432–438

4   Reddi AH (2000) Bone morphogenetic proteins and skeletal development: the kidney-bone connection. *Pediatr Nephrol* 14: 598–601

5   Rueger DC (2002) Biochemistry of bone morphogenetic proteins. In: S Vukicevic, KT Sampath (eds): *Bone morphogenetic proteins. From laboratory to clinical practice.* Birkhauser Verlag, Basel, 289–321

6   Zhao GQ (2003) Consequences of knocking out BMP signalling in the mouse. *Genesis* 35: 43–56

7   Padget RW, St Johnston RD, Gelbart WM (1987) A transcript from a *Drosophila* pattern geen predicts a protein homologous to the transforming growth factor-β family. *Nature (London)* 325: 81–84

8   Sampath TK, Rashka EK, Doctor JS, Tucker RF, Hoffmann FM (1993) *Drosophila* transforming growth factor superfamily proteins induce endochondral bone formation in mammals. *Proc Natl Acad Sci USA* 90: 6004–6008

9   Padget RW, Wozney JM, Gelbart WM (1993) Human BMP sequences can confer normal dorsal-ventral patterning in the *Drosophila* embryo. *Proc Natl Acad Sci USA* 90: 2905–2909

10   Weeks DL, Melton DA (1987) A maternal mRNA localized to the vegetal hemisphere in *Xenopus* eggs codes for a growth factor related to TGF-β. *Cell* 51: 861–867

11 Ozkaynak E, Schnegelsberg PN, Jin DF, Clifford GM, Warren FD, Drier EA, Oppermann H (1992) Osteogenic protein-2. A new member of the transforming growth factor-beta superfamily expressed early in embryogenesis. *J Biol Chem* 267: 25220–25227

12 Wharton KA, Thomsen GH, Gelbart WM (1991) *Drosophila* 60A gene, another transforming growth factor β family member, is closely related to human bone morphogenetic proteins. *Proc Natl Acad Sci USA* 88: 9214–9218

13 Doctor JS, Jackson PD, Rashka KE, Visalli M, Hoffmann FM (1992) Sequence, biochemical caracterization and developmental expression of a new member of the TGF-β superfamily in *Drosophila melanogaster. Dev Biol* 151: 491–505

14 Feng JQ, Harris MA, Ghosh-Choudhury N, Feng M, Mundy GR, Harris SE (1994) Structure and sequence of mouse bone morphogenetic protein-2 gene (BMP-2): comparison of the structures and promoter regions of BMP-2 and BMP-4 genes. *Biochim Biophys Acta* 1218: 221–224

15 Ozkaynak E, Rueger DC, Drier EA, Corbett C, Ridge RJ, Sampath TK, Oppermann H (1990) OP-1 cDNA encodes an osteogenic protein in the TGF-beta family. *EMBO J* 9: 2085–2093

16 Dickinson ME, Kobrin MS, Silan CM, Kingsley DM, Justice MJ, Miller DA, Ceci JD, Lock LF, Lee A, Buchberg AM et al (1990) Chromosomal localization of seven members of the murine TGF-β superfamily suggest close linkage to several morphogenetic mutant loci. *Genomics* 6: 505–520

17 Ceci JD, Kingsley DM, Silan CM, Copeland NG, Jenkins NA (1990) An interspecific backcross linkage map of the proximal half of mouse chromosome 14. *Genomics* 87: 9843–9847

18 Vukicevic S, Helder MN, Luyten FP (1994) Developing human lung and kidney are major sites for synthesis of bone morphogenetic protein-3 (osteogenin). *J Histochem Cytochem* 42: 869–875

19 Vukicevic S, Kopp JB, Luyten FP, Sampath TK (1996) Induction of nephrogenic mesenchyme by osteogenic protein 1 (bone morphogenetic protein 7). *Proc Natl Acad Sci USA* 93: 9021–9026

20 Helder MN, Ozkaynak E, Sampath KT, Luyten FP, Latin V, Oppermann H, Vukicevic S (1995) Expression pattern of osteogenic protein-1 (bone morphogenetic protein-7) in human and mouse development. *J Histochem Cytochem* 43: 1035–1044

21 Ducy P, Karsenty G (2000) The family of bone morphogenetic proteins. *Kidney Int* 57: 2207–2214

22 Dudley AT, Lyons K, Robertson EJ (1995) A requirement for bone morphogenetic protein-7 during development of the mammalian kidney and eye. *Genes Dev* 9: 2795–2807

23 Luo G, Hofmann C, Bronckers AL, Sohocki M, Bradley A, Karsenty G (1995) BMP-7 is an inducer of nephrogenesis, and is also required for eye development and skeletal patterning. *Genes Dev* 9: 2808–2820

24 Letterio JJ, Geiser AG, Kulkarni AB, Roche NS, Sporn MB, Roberts AB (1994) Maternal rescue of transforming growth factor-β1 null mice. *Science* 264: 1936–1938

25 Borovecki F, Grgurevic L, Jelic M, Bosukonda D, Sampath K, Vukicevic S (2004)

Osteogenic protein-1 (bone morphogenetic protein-7) is available to the fetus through placental transfer during early stages of development. *Nephron Exp Nephrol* 97: 26–32

26 Hongbin Z, Bradley A (1996) Mice deficient for BMP-2 are nonviable and have defects in amnion/chorion and cardiac development. *Development* 122: 2977–2986

27 Lyons KM, Pelton RW, Hogan BLM (1990) Organogenesis and pattern formation in the mouse: RNA distribution patterns suggest a role for bone morphogenetic protein-2A (BMP-2A). *Development* 109: 833–844

28 Clement JH, Fettes P, Knochel S, Lef J, Knochel W (1995) Bone morphogenetic protein 2 in early development of *Xenopus laevis*. *Mech Dev* 52: 357–370

29 Tabas JA, Zasloff M, Wasmuth JJ, Emanuel BS, Altherr MR, McPherson JD, Wozney JM, Kaplan FS (1991) Bone morphogenetic protein: chromosomal localization of human genes for BMP1, BMP2A, and BMP3. *Genomics* 9: 283–289

30 Rao VV, Loffler C, Wozney JM, Hansmann I (1992) The gene for bone morphogenetic protein 2A (BMP2A) is localized to human chromosome 20p12 by radioactive and non-radioactive *in situ* hybridization. *Hum Genet* 90: 299–302

31 Luyten FP, Cunningham NS, Ma S, Muthukumaran N, Hammonds RG, Nevins WB, Woods WI, Reddi AH (1989) Purification and partial amino acid sequence of osteo-genin, a protein initiating bone differentiation. *J Biol Chem* 264: 13377–13380

32 Vukicevic S, Helder MN, Luyten FP (1994b) Developing human lung and kidney are major sites for synthesis of bone morphogenetic protein-3 (osteogenin). *J Histochem Cytochem* 42: 869–875

33 Daluiski A, Engstrand T, Bahamonde ME, Gamer LW, Agius E, Stevenson SL, Cox K, Rosen V, Lyons KM (2001) Bone morphogenetic protein-3 is a negative regulator of bone density. *Nat Genet* 27: 84–88

34 Aspenberg P, Basic N, Tagil M, Vukicevic S (2000) Reduced expression of BMP-3 due to mechanical loading: a link between mechanical stimuli and tissue differentiation. *Acta Orthop Scand* 71: 558–562

35 Winnier G, Blessing M, Labosky PA, Hogan BLM (1995) Bone morphogenetic protein-4 is required for mesoderm formation and patterning in the mouse. *Gen Dev* 9: 2105–2116

36 Lawson KA, Pedersen RA (1992) Clonal analysis of cell fate during gastrulation and early neurulation in the mouse. Postimplantation development in the mouse. *CIBA Found* 165: 3–26

37 Duprez D, Bell EJ, Richardson MK, Archer CW, Wolpert L, Bricker PM, Francis-West PH (1996) Overexpression of BMP-2 and BMP-4 alters the size and shape of develop-ing skeletal elements in the chick limb. *Mech Dev* 57: 145–157

38 Shafritz AB, Shore EM, Gannon FH, Zasloff MA, Taub R, Muenke M, Kaplan FS (1996) Overexpression of an osteogenic morphogen in fybrodysplasia ossificans pro-gressiva. *N Engl J Med* 335: 555–561

39 Martinovic S, Mazic S, Kisic V, Basic N, Jakic-Razumovic J, Batinic D, Borovecki F, Simic P, Grgurevic L, Labar B, Vukicevic S (2004) Expression of bone morphogenetic proteins in long-term culture of human bone marrow stromal cells. *J Histoch Cytochem* 52

40 Katoh M, Terada M (1996) Overexpression of bone morphogenetic protein (BMP)-4 mRNA in gastric cancer cel lines of poorly differentiated type. *J Gastroenterol* 31: 137–139

41 Kusafuka K, Yamaguchi A, Kayano T, Fujiwara M, Takemura T (1998) Expression of bone morphogenetic proteins in salivary pleomorphic adenomas. *Virchows Arch* 432: 247–253

42 King JA, Marker PC, Seung KJ, Kingsley DM (1994) BMP5 and the molecular, skeletal, and soft-tissue alterations in short ear mice. *Dev Biol* 166: 112–122

43 Green MC (1968) Mechanism of the pleiotropic effects of the short-ear mutant gene in the mouse. *J Exp Zool* 167: 129–150

44 Kingsley DM, Bland AE, Grubber JM, Marker PC, Russell LB, Copeland NG, Jenkins NA (1992) The mouse short ear skeletal morphogenesis locus is associated with defects in a bone morphogenetic member of the TGFβ superfamily. *Cell* 71: 399–410

45 Hahn GV, Cohen RB, Wozney JM, Levitz CL, Shore EM, Zasloff MA, Kaplan FS (1992) A bone morphogenetic protein subfamily: chromosomal localization of human genes for BMP5, BMP6, and BMP7. *Genomics* 14: 759–762

46 Solloway MJ, Dudley AT, Bikoff EK, Lyons KM, Hogan BL, Robertson EJ (1998) Mice lacking Bmp6 function. *Dev Genet* 22: 321–339

47 Blessing M, Schrimacher P, Kaiser S (1996) Overexpression of bone morphogenetic protein-6 (BMP-6) in the epidermis of transgenic mice: inhibition or stimulation of proliferation depending on the pattern of transgene expression and formation of psoriatic lesions. *J Cell Biol* 135: 227–239

48 Dichmann DS, Miller CP, Jensen J, Heller RS, Serup P (2003) Expression and misexpression of members of the FGF and TGFβ families of growth factors in the developing mouse pancreas. *Dev Dyn* 226: 663–674

49 Perr HA, Ye J-Q, Gitelman SE (1999) Smooth muscle expresses bone morphogenetic protein (Vgr-1/BMP-6) in human fetal intestine. *Biol Neonate* 75: 210–214

50 Zhao GQ, Deng K, Labosky PA, Liaw L, Hogan BL (1996) The gene encoding bone morphogenetic protein 8B is required for the initiation and maintenance of spermatogenesis in the mouse. *Genes Dev* 10: 1657–1669

51 Zhao GQ, Liaw L, Hogan BL (1998) Bone morphogenetic protein 8A plays a role in the maintenance of spermatogenesis and the integrity of the epididymis. *Development* 125: 1103–1112

52 Chen C, Grzegorzewski KJ, Barash S, Zhao Q, Schneider H, Wang Q, Singh M, Pukac L, Bell AC, Duan R et al (2003) An integrated functional genomics screening program reveals a role for BMP-9 in glucose homeostasis. *Nat Biotechnol* 21: 294–301

53 Rankin CT, Bunton T, Lawler AM, Lee SJ (2000) Regulation of left-right patterning in mice by growth/differentiation factor-1. *Nat Genet* 24: 262–265

54 Storm EE, Huynh TV, Copeland NG, Jenkins NA, Kingsley DM, Lee SJ (1994) Limb alterations in brachypodism mice due to mutations in a new member of the TGFβ-superfamily. Nature 368: 639–643

55 Storm EE, Kingsley DM (1996) Joint patterning defects caused by single and double

mutations in members of the bone morphogenetic protein (BMP) family. *Development* 122: 3969–3979

56 Francis-West PH, Abdelfattah A, Chen P, Allen C, Parish J, Ladher R, Allen S, MacPherson S, Luyten FP, Archer CW (1999) Mechanisms of GDF-5 action during skeletal development. *Development* 126: 1305–1315

57 Francis-West PH, Parish J, Lee K, Archer CW (1999) BMP/GDF-signalling interactions during synovial joint development. *Cell Tissue Res* 296: 111–119

58 Chang SC, Hoang B, Thomas JT, Vukicevic S, Luyten FP, Ryba NJ, Kozak CA, Reddi AH, Moos M Jr (1994) Cartilage-derived morphogenetic proteins. New members of the transforming growth factor-beta superfamily predominantly expressed in long bones during human embryonic development. *J Biol Chem* 269: 28227–28234

59 Thomas JT, Lin K, Nandedkar M, Camargo M, Cervenka J, Luyten FP (1996) A human chondrodysplasia due to a mutation in a TGF-β superfamily member. *Nat Gen* 12: 315–318

60 Thomas JT, Kilpatrick MW, Lin K, Erlacher L, Lembessis P, Costa T, Tsipouras P, Luyten FP (1997) Disruption of human limb morphogenesis by a dominant negative mutation in CDMP1. *Nat Genet* 17: 58–64

61 Wolfman NM, Hattersley G, Cox K, Celeste AJ, Nelson R, Yamaji N, Dube JL, DiBlasio-Smith E, Nove J, Song JJ et al (1997) Ectopic induction of tendon and ligament in rats by growth and differentiation factors 5, 6 and 7, members of the TGF-beta gene family. *J Clin Invest* 100: 321–330

62 Lee KJ, Mendelsohn M, Jessell TM (1998) Neuronal patterning by BMPs: a requirement fir GDF7 in the generation of a discrete class of commissural interneurons in the mouse spinal cord. *Genes Dev* 12: 3394–3407

63 McPherron AC, Lawler AM, Lee SJ (1997) Regulation of skeletal muscle mass in mice by a new TGF-beta superfamily member. *Nature* 387: 83–90

64 Elvin JA, Changning Y, Wang P, Nishimori K, Matzuk MM (1999) Molecular characterization of the follicle defects in the growth differentiation factor 9-deficient ovary. Mol Endocrin 6: 1018–1035

65 Elvin JA, Yan C, Matzuk MM (2000) Oocyte-expressed TGF-β superfamily members in female fertility. *Mol Cell Endocrin* 159: 1–5

66 Zhao R, Lawler AM, Lee SJ (1999) Characterization of GDF-10 expression patterns and null mice. *Dev Biol* 212: 68–79

67 McPherron AC, Lawler AM, Lee SJ (1999) Regulation of anterior/posterior patterning of the axial skeleton by growth/differentiation factor 11. *Nat Genet* 22: 260–264

68 Ying Y, Zhao GQ (2001) Cooperation of endoderm-derived BMP2 and extraembryonic ectoderm-derived BMP4 in primordial germ cell generation in the mouse. *Dev Biol* 232: 484–492

69 Katagiri T, Boorla S, Frendo JL, Hogan BL, Karsenty G (1998) Skeletal abnormalities in doubly heterozygous Bmp4 and BMP7 mice. *Dev Genet* 22: 340–348

70 Solloway MJ, Robertson EJ (1999) Early embryonic lethality in Bmp5;Bmp7 double

mutant mice suggests functional redundancy within the 60A subgroup. *Development* 126: 1753–1768

71 Kim RY, Robertson EJ, Solloway MJ (2001) Bmp6 and Bmp7 are required for cushion formation and septation in the developing mouse heart. *Dev Biol* 235:449–466

72 Zhao GQ, Liaw L, Hogan BL (1998) Bone morphogenetic protein 8A plays a role in the maintenance of spermatogenesis and the integrity of the epididymis. *Development* 125: 1103–1112

73 Zhao GQ, Chen YX, Liu XM, Xu Z, Qi X (2001) Mutation in Bmp7 exacerbates the phenotype of Bmp8a mutants in spermatogenesis and epididymis. *Dev Biol* 240: 212–222

74 Smith WC, Harland RM (1992) Expression cloning of Noggin, a new dorsalizing factor localized to the Spemann organizer in *Xenopus* embryos. *Cell* 70: 829–840

75 Reddi AH (2001) Interplay between bone morphogenetic proteins and cognate binding proteins in bone and cartilage development: noggin, chordin and DAN. *Arthritis Res* 3: 1–5

76 Gazzerro E, Gangji V, Canalis E (1998) Bone morphogenetic proteins induce the expression of Noggin, which limits their activity in cultured rat osteoblast. *J Clin Invest* 102: 2106–2114

77 Abe E, Yamamoto M, Taguchi Y, Lecka-Czernik B, O'Brien CA, Economides AN, Stahl N, Jilka RL, Manolagas SC (2000) Essential requirement of BMPs-2/4 for both osteoblast and osteoclast formation in murine bone marrow cultures from adult mice: antagonism by Noggin. *J Bone Miner Res* 5: 663–673

78 Brunet LJ, McMahon JA, McMahon AP, Harland RM. (1998) Noggin, cartilage morphogenesis, and joint formation in the mammalian skeleton. *Science* 280: 1455–1457

79 Piccolo S, Sasai Y, Lu B, De Robertis EM (1996) Dorsoventral patterning in Xenopus: inhibition of ventral signals by direct binding of chordin to BMP-4. *Cell* 86: 589–598

80 Wardle FC, Welch JV, Dale L (1999) Bone morphogenetic protein 1 regulates dorsalventral patterning in early Xenopus embryos by degrading Chordin, a BMP-4 antagonist. *Mech Dev* 86: 75–85

81 Bachiller D, Klingensmith J, Shneyder N, Tran U, Anderson R, Rossant J, De Robertis EM (2003) The role of chordin/Bmp signals in mammalian pharyngeal development and DiGeorge syndrome. *Development*. 130: 3567–3378

82 Sakuta H, Suzuki R, Takahashi H, Kato A, Shintani T, Iemura Si, Yamamoto TS, Ueno N, Noda M (2001) Ventroptin: a BMP-4 antagonist expressed in a double-gradient pattern in the retina. *Science* 293: 111–115

83 Chimal-Monroy J, Rodriguez-Leon J, Montero JA, Ganan Y, Macias D, Merino R, Hurle JM (2003) Analysis of the molecular cascade responsible for mesodermal limb chondrogenesis: Sox genes and BMP signaling. *Dev Biol* 257: 292–301

84 Guo Q, Kumar TR, Woodruff T, Hadsell LA, DeMayo FJ, Matzuk MM (1998) Overexpression of mouse follistatin causes reproductive defects in transgenic mice. *Mol Endocrinol* 12: 96–106

85 Hayette S, Gadoux M, Martel S, Bertrand S, Tigaud I, Magaud JP, Rimokh R (1998) FLRG (follistatin-related gene), a new target of chromosomal rearrangement in malignant blood disorders. *Oncogene* 16: 2949–2954

86   Tsuchida K, Arai KY, Kuramoto Y, Yamakawa N, Hasegawa Y, Sugino H (2000) Identification and characterization of a novel follistatin-like protein as a binding protein for the TGF-beta family. *J Biol Chem* 275: 40788–40796

87   Kim AS, Pleasure SJ (2003) Expression of the BMP antagonist Dan during murine forebrain development. *Brain Res Dev Brain Res* 145: 159–162

88   Nakamura Y, Ozaki T, Nakagawara A, Sakiyama S (1997) A product of DAN, a novel candidate tumour suppressor gene, is secreted into culture medium and suppresses DNA synthesis. *Eur J Cancer* 33: 1986–1990

89   Gerlach-Bank LM, Cleveland AR, Barald KF (2004) DAN directs endolymphatic sac and duct outgrowth in the avian inner ear. *Dev Dyn* 229: 219–230

90   Picollo S, Agius E, Leyns L, Bhattacharyya S, Grunz H, Bouwmeester T, De Robertis EM (1999) The head inducer Cerberus is a multifunctional antagonist of Nodal, BMP and Wnt signals. *Nature* 397: 707–710

91   Silva AC, Filipe M, Kuerner KM, Steinbeisser H, Belo JA (2003) Endogenous Cerberus activity is required for anterior head specification in *Xenopus*. *Development* 130: 4943–4953

92   Belo JA, Bachiller D, Agius E, Kemp C, Borges AC, Marques S, Piccolo S, De Robertis EM (2000) Cerberus-like is a secreted BMP and nodal antagonist not essential for mouse development. *Genesis* 26: 265–270

93   Bell E, Munoz-Sanjuan I, Altmann CR, Vonica A, Brivanlou AH (2003) Cell fate specification and competence by Coco, a maternal BMP, TGFbeta and Wnt inhibitor. *Development* 130: 1381–1389

94   Hsu DR, Economides AN, Wang X, Eimon PM, Harland RM (1998) The Xenopus dorsalizing factor Gremlin identifies a novel family of secreted proteins that antagonize BMP activities. *Mol Cell* 1: 673–683

95   Khokha MK, Hsu D, Brunet LJ, Dionne MS, Harland RM (2003) Gremlin is the BMP antagonist required for maintenance of Shh and Fgf signals during limb patterning. *Nat Genet* 34: 303–307

96   Shi W, Zhao J, Anderson KD, Warburton D (2001) Gremlin negatively modulates BMP-4 induction of embryonic mouse lung branching morphogenesis. *Am J Physiol Lung Cell Mol Physiol* 280: 1030–1039

97   Murphy M, McMahon R, Lappin DW, Brady HR (2002) Gremlins: is this what renal fibrogenesis has come to? *Exp Nephrol* 10: 241–244

98   Sudo S, Avsian-Kretchmer O, Wang LS, Hsueh AJ (2004) Protein related to DAN and cerberus (PRDC) is a BMP antagonist that participates in ovarian paracrine regulation. *J Biol Chem* 279: 23134–23141

99   Van Bezooijen RL, Roelen BA, Visser A, Van Der Wee-Pals L, De Wilt E, Karperien M, Hamersma H, Papapoulos SE, Ten Dijke P, Lowik CW (2004) Sclerostin is an osteocyte-expressed negative regulator of bone formation, but not a classical BMP antagonist. *J Exp Med* 199: 805–814

100 Winkler DG, Sutherland MK, Geoghegan JC, Yu C, Hayes T, Skonier JE, Shpektor D,

Jonas M, Kovacevich BR, Staehling-Hampton K et al (2003) Osteocyte control of bone formation *via* sclerostin, a novel BMP antagonist. *EMBO J* 22: 6267–6276

101 ten Dijke P, Miyazono K, Heldin CH (1996) Signaling *via* hetero-olimeric complexes of type I and type II serine/threonine kinase receptors. *Curr Opin Cell Biol* 8: 139–145

102 Mishina Y, Suzuki A, Ueno N, Behringer RR (1995) Bmpr encodes a type I bone morphogenetic protein receptor that is essential for gastrulation during mouse embryogenesis. *Genes Dev* 9: 3027–3037

103 Ahn K, Mishina Y, Hanks MC, Behringer RR, Crenshaw EB 3rd (2001) BMPR-IA signaling is required for the formation of the apical ctodermal ridge and dorsal-ventral patterning of the limb. *Development* 128: 4449–4461

104 Gaussin V, Van de Putte T, Mishina Y, Hanks MC, Zwijsen A, Huylebroeck D, Behringer R, Schneider MD (2002) Endocardial cushion and myocardial defects after cardiac myocyte-speci.c conditional deletion of the bone morphogenetic protein receptor ALK3. *Proc Natl Acad Sci USA* 99: 2878–2883

105 Oh SP, Seki T, Goss KA, Imamura T, Yi Y, Donahoe PK, Li L, Miyazono K, ten Dijke P, Kim S, Li E (2000) Activin receptor-like kinase 1 modulates transforming growth factor-beta 1 signaling in the regulation of angiogenesis. *Proc Natl Acad Sci USA* 97: 2626–2631

106 Gu Z, Reynolds EM, Song J, Lei H, Feijen A, Yu L, He W, MacLaughlin DT, van den Eijnden-van Raaij J, Donahoe PK, Li E (1999) The type I serine/threonine kinase receptor ActRIA (ALK2) is required for gastrulation of the mouse embryo. *Development* 126: 2551–2561

107 Gu Z, Nomura M, Simpson BB, Lei H, Feijen A, van den Eijnden-van Raaij J, Donahoe PK, Li E. (1998) The type I activin receptor ActRIB is required for egg cylinder organization and gastrulation in the mouse. *Genes Dev* 12: 844–857

108 Yi SE, Daluiski A, Pederson R, Rosen V, Lyons KM (2000) The type I BMP receptor BMPRIB is required for chondrogenesis in the mouse limb. *Development* 127: 621–630

109 Yi SE, LaPolt PS, Yoon BS, Chen JY, Lu JK, Lyons KM (2001) The type I BMP receptor BmprIB is essential for female reproductive function. *Proc Natl Acad Sci USA* 98: 7994–7999

110 Beppu H, Kawabata M, Hamamoto T, Chytil A, Minowa O, Noda T, Miyazono K (2000) BMP type II receptor is required for gastrulation and early development of mouse embryos. *Dev Biol* 221: 249–258

111 Matzuk MM, Kumar TR, Bradley A (1995) Different phenotypes for mice deficient in either activins or activin receptor type II. *Nature* 374: 356–360

112 Oh SP, Li E (1997) The signaling pathway mediated by the type IIB activin receptor controls axial patterning and lateral asymmetry in the mouse. *Genes Dev* 11: 1812–1826

113 Song J, Oh SP, Schrewe H, Nomura M, Lei H, Okano M, Gridley T, Li E (1999) The type II activin receptors are essential for egg cylinder growth, gastrulation, and rostral head development in mice. *Dev Biol* 213: 157–169

114 Lechleider RJ, Ryan JL, Garrett L, Eng C, Deng C, Wynshaw-Boris A, Roberts AB

(2001) Targeted mutagenesis of Smad1 reveals an essential role in chorioallantoic fusion. Dev Biol 240: 157–167

115 Tremblay KD, Dunn NR, Robertson EJ (2001) Mouse embryos lacking Smad1 signals display defects in extra-embryonic tissues and germ cell formation. *Development* 128: 3609–3621

116 Nomura M, Li E (1998) Smad2 role in mesoderm formation, left-right patterning and craniofacial development. *Nature* 393: 786–790

117 Heyer J, Escalante-Alcalde D, Lia M, Boettinger E, Edelmann W, Stewart CL, Kucherlapati R (1999) Postgastrulation Smad2-de.cient embryos show defects in embryo turning and anterior morphogenesis. *Proc Natl Acad Sci USA* 96: 12595–12600

118 Zhu Y, Richardson JA, Parada LF, Graff JM (1998) Smad3 mutant mice develop metastatic colorectal cancer. *Cell* 94: 703–714

119 Ashcroft GS, Yang X, Glick AB, Weinstein M, Letterio JL, Mizel DE, Anzano M, Greenwell-Wild T, Wahl SM, Deng C, Roberts AB (1999) Mice lacking Smad3 show accelerated wound healing and an impaired local inflammatory response. *Nat Cell Biol* 1: 260–266

120 Yang X, Letterio JJ, Lechleider RJ, Chen L, Hayman R, Gu H, Roberts AB, Deng C (1999) Targeted disruption of SMAD3 results in impaired mucosal immunity and diminished T cell responsiveness to TGF-beta. *EMBO J* 18: 1280–1291

121 Yang X, Chen L, Xu X, Li C, Huang C, Deng CX (2001) TGF-beta/Smad3 signals repress chondrocyte hypertrophic differentiation and are required for maintaining articular cartilage. *J Cell Biol* 153: 35–46

122 Sirard C, de la Pompa JL, Elia A, Itie A, Mirtsos C, Cheung A, Hahn S, Wakeham A, Schwartz L, Kern SE et al (1998) The tumor suppressor gene Smad4/Dpc4 is required for gastrulation and later for anterior development of the mouse embryo. *Genes Dev* 12: 107–119

123 Takaku K, Oshima M, Miyoshi H, Matsui M, Seldin MF, Taketo MM (1998) Intestinal tumorigenesis in compound mutant mice of both Dpc4 (Smad4) and Apc genes. *Cell* 92: 645–656

124 Chang H, Matzuk MM (2001) Smad5 is required for mouse primordial germ cell development. *Mech Dev* 104: 61–67

125 Chang H, Huylebroeck D, Verschueren K, Guo Q, Matzuk MM, Zwijsen (1999) Smad5 nockout mice die at mid-gestation due to multiple embryonic and extraembryonic *Development* 126: 1631–1642

126 Chang H, Zwijsen A, Vogel H, Huylebroeck D, Matzuk MM (2000) Smad5 is essential for left right asymmetry in mice. *Dev Biol* 219: 71–78

127 Galvin KM, Donovan MJ, Lynch CA, Meyer RI, Paul RJ, Lorenz JN, Fairchild-Huntress V, Dixon KL, Dunmore JH, Gimbrone MA Jr et al (2000) A role for smad6 in development and homeostasis of the cardiovascular system. *Nat Genet* 24: 171–174

128 Dong C, Zhu S, Wang T, Yoon W, Li Z, Alvarez RJ, ten Dijke P, White B, Wigley FM, Goldschmidt-Clermont PJ (2002) Deficient Smad7 expression: a putative molecular defect in scleroderma. *Proc Natl Acad Sci USA* 99: 3908–3913

Bone morphogenetic proteins in development

Petra Simic and Slobodan Vukicevic

Department of Anatomy, Medical School University of Zagreb, Salata 11, 10000 Zagreb, Croatia

Embryonic development and tissue regeneration is based on cell communications by conserved families of signaling molecules, such as Dpp/BMP, Wnt, FGF, Nodal and Hedgehog. The vertebrate bone morphogenetic proteins (BMPs) are involved in the development of nearly all organs and tissues (Fig. 1), including the nervous system, somites, lung, kidney, skin and gonads, as well as in critical steps in the establishment of the basic embryonic body plan [1].

Embryonic body plan

All vertebrates, despite their many differences, have a very similar basic body plan. Model organisms used for vertebrates include: frog *Xenopus*, mouse, chick and zebrafish. During the first few hours of development, one-cell egg with animal-vegetal polarity and radial symmetry develops into a late gastrula embryo with dorsal-ventral, anterior-posterior and left-right axes (Fig. 2). The next stage in pattern formation in embryos is allocation of cells to different germ layers – the ectoderm, mesoderm and endoderm. Development of germ layers will be described in terms of their dependence along three axes. The mechanism of pattern formation involves localized maternal determinants, external signals and cell–cell interactions. At this early stage, the embryos are capable of considerable regulation emphasizing the essential role of cell–cell interactions in development. Many candidates for the signals in these interactions have been identified and include also members of TGF-β family.

Dorsal-ventral patterning

Unlike mesoderm, it is generally accepted that all the ectoderm and most of endoderm are specified by maternal factors in the egg. The formation of mesoderm in amphibians is totally dependent on inducing signals from the vegetal region of the blastula (Tab. 1). Early patterning of the mesoderm can be accounted for by a four-

Bone Morphogenetic Proteins: Regeneration of Bone and Beyond, edited by Slobodan Vukicevic and Kuber T. Sampath
© 2004 Birkhäuser Verlag Basel/Switzerland

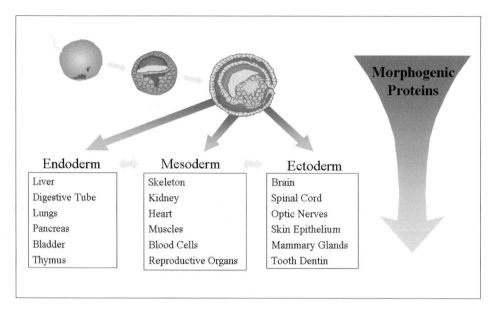

Endoderm	Mesoderm	Ectoderm
Liver	Skeleton	Brain
Digestive Tube	Kidney	Spinal Cord
Lungs	Heart	Optic Nerves
Pancreas	Muscles	Skin Epithelium
Bladder	Blood Cells	Mammary Glands
Thymus	Reproductive Organs	Tooth Dentin

Figure 1
Influence of BMPs on the development of all three germ layers

signal model in *Xenopus* embryo starting at the 32-cell stage (Fig. 3). The first signal (e.g., Vg-1 and activin) is a general mesoderm inducer, specifying a ventral type of mesoderm (mesothelium, mesenchyme and blood). The second signal specifies the dorsal mesoderm, while the third signal (e.g., BMP-4 and Wnt) comes from the ventral side and ventralizes the mesoderm. The fourth signal (e.g., noggin, chordin) originates from the organizer and by interacting with the third signal establishes further pattern within the mesoderm, namely, axial mesoderm (notochord), paraxial mesoderm (somite, muscle) and intermediate mesoderm (kidney) (Fig. 4B) [2]. Then, as described in the next section, dorsalizing signals interfere with endogenous BMP-related signals acting within the ectoderm to specify it as epidermis, resulting in a switch to neural fate.

The most potent ventralizing factor is BMP-4. This role is consistent with the distribution of *bmp-4* transcripts throughout the early gastrula, except in the dorsal lip (Fig. 4A) [1]. Their overexpression in the dorsal territories leads to ventralization of both mesoderm and ectoderm during gastrulation [3]. Loss of BMP signaling leads to severe dorsalization of the embryo, with loss of epidermis, ventral and lateral mesoderm and expansion of dorsal-lateral mesoderm (somites) and anterior neural ectoderm [4].

BMP-2 and BMP-4 share over 90% identity at the amino acid level and are functionally interchangeable in most biological systems. Inactivation of *bmp-2* in the

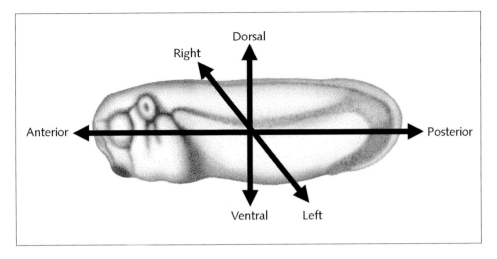

Figure 2
The main axes of a developing embryo (Xenopus laevis tailbud stage embryo)

Table 1 - Signals in early Xenopus development

Factor	Protein family	Effects
Vg-1	TGF-β family	mesoderm induction
activin	TGF-β family	mesoderm induction
BMP-4	TGF-β family	ventral mesoderm patterning
XWnt-8	Wnt family	ventralizes mesoderm
FGF	FGF	ventral mesoderm induction
noggin		dorsalizes-binds BMP-4
chordin		dorsalizes-binds BMP-4
frizbee		dorsalizes-binds BMP-4

mouse results in a phenotype that is somewhat different from that of *bmp-4*, large-ly due to different expression patterns during and before gastrulation. *Bmp-4* is detected by *in situ* hybridization at E3.5 and *bmp-2* is not detectable until E6.0 and is only expressed in visceral endoderm [5]. Without a functional *bmp-2* gene, mutant embryos reveal defects in the development of the amnion and heart, two tissues in which the gene is expressed at high levels.

The *Drosophila* protein Decapentaplegic (*Dpp*) is highly similar in sequence to BMP-4 and BMP-2, and its overexpression in *Xenopus* embryos also results in ven-

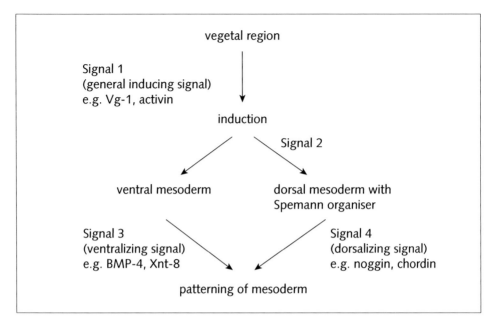

Figure 3
Mesoderm induction in Xenopus

tralization. In *Drosophila*, *dpp* is expressed in the dorsal territories, where it is required for the differentiation of dorsal ectodermal derivatives such as the amnioserosa. The reversed roles of *bmp/dpp* in vertebrates and arthropods support the theory that, during animal evolution, an inversion of the body plan of these two phyla occurred [6].

The activity of BMP signaling is modulated by antagonists such as chordin. The modulation of signaling activity allows a single signal to specify multiple different cell fates. The highest levels of BMP signaling are required for ventral tail fin development and decreasing levels are essential for blood, pronephros and somite development [7]. Similarly, graded BMP signaling is involved in patterning of the neural plate.

Vertebrate chordin and *Drosophila* short gastrulation (Sog) are functional homologues, as ectopic expression of either dorsalizes *Xenopus* embryos and ventralizes *Drosophila* embryos [6]. Chordin and Sog have complementary pattern of expression and an opposite activity to *bmp-4* and *dpp*, respectively. Chordin can bind with high affinity to BMP-4 homodimers, BMP-2 homodimers and BMP-4-BMP-7 heterodimers.

In addition to Chordin, two other extracellular inhibitors of BMP signaling are secreted by the vertebrate organizer – Noggin and Follistatin.

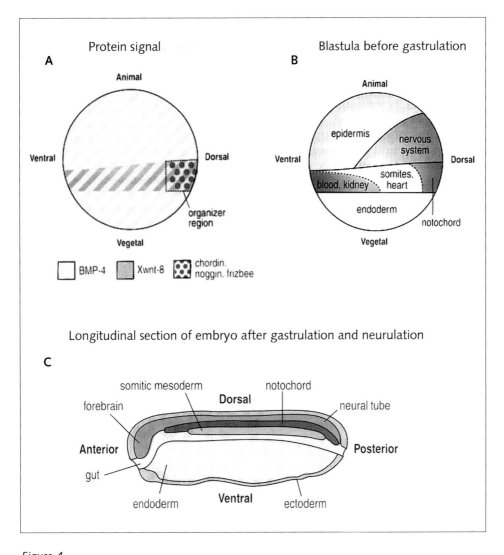

Figure 4
Blastula and gastrula stages of the embryo
A, Distribution of the protein signals in the Xenopus gastrula; B, Fate mape of the Xenopus
blastula; C, Rearrangement of the germ layers during gastrulation and neurulation in Xenopus

Chordin (Chd) and Sog, but not Noggin or Follistatin, are cleaved by *Xenopus* Xolloid and *Drosophila* Tolloid, which are proteins similar to BMP-1 [8]. Xolloid/Tolloid cleaves Chd/Sog – BMP/Dpp complexes, releasing active BMP/Dpp.

The model that emerges from these studies is that antagonistic molecules are secreted by dorsal and ventral cells, leading to a graded distribution of "free" BMP proteins – molecules not complexed to an antagonist – along the dorso-ventral axes of both vertebrate and arthropod embryos (Fig. 5).

Antero-posterior patterning

During gastrulation, the germ layers – mesoderm, endoderm and ectoderm – move to the position in which they will develop into the structure of the adult body. The antero-posterior body axis of the vertebrate embryo emerges, with the head on one end and the future tail on the other (Fig. 4C). After gastrulation, the part of the mesoderm that comes to lie along the dorsal side of the embryo, under the ectoderm, gives rise to notochord and somites and to a small amount of head mesoderm anterior to the notochord [9]. During neurulation, the neural tube (Fig. 4C) is formed from the ectoderm overlying the notochord and develops into the brain and spinal cord.

Several families of secreted signaling molecules have been implicated in the specification and patterning of the antero-posterior axis; in particular, members of the Wnt and TGF-β/BMP family.

Members of the Wnt family of secreted signaling molecules possess axis-inducing and posteriorizing activity when overexpressed. Inhibition of posteriorly localized Wnt signaling by anteriorly localized Wnt antagonists is critical for inducing anterior structures, forebrain and heart, from neural ectoderm and mesoderm, respectively [10]. Negative factors secreted from anterior sources, coupled with posteriorly localized expression of Wnts leads to the establishment of graded Wnt signals along the antero-posterior axis. Interactions between gradients of Wnts and other signaling molecules during gastrulation are critical for the development of tissues in precise locations along the embryonic axes. BMPs also interact with Wnts to specify the locations of various organs. These interactions in neural system, somites and heart development will be described in the following text.

Left-right patterning

Despite an outwardly bilaterally symmetrical appearance, most internal organs of vertebrates display considerable left–right (LR) asymmetry. Mechanistically, the acquisition of LR pattern can be considered to occur in three steps. The first step is, therefore, initial polarization of LR asymmetry with respect to dorsoventral and antero-posterior axes of the embryo. TGF-β superfamily members then function in the second step to reinforce LR asymmetric polarity and to transmit this information to the developing organ primordial. Third, individual organs undergo asym-

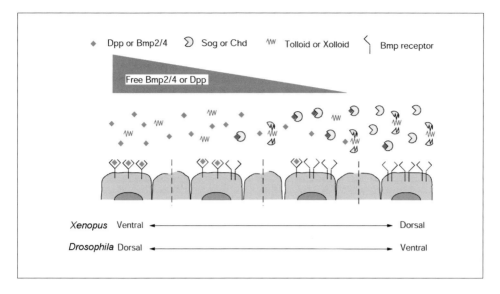

Figure 5
The bmp/dpp genes are expressed uniformly along the dorso-ventral axis
Expression of chd/sog is restricted to the dorsal marginal zone of Xenopus and to the ven-
tral territories of Drosophila. Chd/Sog directly binds Bmp/Dpp and prevents its' association
with receptor, while Xolloid/Tolloid cleaves Chd/Sog–Bmp/Dpp complexes, releasing active
Bmp/Dpp. These opposing activities may fine-tune the level of available Bmp/Dpp along
the dorso–ventral axis.

metric morphogenesis. Disruption of LR pattering, in particular at early points in the programs, cause organ arrangement (heterotaxia) syndromes characterized by striking mirror-image reversals of one or more organs, or in same cases, *situs inversus,* in which the LR asymmetry of the entire viscera is inverted.

The stereotypic orientation of LR asymmetry is ensured by distinct left- and right-side signal transduction pathways that are initiated by divergent members of the TGF-β superfamily of secreted proteins. During early embryogenesis, TGF-β-like protein Nodal is expressed by the left lateral plate mesoderm and provides essential LR signals to the developing organs. BMP signaling is active on the right side of the embryo and must be inhibited on the left in order that Nodal can be expressed. Nodal is set into motion by the inhibition of BMP signaling and then is limited in extent and duration by negative feedback through Lefty, a TGF-β superfamily ligand (Fig. 6) [11]. Left-sided signaling is mediated by transcription factors Smad2 and Smad3, regulated by Nodal, whereas signaling on the right depends on Smad1 and Smad5, regulated by BMPs. It is a confusing problem to understand how similar transduction machinery (e.g., Smad2) can mediate signals that are important for

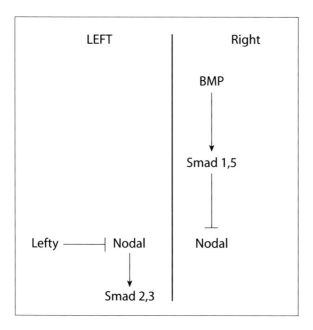

Figure 6
Signaling pathways of LR development
The dashed line indicates midline of the organism.

left–right and dorso-ventral patterning. Presumably, the temporal and special context in which TGF-β superfamily signals are being received prevents cells from confounding mesoderm-inducing, dorso-ventral and left-right patterning signals [12].

Neural patterning

BMPs have been implicated in multiple events during the formation of the central and peripheral nervous system, including primary neural induction, dorso-ventral patterning of the neural tube, regionalization of the brain, eye development, apoptosis and lineage determination in the peripheral nervous system [13].

Neural induction is governed by competitive interactions between epidermalizing and neuralizing factors. BMP-4 and BMP-7 are expressed in the ectoderm and are epidermal inducers, with concurrent inhibition of neurulation [14]. Inhibition of BMP signaling by Chordin, Noggin and Follistatin is required for the conversion of ectoderm to neuroectoderm. In *Xenopus*, overexpression of BMPs or their signaling components alters cell fate from neuronal to epidermal, while expression of dominant negative BMPs or their receptors results in excessive neural tissue [5].

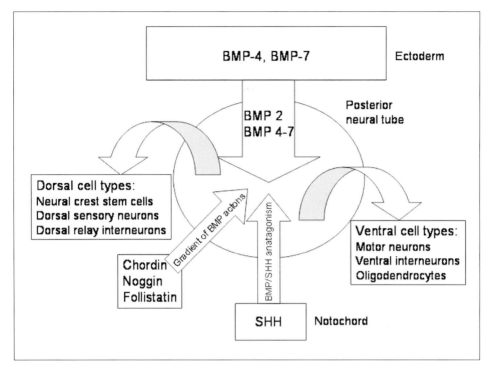

Figure 7
The role of BMPs in dorsoventral patterning of the neural tube
During the development of the neural tube along the antero-posterior axis, BMPs are secreted from the dorsal ectoderm.
BMPs activate certain transcriptional regulators that are associated with the generation of specific dorsal cell types. BMP-binding proteins (noggin, chordin and follistatin), establish a dorsoventral gradient of BMP expression. The combined BMP/SHH expression gradient codes for ventral cell types.

BMPs have also been implicated in dorso-ventral patterning of the neural tube (Fig. 7). In the chick, both *bmp-4* and *bmp-7* are expressed in the non-neural ectoderm bordering the neural plate. As the plate rolls up into a tube, transcript levels of both genes decline in the epidermis, except for *bmp-7* in the forebrain region. However, high levels of BMP-4 transcripts are apparent in dorsal neural folds and in the midline of the neural tube as it closes [15]. By contrast, in murine tissues, BMP-2 rather than BMP-4 is initially expressed in anterior neural folds. Following neural tube closure and maturation, BMP-4 is expressed in the anterior dorsal midline region, BMP-6 along the entire anterior-posterior axis, and BMP-5 and BMP-7 in overlapping domains around the dorsal midline [5, 16]. Dorsalin-1, a BMP relat-

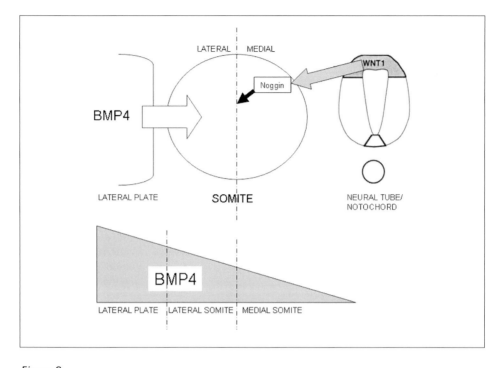

Figure 8
Schematic representation of the interactions between axial and lateral signaling molecules during somite medio-lateral patterning
Lateral plate-derived BMP-4 specifies the lateral somitic compartment. The BMP-4 effect is antagonized by axial factors which include Wnt1. Wnt1 promotes Noggin expression in the medial dermomyotome. This mechanism leads to the establishment of a BMP-4 activity gradient which results in appropriate BMP-4 concentrations to allow medial and lateral somite patterning.

ed molecule, is expressed along the dorsal midline. BMPs act to inhibit the spread of ventralizing signals from the floorplate throughout the neural tube [15]. Sonic Hedgehog (Shh) is a long-range ventralizing factor for the spinal cord.

Bmp-encoding genes are involved in regional neural development. *Bmp-4, bmp-5, bmp-6* and *bmp-7* are expressed in localized regions of the fore- and midbrain in the mouse embryo. *Bmp-7* null embryos have severe defects in eye development [17]. BMP-7 maintains survival of the optic cup and lens.

Neural crest cells, a pluripotent cell population, migrate away from the neural tube and develop into a variety of tissues including brachial arches, a template for construction of skeletal elements of the head, the sensory and autonomic nervous systems and pigment cells. BMP-4 induces regional apoptosis in the nervous system, namely in the dorsal regions of the odd-numbered rhombomeres (brachiogenic neural crest cells) [18]. In the developing peripheral nervous system, BMP-2 and BMP-4 play an instructive role in programming the elaboration of the neuronal lineage from neural crest stem cells. Furthermore, BMP-4 and BMP-7 are expressed by the dorsal aorta at a crucial time during post-migratory neural crest-sympathetic neuronal differentiation. BMP-2, BMP-4 and BMP-7 induce the expression of adrenergic sympathetic neuronal subset.

Somite patterning

Somites are transient embryonic structures found on either side of the neural tube. Somites mature according to a rostro-caudal gradient of differentiation and become subdivided into the dermomyotome dorsally, which gives rise to dermis and muscle and the sclerotome ventrally which yields the axial skeleton [19]. During this process, somites acquire a medio-lateral polarity. Medial dermomyotome gives rise to axial muscles, whereas body wall and limb bud muscles derive from lateral dermomyotome. The polarity along the medio-lateral axis is achieved through the antagonistic influences of the medial neural tube and the lateral plate. Medial patterning is promoted by Wnt1 which is expressed in the dorsal neural tube. Lateral patterning is mediated by BMP-4 which is delivered by the lateral plate and is counteracted locally by Noggin expressed in the medial dermomyotome. Noggin expression in the somite is regulated by Wnt1. Therefore, somite medio-lateral patterning results from a signaling cascade in which Wnt1 produced by the neural tube promotes noggin expression in the medial somite which in turn antagonizes lateral plate-derived BMP-4. This type of BMP-4-Noggin interaction would lead to the establishment of a BMP-4 activity gradient resulting in appropriate BMP-4 signaling to allow medial and lateral somite patterning (Fig. 8).

Limb patterning

The vertebrate limb has a complex arrangement of differentiated cells and tissues including skeleton, muscle, tendon, ligament, skin etc., and this arises from the small bud of apparently homogenous undifferentiated mesenchyme cells encased in ectoderm (Fig. 9). Specification of limb pattern is controlled by cell–cell interactions within the bud. Three different regions signal to undifferentiated mesenchyme at the tip of the limb bud (Fig. 9): (1) the apical ectodermal ridge, which mediates limb bud outgrowth along the proximo-distal axis (shoulder to fingers); (2) ectoderm covering the sides of the limb bud which controls dorso-ventral pattern (knuckle to palm); (3) the polarizing region, a small group of mesenchyme cells at the posterior

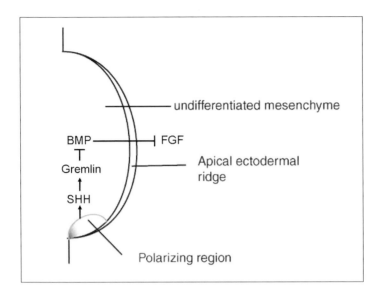

Figure 9
Schematic diagram of an early vertebrate limb
Shh signaling from polarizing region induces expression of several target genes, including Gremlin. Gremlin regulates levels of BMP signaling in the mesenchyme and/or AER. High levels of BMP signaling promote regression of the AER whereas lower levels may positively regulate FGF expression in the AER.

margin of the limb bud, governing antero-posterior pattern (thumb to little finger) [20]. The ectodermal signal from the proximo-distal organizer (the apical ectodermal ridge) is FGF8 and from dorso-ventral organizer is WNT7a, and they both act to position anterior-posterior organizer (the polarizing region) to the distal and dorsal limb mesenchyme [21]. The polarizing region secrets SHH protein. SHH acts to limit the expression of another FGF family member, FGF-4, to the posterior aspect of the apical ectodermal ridge [22]. The path from SHH to FGF-4 is indirect. SHH acts *via* the Formin (limb deformity) gene product to positively regulate expression of Gremlin, an antagonist of BMP signaling molecules. BMP in turn serves to repress expression of *fgf-4* in the apical ectodermal ridge (Fig. 9) [23, 24]. Transcriptional activation of *fgf-4* is dependent on suppression of BMP signaling [25]. Another effector downstream of Formin is BMP-2.

Bmp-2, *bmp-4* and *bmp-7* are all expressed in the apical ectodermal ridge where they control cell proliferation and signaling. In the limb mesenchyme *bmps* have different patterns of expression. High levels of *bmp-4* are seen in the posterior and anterior mesenchyme as well as in the progress zone, and later in joints and ventral footpads [1]. *Bmp-2* is first expressed in regions of the posterior mesenchyme and later in the joints and footpads [1]. *Bmp-7* RNA is first seen diffusely throughout the limb bud mesenchyme and then at high levels around the digit rudiments and in the interdigital mesenchyme, where there is normally a high rate of programmed cell death [26].

Mouse embryos null mutant for *bmp-7* [17, 27] or heterozygous for a null allele of *Bmp-4* [28] display pre-axial (anterior) polydactyly. The absence of webbing indicates that interdigital cell death occurs normally in the mutants while the polydactyly suggests that *Bmp-4* and *bmp-7* normally restrict cell proliferation in the developing limb.

Musculo-skeletal development

Skeletal development

Cartilages and bones develop from multiple embryonic origins (Fig. 10). For example, the craniofacial skeleton in the first branchial arch and other regions of the developing head develops largely from neural crest cells originating from the dorsal neural tube [29]. The ribs and vertebrae form from the sclerotomal part of the somites and appendicular skeleton is derived from the lateral mesoderm. Although the embryonic origin of various skeletal elements is different, the precursor cells will eventually differentiate into three specific cell types that are the same in every skeletal element [30]. The chondrocyte is of mesodermal origin [31], the osteoblast is from neural crest or mesodermal origin, and osteoclast originates from the macrophage monocyte linage.

During vertebrate development, the mesenchymal cells in areas destined to become bones form condensations that have the general shape of the future skeletal elements. This process provides the anlagen or model for the future skeleton. Condensation growth is regulated through signaling pathways involving BMP-2 and BMP-4 [32]. BMP-2 is expressed in mesenchyme surrounding condensations and modulates the condensation size [33]. Growth of condensation ceases when Noggin inhibits BMP signaling, setting the stage for transition to the next stage of skeletal development, namely overt cell differentiation. Recruitment of cells to condensations by BMP is also consistent with the known action of BMP in inducing ectopic bone in adults: implanting BMP outside the skeleton elicits a cascade of events – mesenchymal cell condensation, overt differentiation of cartilage, vascular invasion and replacement of cartilage by bone [34]. This observation reinforces the idea that condensation is an essential initial step in ectopic mineralization, as it is in normal and ectopic skeletogenesis [35]. Several condensations are either abnormal or absent in short ear mice, which harbour various inactivating mutations of BMP-5 gene [36]. The short ear mouse displays numerous skeletal abnormalities such as reduc-

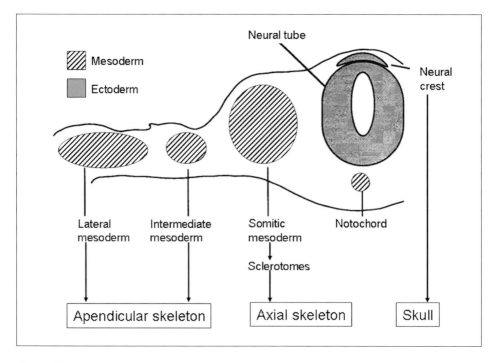

Figure 10
Embryologic structures involved in skeleton formation

tion in body size, absence of xyphoid process, reduction of ventral processes of the cervical vertebrae, deletion of one pair of ribs and a reduced size of the auricle [36]. Mice missing the gene encoding GDF-5 or BMPR1B have abnormalities in digit formation that involve failure of extension of condensations leading to digit formation [37]. Brachypodism is a result of three independent mutations in the GDF-5 gene [38]. Mutations of *cdmp-1*, a human homologue of gdf-5, have been implicated in two recessive chondrodisplasias: the Hunter-Thompson chondrodysplasia [39] and the Grebe type chondrodysplasia [40]. In both cases, the appendicular skeleton is shortened, while the axial skeleton remains largely intact.

BMPs have multiple important roles during cartilage development in the growth plate as well as in joint formation. BMP-2, -3, -4, -5 and -7 are expressed in the perichondrium, BMP-2 and -6 are expressed in hypertrophic chondrocytes and BMP-7 is expressed in proliferating chondrocytes of the growth plate. GDF-5, -6 and -7 are expressed at sites of subsequent joint formation [41]. Addition of BMPs to bone explants increases proliferation of chondrocytes and Noggin blocks chondrocyte proliferation. Furthermore, addition of BMP-2 delays terminal differentiation of

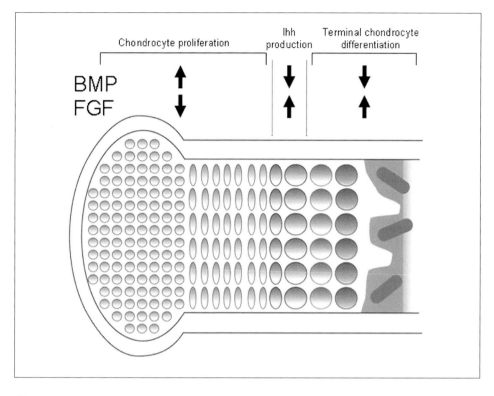

Figure 11

Opposing actions of BMPs and FGFs on the growth plate

FGFs act to decrease chondrocyte proliferation, to increase the production of Ihh, and to accelerate the differentiation of hypertrophic chondrocytes into terminally differentiated chondrocytes. BMPs act at each of these steps in a manner opposite to that of FGFs.

hypertrophic chondrocytes and Noggin hastens terminal hypertrophic differentiation [42]. BMP signaling increases the expression of Indian Hedgehog (Ihh) by pre-hypertrophic chondrocytes [42] and so can increase both the proliferation of chondrocytes and the length of proliferating columns of chondrocytes. In both of these actions, BMPs oppose the effects of FGF signaling to decrease chondrocyte proliferation and to decrease Ihh expression. Because BMPs and FGFs also have opposite effects on terminal hypertrophic chondrocyte differentiation, these pathways can be considered as antagonizing each other at several levels (Fig. 11) [42]. Joint development consists of two main phases: joint patterning and tissue differentiation. CDMP-1/GDF-5 plays a role in the determination of the site of the joint formation. BMP-7 acts as an inhibitory factor for joint formation and the discontinuities in its expression has a permissive role [43] (see chapter by Luyten et al.)

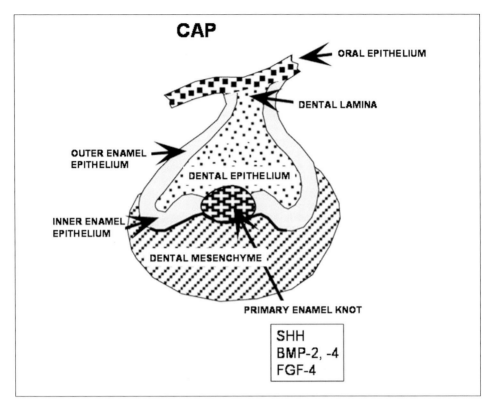

Figure 12
Schematic representation of the sagittal section of cap stage of mouse tooth development

Tooth development

The jaw apparatus is derived from the first branchial arch derivatives. Teeth also develop on the first branchial arch and are derived from both ectoderm and the underlying mesenchyme [40]. Reciprocal signaling interactions between these cell populations also control the odontogenic developmental program, from early patterning of the future dental axis to the initiation of tooth development at specific sites within the ectoderm [44]. The ectodermal bud first forms a cap and then a bell-shaped structure. Subsequently, the mesenchymal papilla differentiates into an epithelial layer (the future odontoblasts) in apposition with the overlying ectoderm (the future ameloblasts) [1].

BMPs act at several different stages in tooth development, in cell proliferation, apoptosis, epithelial–mesenchymal interactions and differentiation [45]. At the bud and cap stages of tooth development, cells which express *Shh*, *bmp-2*, *bmp-4* and *fgf-4* are tightly localized in nested domains in a central region of the ectodermal epithelium known as the enamel knot (Fig. 12) [46]. BMP-4 plays an important role in the earliest epithelial–mesenchymal interactions that initiate tooth development [1]. BMP-2 inhibits the proliferation in the epithelium, whereas BMP-4 may regulate the onset of apoptosis, terminating the signaling function of the knot [46]. At the late bell stage, *Shh*, *bmp-2* and *bmp-4* are expressed in the odontoblast and ameloblast layers, where they are involved in the proliferation and differentiation of these specialized cells [45].

Muscle development

The concentration gradients of BMP and counter gradients of BMP antagonists specify paraxial, intermediate and lateral mesoderm (Fig. 10) [47]. Specification of skeletal myoblasts occurs in the paraxial mesoderm in response to a number of signaling molecules: Shh, Wnts and Noggin as positive activators and BMP-4 as an inhibitor [48]. A BMP inhibitory signal prevents the premature expression of muscle-specific markers before somites are formed. During the maturation of somites, dorsal part forms the dermomyotome. The medial dermomyotome gives rise to axial muscles, whereas body wall and limb bud muscles derive from the lateral dermomyotome.

GDF-8, a member of the TGF-β superfamily is expressed in the myotome compartment of developing somites [49]. GDF-null mice show large and widespread 2–3 time increase in skeletal muscle mass as compared to wild-type animals [49]. The increase in the mass appears to result from a combination of muscle cell hyperplasia and hypertrophy. These results suggest that GDF-8 functions specifically as a negative regulator of skeletal muscle growth [50].

BMPs and organogenesis

Development and differentiation of every cell within our body is regulated by BMP signaling sometime during our life [6]. BMPs mediate inductive interactions between mesenchymal and epithelial cells, control cell proliferation and apoptosis, and in some organs undergoing branching or complex morphogenesis, they are expressed by a small group of cells that behave as organizing centers [2].

Kidney

The development of the kidney begins with the invasion of the ureteric bud into the metanephric mesenchyme. Reciprocal interactions between the two tissues lead to

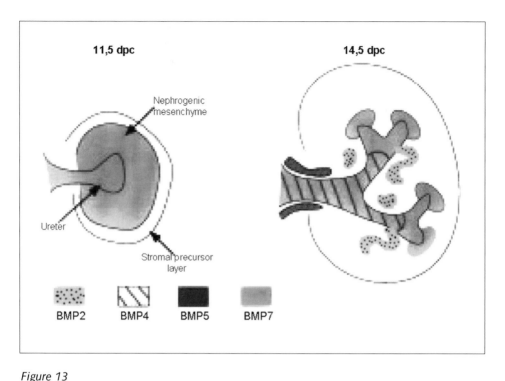

Figure 13
BMP family members are expressed throughout kidney development from the initial inductive interactions at 11.5 dpc through later stages of tubule differentiation
At 11.5 dpc, BMP-7 is expressed in both the ureteric bud and the nephrogenic mesenchyme. No BMP family members have been found to be expressed in the stromal precursor layer. At 14.5 dpc, BMP-4 expression is observed in the maturing tubules along with BMP-2 and BMP-7. BMP-4 and BMP-5 are co-expressed in a population of stromal cells associated with the trunk of the ureter.

the induction of mesenchymal cells near the bud and they convert to an epithelium which goes on to generate the functional filtering unit of the kidney, the nephron [51]. Many BMPs are expressed during organogenesis of the metanephric kidney (Fig. 13). *Bmp-7* is detected in the ureteric bud as it emerges from the mesonephric (Wolffian) duct at 11 dpc and its expression is maintained in derivatives of the bud throughout the development [52] (Fig. 14). *Bmp-3* expression is observed in the trunk of the ureter and the collecting ducts, but not in the tips [52]. *Bmp-2* expression is first detected in the pretubular aggregates and is maintained in the distal part of the early tubules [52]. *Bmp-3, bmp-4* and *bmp-7* are co-expressed in the Bowman's capsule of the developing glomerulus, while *bmp-4* and *bmp-7* are expressed in the presumptive podocyte layer [52]. *Bmp-4* transcripts are detected in a mes-

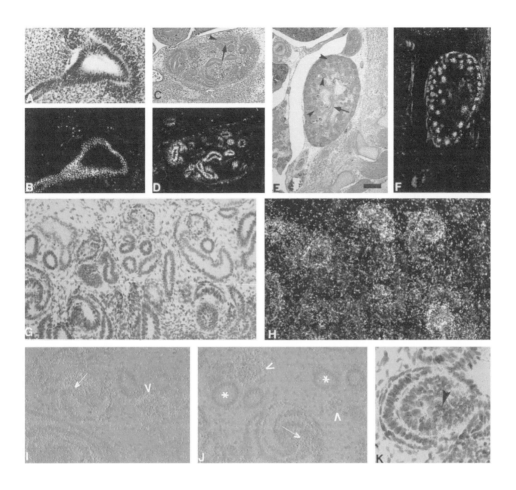

Figure 14
Localization of OP-1 mRNA during mouse kidney development by in situ *hybridization*
In E11.5 embryonic metanephros, BMP-7 transcripts are present in the epithelium of the
ingrowing ureteric bud, but absent from mesenchyme (A and B). At E12.5, BMP-7 transcripts
are present both in ureteric bud epithelium and induced mesenchymal condensates adjacent
(C, arrowhead) to the ureteric bud (C and D, arrow). At E14.5, BMP-7 transcripts are present
in mesenchymal condensates (I and J, arrowheads), comma (I, arrow), and S-shaped (J,
arrow) bodies and in vascularized glomeruli (G and H, arrowheads; E and K), and they
remain absent in tubules derived from the ureteric bud (E and F, arrow). No BMP-7 tran-
scripts were detected in convoluted tubules and collecting ducts (G and H; J, asterisk). A, C,
and E, brightfield; B, D, F, and H, darkfield; and G, I, J, and K, epipolarization. (E, bar =
200 μm.)

enchymal cell population adjacent to the ureter. Later this cell population also expresses *bmp-5*, while loose stromal cells, at some distance from the ureteric tree, express *bmp-6*. In most tissues *bmp* mRNA expression and BMP protein are found colocalized [53]. Restricted diffusion of BMP molecules may result from strong binding to basement membrane-binding proteins [53] or heparin sulfate containing proteoglycans [54].

Homozygous mice carrying a disrupted BMP-7 gene die soon after birth and possess dysplastic kidneys that appear arrested in development [17]. Kidney development is essentially normal until 14.5 dpc, but then branching of the ureteric epithelium, formation of the mesenchymal condensations and differentiation of epithelial structures, all tissues in which *bmp-7* is expressed, all cease. The mutant kidneys suffer massive apoptosis of the disorganized, uninduced mesenchymal cells [55] showing that BMP-7 is necessary for their continued proliferation, differentiation and survival. BMP-2 and BMP-4-null embryos die early during gastrulation and, therefore, preclude analysis of their role during kidney development [56, 57]. Both targeted mutation and kidney culture studies show that BMP-4 inhibits the ureteric budding process, but promotes the growth or elongation of the stalk of ureteric bud, thus exhibiting a unique function in kidney branching morphogenesis [58]. BMP-2 and BMP-7 have distinct and opposite effects on renal branching morphogenesis, which might be caused by competition for the same receptor complexes in ureteric budding [59]. BMP-5-null animals have the ureter increased in length and convoluted though not dilated [60].

Recombinant BMP-7 can induce differentiation of the metanephric mesenchyme in *in vitro* culture experiments [61]. When *in vivo bmp-7* activity is lost, mesenchymal development proceeds for a while before ceasing. It is possible that other BMPs expressed in the developing kidney compensate for the absence of BMP-7 early in the development, but cannot suffice as the number of end buds increases [1]. BMP-7 promotes survival of explants of the metanephric mesenchyme [62]. BMP-7 also functions to inhibit differentiation of the metanephric mesenchyme [62]. BMP-7 signaling in conjunction with FGF-2 promotes expansion of the stromal progenitor cell population adjacent to the nephrogenic mesenchyme [62]. This growth occurs simultaneously with inhibition of tubulogenesis.

As the kidney matures, most of the functional units are epithelial tissues derived from either the ureteric bud or from mesenchyme to epithelial transformation [52]. At low concentrations (< 0.5 nM), BMP-7 signaling promotes general growth of the explant and increased branching of the ureter while at higher doses, there is a decrease in both parameters [59]. In contrast, BMP-2 is inhibitory to growth and branching [52].

Mature epithelium is able to replace cells that are damaged and BMP-7 is an important component of this process. Intravenous administration of BMP-7 post ischemic injury improved both the function and histology of the kidney [63].

Gastro-intestinal tract

Gut

The gut develops ventrally from an endodermal tube surrounded by splanchnic mesoderm. Signaling between the endoderm and mesoderm establishes the normal antero-posterior axis and tissue-specific differentiation. Factors implicated in positional specification of the anterior-posterior regions of the gut (Fig. 15) include endodermally expressed *Shh*, mesodermally expressed *bmp-4* and members of the Hox gene family [64]. *Shh* can induce *bmp-4* in the mesoderm of the hindgut and midgut, but *bmp-4* is not induced in the stomach region of the foregut, the gizzard [65]. BMP-4 has several roles in the gut (Fig. 15). First, BMP signaling is responsible for mediating the thickness of the mesoderm in the small intestine [64, 66]. BMP-4 mediates both proliferation and differentiation of the splanchnic mesoderm. Ectopic BMP-4 in the gizzard results in a thinning of the mesoderm [64]. Second, BMP signaling plays a role in regulating smooth muscle differentiation [65]. The mesoderm of the gut is organized into two major layers, an inner submucosa and an outer muscularis layer of the smooth muscle. In the gizzards with ectopic BMP signaling, smooth muscle is disorganized and regions of the mesoderm are undifferentiated [65]. Third, BMP signaling patterns the pyloric sphincter at the border between the gizzard and the small intestine [65]. Hox gene mesodermal expression is restricted to the most posterior part of the hindgut (cloaca) and leads to regionally specific differentiation of the gut endoderm [67].

Salivary gland

Embryonic salivary glands initiation and branching morphogenesis are dependent on cell–cell communications between and within the epithelium and mesenchyme. BMPs and FGFs play reciprocal roles in regulating branching morphogenesis [68]. FGF-7 and FGF-10 increase the size of the buds and the width of the ducts. BMP-4 inhibits the number of buds and further branching, while BMP-7 increases the number of the buds [68].

Liver

The embryonic hepatocytes normally arise by diversion of an endodermal cell population that would otherwise default to a pancreatic fate [69]. Convergent FGF and BMP signals from distinct mesoderm cell types, the cardiac mesoderm and septum transversum mesenchyme, control this transition [69]. The early septum transver-

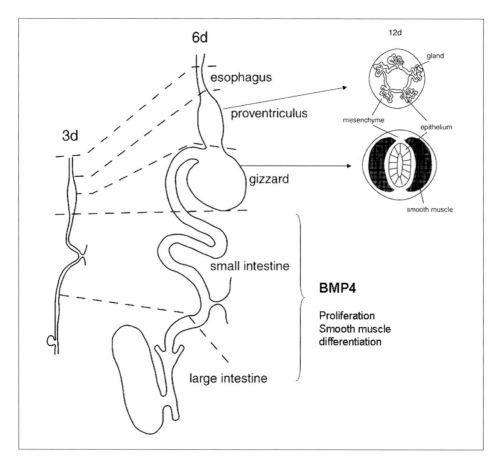

Figure 15
Schematic figure of development of proventriculus and gizzard in chicken embryo
On embryonic day 3 (3d), the gut is a simple tube. On 6d, digestive organs can be distin-
guished. On 12d, the epithelium has formed many compound glands in the proventriculus.
BMP-4 regulates proliferation and subsequent mesodermal thickness within the small intestine.

sum mesenchyme cells produce BMPs -2, -4 and -7 [70]. BMP-9 is an autocrine/
paracrine mediator in the hepatic reticuloendothelial system [71].

Pancreas

The pancreas originates from a ventral and dorsal evagination from the endoderm,
which later fuse to give rise to the adult organ. Correct morphology is obtained

through growth and iterative branching of the epithelia and is dependent on interactions with the surrounding mesenchyme. FGFs and BMPs are crucial mediators of signals between epithelia and mesenchyme [72]. Several FGFs are able to promote proliferation and growth of pancreatic epithelia [73]. BMP-4, BMP-5, BMP-6 and BMP-7 are expressed at the time of endocrine specification [72]. BMP-6 misexpression assay results in complete agenesis of the pancreas [72].

Lung

The lung develops from a bud of the foregut endoderm surrounded by splanchnic mesoderm. Reciprocal interactions between the epithelial endoderm and surrounding mesenchyme are critical for the branching and differentiation of the lung [74]. The proximal–distal (P-D) axis is delineated by two distinct epithelial subpopulations: the proximal bronchiolar epithelium and the distal respiratory epithelium. BMPs are important signaling molecules that pattern the lung along the proximal-distal axis. *Bmp-3*, *-4*, *-5* and *-7* exhibit complex expression patterns within the developing lung with *bmp-3*, *bmp-4* and *bmp-7* expression observed in the developing epithelium and *bmp-5* expression observed in the lung mesenchyme [75, 76]. *Bmp-4* is expressed in the epithelial cells in the tips of the growing limb buds [77]. Endodermal cells at the periphery of the lung, which are exposed to high levels of *bmp-4*, maintain or adopt a distal character, while cells receiving little or no *bmp-4* signal initiate a proximal differentiation program [70]. Inhibiting *Bmp* signaling results in a severe reduction in distal epithelial cell types and a concurrent increase in proximal cell types [78].

Once a lung bud has been initiated by the influence of *Shh* [74], *Fgf-10* in the mesenchyme directly promotes both the proliferation and chemotaxis of the underlying epithelium and the bud extends. As the bud outgrowth continues, endodermal *bmp-4* expression increases. *Fgf-10* and *bmp-4* play opposing roles during branching morphogenesis [79]. The BMP-4 produced in the tip acts as a lateral inhibitor of budding, thus ensuring a single extending bud, rather than a cluster of buds, in response to mesenchymal FGF-10. Only at discrete distances from the tip, where *bmp-4* levels are low but *fgf-10* expression is maintained, can a new bud form [79]. In this way a combination of *bmp-4* and *fgf-10* signaling contributes to the patterning of the early lung. Additionally, BMP signaling is regulated by specific inhibitors such as gremlin and noggin (Fig. 16). This regulation involves both spatial and temporal restriction of BMP activity, since gremlin is expressed in the proximal epithelium of the lung and in the later stages of lung development, whereas noggin is expressed in the distal mesenchyme of the lung and early in the lung development [75]. A fine tuned balance of BMP activity and inhibition is required for proper proximal-distal patterning in the lung during embryonic development.

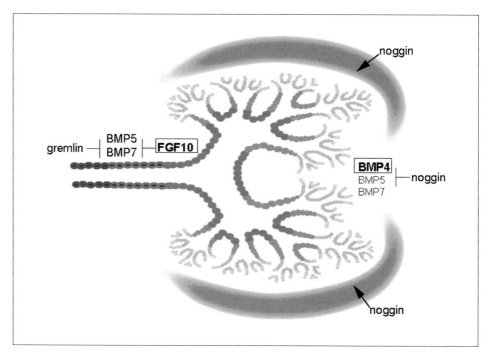

Figure 16
Regulation of BMP signaling in the developing lung
The distal epithelium is exposed to high levels of BMP-4, whereas the proximal epithelium is exposed to high levels of FGF-10. The activity of BMPs is regulated by gremlin in the proximal regions of the lung, whereas noggin regulates BMP activity in the distal regions.

Cardio-vascular system

Heart

Heart muscle cell specification (cardiac myogenesis) and creating the four-chambered heart (cardiac morphogenesis) are subject to regulation by BMPs and their receptors [80]. At least six BMPs are expressed in the heart: BMP-2, -4, -5, -6, -7 and -10, with distinct but partially overlapping distributions [80]. *Bmp-2* is first detected in the promyocardium and its surrounding mesoderm and subsequently in the myocardium of the common atrium and atrio-ventricular canal [5]. Mice deficient for BMP-2 display abnormal placement and poor development of the heart [57]. *Bmp-4* is expressed predominantly on the left side of the heart tube [81]. *Bmp-4* knockout mice fail to form mesoderm and cardiac development is impaired [56].

Bmp-7 is expressed widely in the myocardium while *bmp-5* is mainly in the ventricular myocardium and *bmp-6* in dorsal myocardium, where the level of *bmp-7* expression is relatively low [5]. Single mutants of either *bmp-6* or *bmp-7* do not reveal any heart defects, while their double mutants exhibit defects in heart cushion formation and separation [82]. Delayed heart development is also observed in *bmp-5* and *bmp-7* double mutants [83].

The heart forms soon after gastrulation in a specific region of the anterior mesoderm adjacent to the endoderm, while blood cells arise from the posterior mesoderm [84]. BMP-2 secreted from anterior endoderm plays a key role in cardiac induction [85]. BMP-2 and BMP-4 can induce ectopic expression of the heart-specific genes in non-precardiac mesoderm and its differentiation to beating cardiomyocytes, whereas inhibition of BMP-2, -4 and/or -7 signaling blocks late expression of heart-specific genes and cardiac differentiation [86, 87]. In addition to BMPs, the Wnt proteins also influence the determination of early mesodermal cells to become heart [84]. Expression of BMPs is detected in the endoderm along the entire anterior–posterior axis, whereas the members of the Wnt family are highly expressed in the neural tube located at the posterior region of the embryo [88]. Wnt3a and Wnt8 inhibit cardiogenesis when misexpressed in the anterior heart-forming regions of chick and frog embryos [89]. Wnt antagonists, like crescent and Dkk-1, are expressed in the region of anterior endoderm permissive for heart formation where it can interfere with Wnt signaling and promote cardiogenesis [84] (Fig. 17). BMP signaling can also be blocked by BMP antagonists noggin and chordin, which are secreted from the notochord and cooperate with Wnts to prevent cardiogenesis [90]. Two orthogonal gradients, one of Wnt activity along the anterior–posterior axis and the other of BMP signals along the dorsal–ventral axis, intersect in the heart forming region to induce cardiogenesis in a region of high BMP and low Wnt activity [89].

After the induction, progenitor cells migrate anteriorly over the foregut lip to form the cardiac crescent and fuse at the midline to create the linear heart tube, which consists of an outer myocardium and inner endocardium (Fig. 17). As development proceeds, the heart tube generates a right-sided bend, the d-loop, and primitive cardiac chambers become evident as a pattern of swellings and constrictions along the heart tube [85]. Looping is guided by an embryonic left–right axial pathway that determines the rightward direction of ventricular bending and distinct morphological identities of left and right atria. In zebrafish, BMP-4 is expressed predominantly on the left side of the heart tube and may impart laterality of the developing heart tube [91]. After the completion of the d-loop, some of the endocardial cells change their phenotype to that of mesenchymal cells, and this embryonic phenomenon is called endothelial–mesenchymal transformation [92]. Endothelially expressed TGF-β3 and myocardially expressed BMP-2 act synergistically in the initiation of endocardial cushion endothelial-to-mesenchymal transformation and consequently septation and valvulogenesis completing the separation and connectivity of the chambers [93].

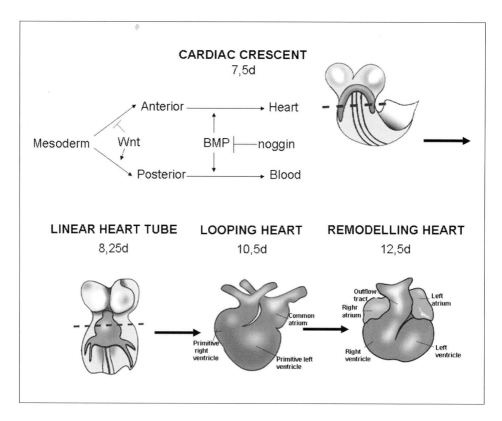

Figure 17

Heart development

Pictures show the major transitions in early mammalian heart development. BMP and Wnt signaling pathways influence the decision of early mesodermal cells to become heart or blood. BMP signaling plays a permissive role in heart and blood formation, which are localized to the anterior and posterior mesoderm. Wnt signaling promotes development of blood and inhibits cardiogenesis.

Hematopoiesis and vasculogenesis

"Blood islands" of primitive hematopoietic cell clusters, surrounded by a layer of endothelial cells, form in the yolk sac [94]. Formation of embryonic hematopoietic and endothelial (angioblastic) stem/progenitor cells from nascent mesoderm requires signals from the adjacent outer layer of visceral endoderm. Indian hedgehog (Ihh) is secreted by visceral endoderm and alone is sufficient to induce hematopoiesis and vasculogenesis in explanted embryos [95]. Ihh functions through upregulation of genes encoding *bmp-4* in target tissue [95, 96].

Gametogenesis

Oogenesis

Female germ cells enter meiosis I during embryonic development. As the oocyte develops, a group of surrounding somatic cells start to proliferate and together they form a primordial follicle which further matures to become a fully grown follicle. Folliculogenesis is regulated by local ovarian factors and systemic growth factors or hormones. Among the local ovarian factors, TGF-β superfamily members have a biological role in folliculogenesis. These members include BMP-4, BMP-8b, GDF-9, BMP-15, BMP-6, activin, inhibin, TGF-β and Mullerian inhibitory substance (MIS) [97].

BMP-4 and BMP-8b produced by the extraembryonic ectoderm are essential for the generation of primordial germ cells in the mouse [98]. *Bmp-4* and *bmp-7* are also expressed in the adult ovary, especially in thecal cells of Graafian follicles [99]. GDF-9 is produced by the oocytes and targeted deletion of the GDF-9 gene resulted in failure of theca cells to organize, abnormal oocyte growth and decreased granulosa cell proliferation which resulted in a block of folliculogenesis at the primary follicle stage [100]. BMP-15 (also known as GDF-9b) is an oocyte–derived factor which is expressed from the primary phase follicle onward [97]. A natural X-linked mutation in sheep identified BMP-15 to be essential for female fertility and that natural mutations can cause both infertility phenotypes in homozygous mutants and increased ovulation rate in heterozygous individuals, showing dosage-sensitive effect [101]. BMP-6 is also expressed in the oocyte. Based on knockout mouse model BMP-6 does not play an essential role in ovarian function [102]. Activin is produced by the granulosa cells and promotes its proliferation [103]. Inhibin is also produced by the granulose cells, but it has an endocrine effect as it is released into the circulation and suppresses FSH secretion [103]. TGF-β is expressed in follicular cells and has proliferative and cytodifferentiation actions on granulosa cells [97]. MIS, produced in granulosa cells, plays an important role in regulating early follicle development [104].

BMP-7 is expressed at high levels in the endometrium of the uterus of non-pregnant mice. During pregnancy, the BMP-7 mRNA in the endometrium rapidly declines at 4 dpc. Thereafter, BMP-7 transcripts are detected in the trophoblastic giant cells of the placenta and the fetal tissues [105, 106].

Spermatogenesis

Spermatogenesis is regulated systemically and locally by cell–cell interactions partially mediated by BMPs. *Bmp-2* is mainly expressed in the interstitial compartment, *bmp-4* mainly in spermatocytes, *bmp-7*, *bmp-8a* and *bmp-8b* in germ cells of several stages [5]. *Bmp-7*, *bmp-8a* and *bmp-8b* have overlapping functions in supporting meiotic germ cell survival. Targeted mutagenesis of *bmp-8b* causes male infertility due to germ cell degeneration [107]. Homozygous *bmp-8a* mutants do not show obvious germ cell defects during the initiation of spermatogenesis, but 47% of adult males show germ cell degeneration, establishing a role of *bmp-8a* in the maintenance of spermatogenesis [107]. *Bmp-4* heterozygotes alone on C57BL/6 background show a decrease in testis weight and an increase in the degeneration of meiotic germ cells [5]. Meiotic germ cells are prime targets of BMPs.

Skin

BMP plays essential roles in the development of skin and cutaneous appendages. During gastrulation, when ectodermal cells choose between neural and epidermal fates, BMP-4 induces epidermal differentiation and inhibits neural development [108]. During later steps of embryonic development, BMPs show strict spatiotemporal expression patterns in the skin epithelium and mesenchyme as well as in neural crest cells that differentiate into melanocytes and components of cutaneous innervation [109].

Epidermis

Bmp-6 is expressed in suprabasal layers of murine embryonic epidermis [110], while *bmp-7* is expressed in the basal epidermal layer during the last stages of development [111]. Both BMP-6 and BMP-7 are involved in the control of cell proliferation and differentiation in the epidermis. Noggin is seen in mesenchymal cells under the basal membrane of the epidermis, whereas later, noggin is restricted to the dermal papilla and connective tissue of the hair follicle [109]. High expression of BMP-6 transgene in the suprabasal epidermal layer inhibits epidermal proliferation; however moderate BMP-6 expression stimulates proliferation of basal epidermal keratinocytes [112].

Skin appendages

During hair follicle induction in mice, BMP-2 is expressed in hair placode, whereas BMP-4 and noggin expression is seen in cells of mesenchymal condensations beneath the placode [110, 113]. As shown in the chicken model, increased BMP signaling inhibits the initiation phase during feather development [114, 115]. In contrast, downregulation of BMP activity by noggin stimulates feather placode induction [115]. BMP signaling is also involved in the control of cell differentiation during hair follicle development. The inductive interactions between keratinocytes of the hair placode and fibroblasts of the dermal papilla lead to the construction of the hair bulb [109]. BMP-2 and BMP-4 are expressed in the epithelial and mesenchymal cells of the developing hair bulb, whereas the expression of noggin is restricted to the dermal papilla and follicular connective tissue sheath cells [113].

References

1   Hogan BLM (1996) Bone morphogenetic proteins-multifunctional regulators of vertebrate development. *Gen Develop* 10: 1580–1594

2   Wolpert L, Beddington R, Brockes J, Jessell T, Lawrence P, Meyerowitz E (1998) Patterning of the vertebrate body plan I: axes and germ layers. In: L Wolpert (ed): *Principles of development*. Current Biology Ltd, London, 61–96

3   Lemaire P, Yasuo H (1998) Developmental signalling: A careful balancing act. *Curr Biol* 8: 228–231

4   Schier AF (2001) Axis formation and pattering in zebrafish. *Curr Op Gen Develop* 11: 393–404

5   Zhao GQ (2003) Consequences of knocking out BMP signalling in the mouse. *Genesis* 35: 43–56

6   DeRobertis EM, Sasai Y (1996) A common plan for dorsoventral patterning in *Bilateria*. *Nature* 380: 37–40

7   Mullins MC, Hammerschmidt M, Kane DA, Odenthal J, Brand M, van Eeden FJ, Furutani-Seiki M, Granato M, Haffter P, Heisenberg CP et al (1996) Genes establishing dorsoventral pattern formation in the zebrafish embryo: the ventral specifying genes. *Development* 123: 81–93

8   Blader P, Rastegar S, Fischer N, Strahle U (1997) Cleavage of the BMP-4 antagonist chordin by zebrafish tolloid. *Science* 278: 1937–1940

9   Wolpert L, Beddington R, Brockes J, Jessell T, Lawrence P, Meyerowitz E (1998) Patterning of the vertebrate body plan II: the mesoderm and early nervous system. In: L Wolpert (ed): *Principles of development*. Current Biology Ltd, London, 97–124

10 Yamaguchi TP (2001) Heads or tails: Wnts and anterior-posterior patterning. *Curr Biol* 11: 713–724

11 Meno C, Saijoh Y, Fujii H, Ikeda M, Yokoyam M, Toyoda Y, Hamada H (1996) Left-right asymmetric expression of the TGFβ-family member lefty in mouse embryos. *Nature* 381: 151–155

12 Whitman M, Mercola M (2001) TGF-β superfamily signalling and left-right asymmetry. *Sci STKE* 64: 1–7

13 Mehler MF, Mabie PC, Zhang D, Kessler JA (1997). Bone morphogenetic proteins in the nervous system. *Trends Neurosci* 20: 309–317

14 Wilson PA, Hemmati-Brivanlou A (1995) Induction of epidermis and inhibition of neural fate by BMP-4. *Nature* 376: 331–333

15   Liem JKF, tremmal G, Roelink H, Jessel TM (1995) Dorsal differentiation of neural plate cells induced by Bmp-mediated signals from epidermal ectoderm. *Cell* 82: 969–976

16   Lein P, Drahushuk KM, Higgins D (2002) Effects of bone morphogenetic proteins on neural tissues. In: S Vukicevic, KT Sampath (eds): Bone morphogenetic proteins. From laboratory to clinical practice. Birkhäuser Verlag, Basel, Switzerland, 289–321

17   Dudley AT, Lyons KM, Robertson EJ (1995) A requirement for bone morphogenetic protein-7 during development of the mammalian kidney and eye. *Genes Dev* 9: 2795–2807

18   Graham A, Heyman I, Lumsden A (1993) Even numbered rhombomeres control the apoptotic elimination of neural crest cells from odd-numbered rhombomeres in the chick hindbrain. *Development* 119: 233–245

19   Hirsinger E, Jouve C, Malapert P, Pourquie O (1998) Role of growth factors in shaping the developing somite. *Mol Cell Endocrin* 140: 83–87

20   Sanz-Ezquerro JJ, Tickle C (2001) "Fingering" the vertebrate limb. *Differentiation* 69: 91–99

21   Niswander L (2002) Interplay between the molecular signals that control vertebrate limb development. *Int J Dev Biol* 46: 877–881

22   Laufer E, Nelson C, Johnson RL, Morgan BA, Tabin C (1994) Sonic hedgehog and Fgf-4 act through a signalling cascade and feedback loop to integrate growth and patterning of the developing limb bud. *Cell* 79: 993–1003

23   Capdevila J, Tsukui T, Rodriquez Esteban C, Zappavigna V, Izpisua Belmonte JC (1999) Control of vertebrate limb outgrowth by the proximal factor Meis2 and distal antagonism of BMPs by Gremlin. *Mol Cell* 4: 893–849

24   Merino R, Rodriguez-Leon J, Macias D, Ganan Y, Economides AN, Hurle JM (1999) The BMP antagonist Gremlin regulates outgrowth, chondrogenesis and programmed cell death in the developing limb. *Development* 127: 5515–5522

25   Dudley AT, Tabin CJ (2000) Constructive antagonism in limb development. *Curr Op Gen Develop* 10: 387–392

26   Lyons KM, Hogan BLM, Robertson EJ. (1995) Colocalization of Bmp 7 and Bmp 2 RNAs suggest that these factors cooperatively mediate tissue interactions during murine development. *Mech Dev* 50: 71–83

27   Hofmann C, Luo G, Balling R, Karsenty G (1996) Analysis of limb patterning in BMP-7 deficient mice. *Dev Gen* 19: 43–50

28   Dunn Nr, Winnier GE, Hargett LK, Schrick JJ, Fogo AB, Hogan BLM (1997) Haploinsufficient phenotypes in Bmp4 heterozygous null mice and modification by mutations in Gli3 and Alx4. *Dev Biol* 188: 235–247

29   Köntges G, Lumsden A (1996) Rhombocephalic neural crest segmentation is preserved throughout craniofacial ontogeny. *Development* 122: 3229–3242

30   Karsenty G (1998) Genetics of skeletogenesis. *Dev Gen* 22: 301–313

31   Reddi AH (1994) Bone and cartilage differentiation. *Curr Opin Genet Dev* 4: 737–744

32   Hall BK, Miyake T (1995) Divide, accumulate, differentiate: cell condensation in skeletal development revisited. *Int J Dev Biol* 39: 881–893

33 Duprez D, Bell EJdeH, Richardson MK, Archer CW, Wolpert L, Brickell PM, Francis-West PH (1996) Overexpression of BMP-2 and BMP-4 alters the size and shape of developing skeletal elements in the chick limb. *Mech Dev* 57: 145–157

34 Urist MR (1965) Bone: formation by autoinduction. *Science* 150: 893–899

35 Hall BK, Miyake T (2000) All for one and one for all: condensations and the initiation of skeletal development. *BioEssays* 22: 138–147

36 Kingsley DM, Bland AE, Grubber JM, Marker PC, Russel LB, Copeland NG, Jenkins NA (1992) The mouse short ear skeletal morphogenesis locus is associated with defects in a bone morphogenetic member of the TGF beta superfamily. *Cell* 71: 399–410

37 Baur ST, Mai JJ, Dymecki SM (2000) Combinatorial signaling through BMP receptor IB and GDF5; shaping of the distal mouse limb and the genetics of distal limb diversity. *Development* 127: 605–619

38 Storm EE, Huynh TV, Copeland NG, Jenkins NA, Kingsley DM, Lee SJ (1994) Limb alterations in brachypodism mice due to the mutations in a new member of the TGFβ-superfamily. *Nature* 368: 639–643

39 Thomas JT, Lin K, Nandedkar M, Camargo M, Cervenka J, Luyten FP (1996) A human chondrodysplasia due to the mutation in a TGF-β superfamily member. *Nat Gen* 12: 315–318

40 Thomas JT, Kilpatrick MW, Lin K, Erlacher L, Lembessis P, Costa T, Tsipouras P, Luyten FP (1997) Disruption of human limb morphogenesis by a dominant negative mutation in CDMP1. *Nat Gen* 17: 58–64

41 Kronenberg HM (2003) Development regulation of the growth plate. *Nature* 423: 332–336

42 Minina E, Kreschel C, Naski MC, Ornitz DM, Vortkamp A (2002) Interaction of FGF, Ihh/PthIh, and BMP signaling integrates chondrocyte proliferation and hypertophic differentiation. *Dev Cell* 3: 439–449

43 Macias D, Ganan Y, Sampath TK, Piedra ME, Ros MA, Hurle JM (1997) Role of BMP-2 and OP-1 (BMP-7) in programmed cell death and skeletogenesis during chick limb development. *Development* 124: 1109–1117

44 Cobourne MT, Sharpe PT (2003) Tooth and jaw: molecular mechanisms of pattering in the first branchial arch. *Arch Oral Biol* 48: 1–14

45 Vainio S, Karavanova I, Jowett A, Thesleff I (1993) Identification of Bmp-4 as signal mediating secondary induction between epithelial and mesenchymal tissues during early tooth development. *Cell* 75: 45–58

46 Vaahtokari A, Aberg T, Thesleff I (1996) Apoptosis in the developing tooth: association with an embryonic signaling center and supression by EGF and FGF. *Development* 122: 121–129

47 Naeve B, Holder N, Patient R (1997) A graded response to BMP-4 spatially coordinates patterning of the mesoderm and ectoderm in the zebrafish. *Mech Dev* 62: 183–195

48 Cossu G, Borello U (1999) Wnt signalling and the activation of myogenesis in mammals. *EMBO J* 18: 6867–6872

49   McPherron AC, Lawler AM, Lee S-J (1997) Regulation of skeletal muscle mass in mice by a new TGF-β superfamily member. *Nature* 387: 83–90

50   Paralkar VM, Grasser WA, Baumann AP, Castleberry TA, Owen TA, Vukicevic S (2002) Prostate-derived factor and growth and differentiation factor-8: newly discovered members of TGF-β superfamily. In: S Vukicevic, KT Sampath (eds): *Bone morphogenetic proteins. From laboratory to clinical practice*. Birkhäuser Verlag, Basel, Switzerland, 19–31

51   Lechner MS, Dressler GR (1997) The molecular basis of embryonic kidney development. *Mech Dev* 62: 105–120

52   Godin RE, Robertson EJ, Dudley AT (1999) Role of BMP family members during kidney development. *Int J Dev Biol* 43: 405–411

53   Vukicevic S, Latin V, Chen P, Batorsky R, Reddi AH, Sampath K (1994) Localization of osteogenic protein-1 (Bone morphogenetic protein 7) during human embryonic development: high affinity binding to basement membranes. *Biochem Biophys Res Commun* 198: 693–700

54   Ruppert R, Hoffmann E, Sebald W (1996) Human bone morphogenetic protein 2 contains a heparin binding site which modifies its biological activity. *Eur J Biochem* 237: 295–302

55   Luo G, Hofmann C, Bronckers ALJJ, Sohocki M, Bradley A, Karsenty G (1995) Bmp7 is an inducer of nephrogenesis, and is also required for eye development and skeletal patterning. *Genes Dev* 9: 2808–2820

56   Winnier G, Blessing M, Labosky PA, Hogan BLM (1995) Bone morphogenetic protein-4 is required for mesoderm formation and patterning in the mouse. *Genes Dev* 9: 2105–2116

57   Zhang H, Bradley A (1996) Mice deficient for BMP2 are nonviable and have defects in amnion/chorion and cardiac development. *Development* 122: 2977–2986

58   Miyazaki Y, Oshima K, Fogo A, Hogan BL, Ichikawa I (2000) Bone morphogenetic protein 4 regulates the budding site and elongation of the mouse ureter. *J Clin Invest* 105: 863–873

59   Piscione TD, Yager TD, Gupta IR, Grinfeld B, Pei Y, Attisano L, Wrana JL, Rosenblum ND (1997) BMP-2 and OP-1 exert direct and opposite effects on renal branching morphogenesis. *Am J Physiol* 273: 961–975

60   King JA, Marker PC, Seung KS, Kingsley DM (1994) BMP5 and the molecular, skeletal and soft tissue alterations in short ear mice. *Dev Biol* 166: 112–122

61   Vukicevic S, Kopp JB, Luyten FP, Sampath TK (1996) Induction of nephrogenic mesenchyme by osteogenic protein 1 (bone morphogenetic protein 7). *Proc Natl Acad Sci USA* 93: 9021–9026

62   Dudley AT, Godin RE, Robertson EJ (1999) Interaction between FGF and BMP signalling pathways regulates development of metanephric mesenchyme. *Genes Dev* 13: 1601–1613

63   Vukicevic S, Basic V, Rogic D, Basic N, Shih MS, Shepard A, Jin D, Dattatreyamurty B, Jones W, Dorai H et al (1998) Osteogenic protein-1 (bone morphogenetic protein-7)

reduces severity of injury after ischemic acute renal failure in rat. *J Clin Invest* 102: 202–214

64   Roberts DJ, Smith DM, Goff DJ, Tabin CJ (1998) Epithelial-mesenchymal signaling during the regionalization of the chick gut. *Development* 125: 2791–2801

65   Smith DM; Nielsen C, Tabin CJ, Roberts DJ (2000) Roles of BMP signaling and Nkx2.5 in patterning at the chick midgut-foregut boundary. *Development* 127: 3671–3681

66   Maric I, Poljak L, Zoricic S, Bobinac D, Bosukonda D, Sampath KT, Vukicevic S (2003) Bone morphogenetic protein-7 reduces the severity of colon tissue damage and accelerates the healing of inflammatory bowel disease in rats. *J Cell Physiol* 196: 258–264

67   Roberts DJ, Johnson RL, Burke AC, Nelson CE, Morgan BA, Tabin CT (1995) Sonic hedgehog is an endodermal signal inducing Bmp-4 and Hox genes during induction and regionalization of the chick hindgut. *Development* 121: 3163–3174

68   Hoffman MP, Kidder BL, Steinberg ZL, Lakhani S, Ho S, Kleinman HK, Larsen M (2002) Gene expression profiles of mouse submandibular gland development: FGFR1 regulates branching morphogenesis *in vitro* through BMP- and FGF-dependent mechanisms. *Development* 129: 5767–5778

69   Zaret KS (2001) Hepatocyte differentiation: from the endoderm and beyond. *Curr Opin Gen Develop* 11: 568–574

70   Furuta Y, Piston DW, Hogan BL (1977) Bone morphogenetic proteins (BMPs) as regulators of dorsal forebrain development. *Development* 124: 2203–2212

71   Miller AF, Harvey SA, Thies RS, Olson MS (2000) Bone morphogenetic protein-9. An autocrine/paracrine cytokine in the liver. *J Biol Chem* 275: 17937–17945

72   Dichmann DS, Miller CP, Jensen J, Scott Heller R, Serup P (2003) Expression and misexpression of members of the FGF and TGFβ families of growth factors in the developing mouse embryos. *Dev Dyn* 226: 663–674

73   Miralles F, Czernichow P, Ozaki K, Itah N, Scharfmann R (1999) Signaling through fibroblast growth factor receptor 2b plays a key role in the development of the exocrine pancreas. *Proc Natl Acad Sci USA* 96: 6267–6272

74   Warburton D, Schwarz M, Tefft D, Flores-Delgado G, Anderson KD, Cardoso WV (2000) The molecular basis of lung morphogenesis. *Mech Dev* 92: 55–81

75   Lu MM, Yang H, Zhang L, Shu W, Blair DG, Morrisey EE (2001) The bone morphogenetic protein antagonist gremlin regulates proximal-distal patterning of the lung. *Dev Dyn* 222: 667–680

76   Vukicevic S, Helder MN, Luyten FP (1994) Developing human lung and kidney are major sites for synthesis of bone morphogenetic protein-3 (osteogenin). *J Histochem Cytochem* 42: 869–875

77   Bellusci S, Henderson R, Winnier G, Oikawa T, Hogan BLM (1996) Evidence from normal expression and targeted misexpression that bone morphogenetic protein-4 (Bmp-4) plays a role in mouse embryonic lung morphogenesis. *Development* 122: 1693–1702

78   Weaver M, Yingling JM, Dunn NR, Bellusci S, Hogan BLM (1999) Bmp signalling regulates proximal-distal differentiation of endoderm in mouse lung development. *Development* 126: 4005–4015

79   Weaver M, Dunn NR, Hogan BLM (2000) Bmp4 and Fgf10 play opposing roles during lung bud morphogenesis. *Development* 127: 2695–2704

80   Schneider MD, Gaussin V, Lyons KM (2003) Tempting fate: BMP signals for cardiac morphogenesis. *Cytokine Growth Fact Rev* 14: 1–4

81   Walters MJ, Wayman GA, Christian JL (2001) Bone morphogenetic function is required for terminal differentiation of the heart but not for early expression of cardiac marker genes. *Mech Dev* 100: 263–273

82   Kim RY, Robertson EJ, Solloway MJ (2001) Bmp6 and Bmp7 are required for cushion formation and septation in the developing mouse heart. *Dev Biol* 235: 449–466

83   Solloway MJ, Robertson EJ (1999) Early embryonic lethality in Bmp5 Bmp7 double mutant mice suggests functional redundancy within the 60A subgroup. *Development* 126: 1753–1768

84   Olson EN (2001) The path to the heart and the road not taken. *Science* 291: 2327–2328

85   Solloway MJ, Harvey RP (2003) Molecular pathways in myocardial development: a stem cell perspective. *Cardiovasc Res* 58: 264–277

86   Schultheiss TM, Burch JB, Lassar AB (1997) A role of bone morphogenetic proteins in the induction of cardiac myogenesis. *Genes Dev* 11: 451–462

87   Andree B, Duprez D, Vorbusch B, Arnold HH, Brand T (1998) BMP-2 induces ectopic expression of cardiac lineage markers and interferes with somite formation in chicken embryos. *Mech Dev* 70: 119–131

88   Monzen K, Nagai R, Komuro I (2002) A role for bone morphogenetic protein signalling in cardiomyocyte differentiation. *Trends Cardiovasc Med* 12: 263–269

89   Marvin MJ, Di Rocco G, Gardiner A, Bush SM, Lassar AB (2001) Inhibition of Wnt activity induces heart formation from posterior mesoderm. *Genes Dev* 15: 316–327

90   Schneider VA, Mercola M (2001) Wnt antagonism initiates cardiogenesis in *Xenopus laevis*. *Genes Dev* 15: 304–315

91   Breckenridge RA, Mohun TJ, Amaya E (2001) A role for BMP signalling in heart looping morphogenesis in *Xenopus*. *Dev Biol* 232: 191–203

92   Nakajima Y, Yamagishi T, Hokari S, Nakamura H (2000) Mechanisms involved in valvuloseptal endocardial cushion formation in early cardiogenesis: roles of transforming growth factor (TGF)-β and bone morphogenetic protein (BMP). *Anat Rec* 258: 119–127

93   Delot EC (2003) Control of endocardial cushion and cardiac valve maturation by BMP signalling pathways. *Mol Gen Met* 80: 27–35

94   Baron MH (2001) Induction of embryonic hematopoietic and endothelial stem/progenitor cells by hedgehog-mediated signals. *Differentiation* 68: 175–185

95   Baron MH (2003) Embryonic origins of mammalian hematopoiesis. *Exp Hemat* 31: 1160–1169

96   Martinovic S, Mazic S, Kisic V, Basic N, Jakic-Razumovici J, Borovecki F, Batinic D, Simic P, Grgurevic L, Labar B et al (2004) Expression of bone morphogenetic proteins in stromal cells from human bone marrow long-term culture. *J Histochem Cytochem; in press*

97 Findlay JK, Drummond AE, Dyson ML, Baillie AJ, Robertson DM, Ethier JF (2002) Recruitment and development of the follicle; the roles of the transforming growth factor-β superfamily. *Mol Cell Endocrin* 191: 35–43

98 Ying Y, Qi X, Zhao GQ (2001) Induction of primordial germ cells from murine epiblast by synergistic action of BMP4 and BMP8B signalling pathways. *Proc Natl Acad Sci USA* 98: 7858–7862

99 Shimasaki S, Zachow RJ, Li DM, Kim H, Iemura S, Ueno N, Sampath K, Chang RJ, Erickson GF (1999) A functional bone morphogenetic protein system in the ovary. *Proc Natl Acad Sci USA* 96: 7282–7287

100 Dong J, Albertini D, Nishimori K, Lu N, Matzuk M (1996) Growth differentiation factor-9 is required during early ovarian folliculogenesis. *Nature* 383: 531–535

101 Galloway SM, McNatty KP, Cambridge LM, Laitinen MPE, Juengel JL, Jokiranta TS, McLaren RJ, Luiro K, Dodds KG, Montgomery GW et al (2000) Mutations in an oocyte-derived growth factor gene (BMP15) cause increased ovulation rate and infertility in a dosage-sensitive manner. *Nature* 25: 279–283

102 Solloway MJ, Dudley AT, Bikoff EK, Lyons KM, Hogan BLM, Robertson EJ (1998) Mice lacking Bmp6 function. *Dev Gen* 22: 321–339

103 Knight PG, Glister C (2001) Potential local regulatory functions of inhibins, activins and follistatin in the ovary. *Reproduction* 121: 503–512

104 Durlinger AL, Kramer P, Karels B, de Jong FG, Uilenbroek JT, Grootegoed JA, Themmen AP (1999) Control of primordial follicle recruitment by anti-Mullerian hormone in the mouse ovary. *Endocrinology* 140: 5789–5796

105 Martinovic S, Latin V, Suchanek E, Stavljenic-Rukavina A, Sampath KI, Vukicevic S. (1996) Osteogenic protein-1 is produced by human fetal trophoblasts *in vivo* and regulates the synthesis of chorionic gonadotropin and progesterone by trophoblasts *in vitro*. *Eur J Clin Chem Clin Biochem* 34: 103–109

106 Ozkaynak E, Jin DF, Jelic M, Vukicevic S, Oppermann (1997) Osteogenic protein-1 mRNA in the uterine endometrium. *Biochem Biophys Res Commun* 234: 242–246

107 Zhao GQ, Liaw L, Hogan BLM (1998) Bone morphogenetic protein 8A plays a role in the maintenance of spermatogenesis and the integrity of epididymis. *Development* 125: 1103–1112

108 Wilson PA, Hemmati-Brivanlou A (1995) Induction of epidermis and inhibition of neural fate by BMP-4. *Nature* 376: 331–333

109 Botchkarev VA (2003) Bone morphogenetic proteins and their antagonists in skin and hair follicle biology. *J Invest Dermatol* 120: 36–47

110 Lyons KM, Peltron RW, Hogan BL (1989) Patterns of expression of murine Vgr-1 and BMP-2a RNA suggest that transforming growth factor-beta-like genes coordinately regulate aspects of embryonic development. *Genes Dev* 3: 1657–1668

111 Takahashi H, Ikeda T (1996) Transcripts for two members of the transforming growth factor-beta superfamily BMP-3 and BMP-7 are expressed in developing rat embryos. *Dev Dyn* 207: 439–449

112 Blessing M, Schirmacher P, Kaiser S (1996) Overexpression of bone morphogenetic pro-

tein-6 in the epidermis of transgenic mice: inhibition or stimulation of proliferation depending on the pattern of transgene expression and formation of psoriatic lesions. *J Cell Biol* 135: 227–239

113 Botchkarev VA, Botchkareva NV, Roth W, Nakamura M, Chen LH, Herzog W, Lindner G, McMahon JA, Peters C, Lauster R et al (1999) Noggin is mesenchymally-derived stimulator of hair follicle induction. *Nature Cell Biol* 1: 158–164

114 Jung HS, Francis West PH, Widelitz RB (1998) Local inhibitory action of BMP and their relationship with activators in feather formation: implications for periodic patterning. *Dev Biol* 196: 11–23

115 Noramly S, Morgan BA (1998) BMPs mediate lateral inhibition at successive stages in feather tract development. *Development* 125: 3775–3787

Bone morphogenetic proteins in articular cartilage repair

David C. Rueger[1] and Susan Chubinskaya[2]

[1]Stryker, 35 South Street, Hopkinton, MA 01748, USA; [2]Department of Biochemistry and Section of Rheumatology, Rush University Medical Center, Chicago, IL 60612, USA

Introduction

Cartilage regeneration and repair is one of the major obstacles in current orthopedics. The importance is enormous since osteoarthritis (OA) is a major cause of disability among the adult population in the United States. OA is considered a process of attempted, but gradually failing, repair of damaged cartilage extracellular matrix, as the balance between synthesis and breakdown of matrix components is disturbed and shifted towards catabolism. In recent times, members of the bone morphogenetic proteins (BMP) family of proteins have received the highest attention among potential anabolic factors for cartilage repair because of their ability to induce matrix synthesis and promote repair in different connective tissues, including cartilage. This Chapter will review the information accumulated on BMPs from *in vitro* studies as well as from studies of repair in various animal models. The data show significant promise for BMPs in cartilage repair and suggest that this indication could become the most important application for BMPs in orthopedics.

In vitro studies

The *in vitro* studies covering BMPs in cartilage repair are reviewed in two parts. First, the studies which address the anabolic activity of recombinant BMPs on chondrocytes are described. In this section, the activity of exogenous BMP preparations is characterized using cells either embedded in native cartilage matrix and cultured as explants or isolated from the extracellular matrix and cultured as monolayers, pellets, or embedded in different scaffolds and polymeric matrices. Secondly, studies which investigate endogenous BMPs expressed by chondrocytes are described.

Bone Morphogenetic Proteins: Regeneration of Bone and Beyond, edited by Slobodan Vukicevic and Kuber T. Sampath
© 2004 Birkhäuser Verlag Basel/Switzerland

Anabolic activity of BMPs

The ability of BMPs to induce an anabolic response in cartilage *in vitro* has been documented using different BMP in multiple species, including human, bovine, rat, rabbit, mouse, and a variety of culture conditions. Among the BMPs, OP-1 (BMP-7) has been by far the most extensively studied *in vitro*. Its anabolic effects have been well documented in cultures of bovine [1, 2] and porcine articular chondrocytes [3], where OP-1 (BMP-7) maintained chondrocyte phenotype and induced matrix synthesis above serum levels. Studies using either bovine or human cells demonstrated that OP-1 (BMP-7) does not induce chondrocyte proliferation or type X collagen synthesis and the cells do not undergo hypertrophy [5–10]. In a detailed analysis OP-1 (BMP-7) treatment resulted in synthesis of all of the major cartilage matrix components: proteoglycans (PGs), collagen type II, and hyaluronan [2–5]. Furthermore, in porcine cartilage explants, a dose-dependent stimulation of PG synthesis and a decrease in PG release were demonstrated with OP-1 (BMP-7) and the induction generated normal, functional PGs, with a hydrodynamic size unaltered by the treatment [3]. In regard to other BMPs, BMP-4 has been shown to induce PG synthesis in bovine explant cultures in a dose-dependent manner under short-term conditions; longer cultures led to a decrease in PG synthesis and collagen metabolism [11–13]. BMP-2 and BMP-9 were shown to induce a similar anabolic response in the absence of serum, while in the presence of serum the bovine chondrocytes became unresponsive [14, 15]. In addition, BMP-2 has been reported to induce PG synthesis and maintain bovine chondrocytic phenotype through induction of the genes for aggrecan and type II collagen and by depressing the type X collagen genes in long-term monolayers [16, 17].

The anabolic activity of OP-1 (BMP-7) was observed to extend beyond matrix synthesis. OP-1 (BMP-7) is capable of modulating expression of receptors for certain matrix components, for instance CD-44, enzymes responsible for their synthesis (hyaluronan synthase [10]), as well as components of the chondrocyte cytoskeleton. In regard to the cytoskeleton, stimulation of four cytoskeletal proteins was observed in bovine and murine chondrocytes: tensin, talin, paxillin, and focal adhesion kinase [18].

Since a primary objective of studies on BMPs is to develop new therapeutic tools for cartilage repair in humans, a special emphasis in this review will be given to studies on human chondrocytes. The ability of BMP-6 to induce the synthesis of cartilage matrix components by human normal and OA chondrocytes was recently investigated [19] and it was shown that BMP-6 is capable of increasing PG synthesis and this response to BMP-6 declined with age. BMP-6 had no effect on proliferation of human chondrocytes. These data were similar to that observed with OP-1 (BMP-7) treatment of human cells [9]. In our laboratory, we recently compared the response of human normal adult articular chondrocytes to recombinant BMP-2, 4,

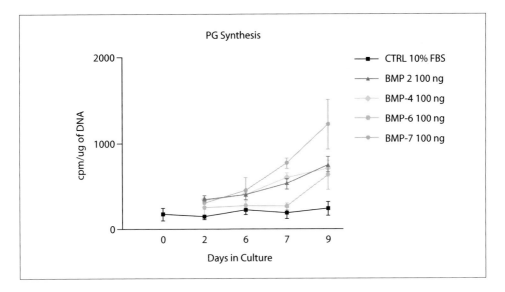

Figure 1
The levels of PG synthesis in human normal chondrocytes cultured in alginate beads in the presence of 10% fetal bovine serum and stimulated with four different BMPs: BMP-2,-4, -6, and 7
All BMPs were added at the concentration 100 ng/ml with media changes every other day. The results are measured in triplicates and represent the average of three different experiments. The values of PG synthesis normalized to the DNA content.

6, and 7 with regard to PG synthesis (Fig. 1). Isolated cells were cultured in alginate for nine days in the presence of 10% fetal bovine serum. Each growth factor was added at the same concentration (100 ng/ml) every other day and the values were normalized to the DNA content. Within the first five days of treatment, all BMPs studied induced a similar response, however by day nine, the highest levels of PG synthesis were identified in the OP-1 (BMP-7)-treated group. Importantly, all of these BMPs were able to stimulate an anabolic response in human chondrocytes above serum levels. These findings were similar to previous data where OP-1 (BMP-7) was found to be more potent than TGF-β and activin [5].

To date, a number of additional anabolic effects of OP-1 (BMP-7) have been observed for human chondrocytes under a variety of culture conditions. OP-1 (BMP-7) has been shown in different types of cultures (organ culture, high-density monolayers, alginate beads) to induce the synthesis of the major components of the cartilage matrix: aggrecan, decorin, collagen type II, collagen type VI, hyaluronan, fibronectin [5, 7–8, 10]. OP-1 (BMP-7) has also been shown to enhance gene expression of the anabolic molecule tissue inhibitor of metalloproteinase (TIMP) [20], an effect which was recently confirmed in a study exploring different anabolic and catabolic genes in normal and OA chondrocytes [21]. In this study, the expression of anabolic genes in OA chondrocytes was significantly upregulated by OP-1 (BMP-7) when compared to normal chondrocytes and that in neither normal nor OA cells was a stimulating effect on matrix-degrading enzymes (MMP-1, 3, 13 and ADAMTS-4) observed. Interestingly, on the protein level, OA chondrocytes also showed a greater ability to synthesize PGs upon response to OP-1 (BMP-7) when compared to normal chondrocytes.

OP-1 (BMP-7) has also been observed to have anti-catabolic activity. Interleukin-1 and fibronectin-fragments-induced inhibition of PG synthesis [6, 22] as well as cytokine-induced MMP-1, 3 and MMP-13 expression and activity can be overcome by OP-1 (BMP-7) [20, 23]. Detailed analyses suggest that the inhibition of transcriptional factors NF-κB and AP-1 by OP-1 (BMP-7) [23] or upregulation of inhibitors of matrix proteinases [20] may be part of the underlying mechanisms responsible for the anti-catabolic activity of this BMP. In regard to other BMPs, BMP-2 was not able to counteract IL-1-induced inhibition of PG synthesis in an *in vivo* model using mice [24]. In this study, co-injection of IL-1 and BMP-2 into normal murine knees led to the abrogation of an anabolic activity of BMP-2 while TNF-α at the same concentration had no effect on the BMP-2-mediated PG synthesis [25, 26]. However, BMP-2 has not been evaluated in any *in vitro* cell culture study nor has OP-1 (BMP-7) been evaluated in this *in vivo* injection model. There is clearly a need for a direct comparison of BMPs in the same model to determine if such an important difference exists between these BMPs.

Endogenous BMPs in cartilage

As was stated at the beginning of this review, cartilage regeneration and repair in adults is one of the most important tasks in current orthopedics. Development of new methodology (real-time PCR, gene and protein array, proteomics, transfections, point mutations, siRNA, etc.) have allowed more powerful approaches in studying cartilage autocrine growth factors. Although the application of recombinant BMPs still remains the primary focus in the BMP field, the understanding of the regulation and function of the BMPs that are endogenously expressed in adult articular cartilage offers significant supporting data. Knowledge of the mechanisms that control BMP synthesis, activation, induction, and signaling in articular cartilage will provide critical missing information which is necessary to develop correct strategies for the application of recombinant BMPs.

Unfortunately, there are few studies that have focused on endogenous expression of BMPs and their role in human cartilage homeostasis. Fukui et al. [27] reported the expression of BMP-2, 4, 6, and CDMP-1 (GDF-5) in normal and OA cartilage.

They showed that BMP-2 is expressed in both normal and OA tissue and that it is greatly induced by inflammatory cytokines, interleukin-1β and TNF-α. No differences were found for the other BMPs studied. Bobacz et al. [19] recently documented the expression of BMP-6 in normal and OA cartilage and found no changes in its expression levels with aging or the presence of OA. The authors concluded that it is unlikely that this BMP is involved in the pathogenesis of OA but may be important for the maintenance of tissue homeostasis in articular cartilage. The long-term focus of our laboratory is autocrine and paracrine OP-1 (BMP-7) and its function in adult articular chondrocytes and the availability of normal human tissue has allowed us to evaluate not only normal cartilage but cartilage that has undergone a variety of degenerative changes. For our studies we have used cartilage from femurs of the tibia-femoral joint and talar domes from the talocrural joint. We detected that OP-1 (BMP-7) is expressed in fetal, newborn, young and adult normal and OA articular cartilage at both gene and protein levels [28]. In addition, through the use of antibodies to the pro- and mature domains of the OP-1 (BMP-7) molecule, we found by immunohistochemistry that OP-1 (BMP-7) is present in normal and OA cartilage in two forms: the pro-form (inactive) and the mature (active) form. These forms of OP-1 (BMP-7) have a distinct localization in cartilage tissue: mature OP-1 (BMP-7) is primarily localized in the upper, superficial zone, while the pro-form is primarily identified in the middle and deep cartilage zones. In OA cartilage mature OP-1 (BMP-7) is selectively localized in cell clusters. Preferential distribution of mature OP-1 (BMP-7) in the upper cartilage zone was similar to the previously published results by Erlacher et al. [29] for cartilage-derived morphogenetic proteins (CDMPs), CDMP-1 (GDF-5) and CDMP-2 (GDF-6). Both of these proteins were also primarily detected in the superficial layer of normal cartilage and in chondrocyte clusters of OA cartilage.

For use in other studies we developed a quantitative sandwich ELISA method that allowed the detection of the OP-1 (BMP-7) protein in extracts of connective tissues and in synovial fluid at picogram quantities [30]. Applying quantitative ELISA and PCR methods, we determined the levels of OP-1 (BMP-7) in normal, degenerated and OA cartilage. We found that in normal cartilage the concentration of OP-1 (BMP-7) protein is around 50 ng/g dry tissue, which is within the physiological range of the anabolic activity of recombinant human OP-1 (BMP-7) (50–200 ng/ml). However, with aging and the progression of cartilage degeneration the levels of endogenous OP-1 (BMP-7) expression dramatically decrease and this decrease is primary due to a reduction in the level of the mature (active) form of the protein [30–32]. We also detected OP-1 (BMP-7) in synovial fluid from normal joints and from patients with OA and rheumatoid arthritis (RA) [33]. Interestingly, in synovial fluid from RA patients, the levels of OP-1 (BMP-7) protein were at least two-fold higher than in synovial fluid from OA patients and normal donors. Our very recent data indicate that in normal synovial fluid OP-1 (BMP-7) is primarily present in the pro form (inactive), while in OA and RA samples a portion of the OP-1 (BMP-7) was identified to be in the mature (active) form. The origin of OP-1 (BMP-7) in the synovial cavity is unknown; however, recent detection of OP-1 (BMP-7) in cartilage, synovium, ligament, tendon and menisci (Fig. 2) suggests that it could be synthesized and released by all these tissues.

Taken together with our previous results [34–36] these data suggest that the increased levels of OP-1 (BMP-7) in RA synovial fluid and the presence of mature OP-1 (BMP-7) in the diseased samples could be explained, at least partially, by the response to catabolic cytokines/inflammatory processes that accompanied RA and late stages of OA. We found that OP-1 (BMP-7) is an inducible protein that can be upregulated in response to interleukin-1 or TNF-α [34], to biomechanical loading [36] or to the treatment with the catabolic enzyme chymopapain used to create cartilage degeneration in animal models of OA [35]. Upregulation of endogenous BMP expression in response to catabolic cytokines as an attempt to repair appears to be a common feature for these factors, since similar findings were reported by Fukui et al. [27] for BMP-2.

In addition to human adult cartilage, endogenous expression of OP-1 (BMP-7) has been detected in adult bovine, rabbit, and goat tissues [29, 35, 37–39]. The distribution of mature (active) OP-1 (BMP-7) protein in normal goat and rabbit articular cartilage is similar to human tissues, while in bovine cartilage mature OP-1 (BMP-7) is strongly detected throughout the entire cartilage thickness. In rat cartilage, the mature form of OP-1 (BMP-7) and a number of other BMPs (BMP-2, -3, -4, -5, and -6) was detected [38] in the middle and deep cartilage zones. Noteworthy, the use of different antibodies and different procedures for treatment of tissues may account for the conflicting results we obtained for OP-1 (BMP-7), but it is also possible that in rat cartilage the localization does indeed differ from cartilage of other species.

The detection of BMPs in adult articular cartilage of different species provides support for the concept that these growth factors are not only important for skeletal development and bone healing, but they may also have a functional role in the maintenance of normal cartilage homeostasis. This was confirmed by our recent OP-1 (BMP-7) antisense studies [40], where transfection of human adult articular chondrocytes with OP-1 (BMP-7) antisense oligos led to about 70% inhibition in OP-1 (BMP-7) gene expression. Downregulation of OP-1 (BMP-7) mRNA induced significant inhibition of aggrecan mRNA expression and about a 50% decrease in newly PG synthesis as detected by sulfate incorporation. Recovery (add-back) experiments with recombinant OP-1 (BMP-7) were able to restore, at least partially, PG synthesis indicating a direct role of autocrine OP-1 (BMP-7) in the regulation of PG metabolism. Histological evaluation of cartilage explants cultured in the presence of OP-1 (BMP-7) antisense oligonucleotides revealed a remarkable depletion of PGs as identified by safranin O stain. Antisense-treated tissue showed a paucity of chondrocytes, initial fibrillation of cartilage surface and a decrease in safranin O staining in the upper and middle cartilage zones. These data together with previous

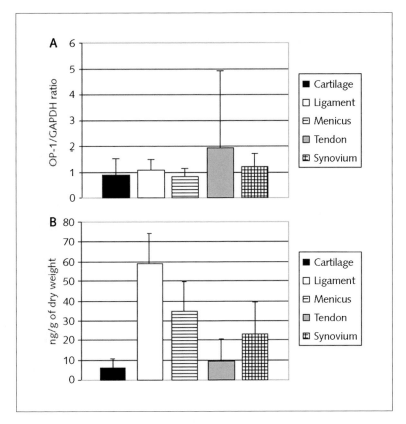

Figure 2
The levels of endogenous OP-1 gene expression (A) and protein concentration (B) in con-
nective tissues of the knee joints obtained from three normal adult human donors
OP-1 PCR results normalized to GAPDH levels; concentration of OP-1 protein is detected by
sandwich ELISA and the results normalized to tissue dry weight.

data, provide strong evidence for endogenous OP-1 (BMP-7) being a critical factor
that controls cartilage matrix integrity and is involved in the maintenance of normal cartilage homeostasis. In addition, the data further suggest that the lack of
endogenous OP-1 (BMP-7) could predispose cartilage to degenerative processes and
make tissue more susceptible to the influence of catabolic agents. Our future studies are aimed to elucidate the mechanisms of OP-1 (BMP-7) signaling in cartilage
and the relationship between the OP-1 (BMP-7) signaling pathway and other anabolic signaling pathways that are important in normal cartilage homeostasis. Furthermore, more will be done investigating the role of other BMPs in cartilage
homeostasis.

Animal studies

Repair of osteochondral defects, which involve both the cartilage tissue and the underlying bone, is known to occur to a limited extent promoted by the presence of both stem cells and growth and differentiation factors brought into the defect by the blood/marrow. In animal studies these defects undergo some repair with formation of a new layer of bone and cartilage, but the macromolecular organization and the bio-chemical characteristics of the cartilage matrix are imperfect. High levels of type I instead of type II collagen and proteoglycans that are not cartilage specific, such as dermatan sulphate containing proteoglycans, make up the repair tissue and result in fibrillations and degenerative changes over time. In contrast to the osteochondral defects, the repair of cartilage defects that do not penetrate into the subchondral bone (chondral defects) is not believed to occur even to a limited extent.

The pivotal role of BMPs in the development and regeneration process of the skeleton had originally suggested a role in articular cartilage repair. Furthermore, the accumulation of data from numerous *in vitro* studies as reviewed in the first part of this Chapter have clearly demonstrated that certain BMPs have an important role in chondrocyte differentiation and extracellular matrix production as well as the maintenance of adult chondrocyte phenotype. In light of these data the testing of BMPs in animal models of cartilage repair was begun. For the most part two BMPs, OP-1 (BMP-7) and BMP-2, have been the subject of these investigations with the objective to determine if a BMP could improve the healing in osteochondral defects or promote repair in chondral defects. The earliest studies were done using deep osteochondral defects with the BMPs delivered locally into the defect site on a variety of scaffold materials that were press fitted into the defect site. More recent studies were done using chondral defects that did not penetrate the calcified cartilage layer and where the BMPs were delivered by a variety of means including *via* a minipump to the synovial fluid. This section summarizes the results from these studies involving delivery of recombinant BMPs as well as BMP genes to defect sites and contemplates the future of BMPs in articular cartilage repair.

Osteochondral defect repair studies

Most investigations evaluating articular cartilage healing using BMPs have involved osteochondral defect repair models. These studies have demonstrated that certain BMPs can improve the healing response in both the bone and cartilage tissue in comparison to the untreated controls. The models have involved deep as well as shallow defects of the knee and have utilized a variety of animal species, including rabbit, dog, sheep or goat models. For the most part type I collagen has been used to locally deliver the BMPs into the defect sites.

Rabbit studies

The earliest studies were done using various models in the rabbit knee. The most comprehensive study was conducted using recombinant BMP-2 delivered in a type I collagen sponge to defects created in the trochlear groove [41, 42]. Repair of defects (3 mm diameter, 3 mm deep) was evaluated at 4, 8, 24 and 52 weeks postimplantation. In comparison with control defects, treatment with BMP-2 greatly accelerated the formation of new subchondral bone and improved the articular cartilage with the formulation of a tidemark between the tissues. At six months, the thickness of the cartilage was 70% that of the normal adjacent cartilage and this was maintained to 12 months. In this regard it is interesting that this repair did not require the presence of BMP-2 after the initial induction process, since the residence time of the BMP-2 was demonstrated to be 14 days in the defect site. The BMP-2 treated defects showed an improvement in the integration of the repair cartilage with the adjacent cartilage, an improvement in cellular morphology and significantly less type I collagen when compared with untreated defects. In conclusion, the authors clearly stated that the repair cartilage was not identical to normal articular cartilage, but BMP-2 stimulated a faster repair with more abundant and more hyalin-like cartilage over that seen in the control defects.

A second rabbit study has been reported evaluating recombinant BMP-2, but in the presence and absence of allogenic chondrocytes [43]. In this study, BMP-2 was delivered on a type I collagen bilayer sponge to femoral trochlea defects (3 mm diameter, 2 mm deep). Test groups included sponge seeded with chondrocytes with and without BMP-2 and sponge with and without BMP-2. At six weeks postoperatively, histological characterization showed a significant and similar cartilage repair in the three groups containing BMP-2 or chondrocytes alone or the combination, but not in the group that received sponge alone. The defects receiving BMP-2 alone induced a repair quantitatively and qualitatively as good as that achieved with the chondrocytes alone, but the combination did not improve the outcome. In each of these three groups, there was good restoration of subchondral bone and significant cartilage restoration with a high percentage of hyaline cartilage. The matrix alone repair was disorganized, hypercellular and fibrillated. Thus, both the BMP-2 and cell treatments were successful in this model and the authors suggested that the use of BMP-2 may be an attractive alternative to autologous chondrocyte transplantation.

Two rabbit studies have been reported using recombinant OP-1 (BMP-7) delivered on bone-derived type I collagen particles [44, 45]. The results were similar to that seen in the BMP-2 studies. In one study, defects were created in femoral chondyles (6 mm long, 3 mm wide and 3 mm deep) and histological examination of the repair at two months showed a significant difference in healing of the defects treated with OP-1 (BMP-7) compared to those left empty or treated with the collagen only [44]. Defects that were not treated with OP-1 (BMP-7) were filled with, primarily fibrous tissue and fibrocartilage. Defects treated with OP-1 (BMP-7) showed extensive regeneration of both the subchondral bone and a hyaline-like cartilage layer on top. These defects were completely bridged with abundant tissue that consisted of small rounded cells organized in columns and embedded in compact extracellular matrix. In the second study, OP-1 (BMP-7) treatment was evaluated in larger osteochondral defects (4 mm diameter, 5 mm deep) made in the femoral patellar groove and histological evaluation was done at three months [45]. The results were similar to that seen in the eight week study. In control defects, primarily fibrous tissue or fibrocartilage was observed that was discontinuous with the surrounding cartilage and was opaque and inhomogeneous. In OP-1 (BMP-7) treated sites, maturing hyaline-like cartilage was present and of similar thickness to the surrounding cartilage and the integration of newly formed cartilage with old, intact cartilage was reported to be satisfactory. In addition, the subchondral bone had for the most part completely regenerated with a tidemark between the tissues.

One rabbit study has also been described using a non-collagen carrier material [46]. In this study, N,N-dicarboxymethyl chitosan was used to deliver OP-1 (BMP-7) to 2.5 mm diameter, 2.5 mm deep defects in the femoral chondyles and healing evaluated at 7, 14 and 60 days post surgery. By 60 days, similar incomplete scar-like defect fill was observed in both OP-1 (BMP-7) and control defects, although the OP-1 (BMP-7) group showed a higher number of cells with features of chondrocytes. However, the defect surfaces in both OP-1 (BMP-7) and control groups were covered with synovial membrane that had migrated to the site and the authors attributed this activity to the chitosan carrier material.

More recently, a rabbit study was reported evaluating the delivery of the OP-1 (BMP-7) gene to repair osteochondral defects [47]. The study was designed to see if a BMP gene could be used to augment repair with stem cells. In this study, the OP-1 (BMP-7) gene was introduced into periosteal-derived allogenic mesenchymal stem cells which were then seeded onto a polyglycolic acid scaffold and implanted into defects (3 mm diameter, 2 mm deep) in the femoropatellar groove. Regeneration of the bone and cartilage was evaluated at 4, 8 and 12 weeks. The data showed that defects treated with the OP-1 (BMP-7) gene modified cells showed complete or near complete bone and cartilage regeneration at 8 and 12 weeks while the controls with empty defects or allogenic cells only had poor repair of both bone and cartilage. It was concluded that the cell-based approach was not successful in this model without the introduction of the OP-1 (BMP-7) gene.

Large animal studies

The encouraging results from multiple rabbit studies provided the impetuous to extend the studies on osteochondral defect repair to larger animal models. The most extensive evaluation has been done in dogs using recombinant OP-1 (BMP-7)

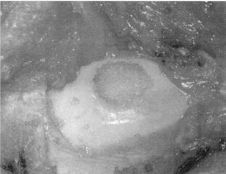

Figure 3
Canine osteochondral defect model
Photographs of the defect being created on the left side and of the defect after filling with the OP-1(BMP-7) collagen particles on the right.

and the data has demonstrated that significant repair could indeed be observed in a large animal model [48]. Variables such as OP-1 (BMP-7) dose, delivery materials and containment methods have been investigated in defects (5 mm diameter, 6 mm deep) created in the femoral condyles. OP-1 (BMP-7) was delivered on bone-derived type I collagen particles with carboxymethylcellulose as an additive to make a putty-like formulation and compared to defects left empty or treated with the delivery material alone (Fig. 3). The repair has been evaluated at up to 52 weeks postoperatively, but most animals have been sacrificed at 16 weeks or less. Similar to the rabbit data, there were marked differences in the gross and histologic appearance of osteochondral defect sites treated with OP-1 (BMP-7) when compared to the control sites (Fig. 4). Both the subchondral bone and cartilage showed statistically significant improvement in OP-1 (BMP-7) treated sites although there was a high degree of animal to animal variability. The differences between OP-1 (BMP-7) treated and control defects were most pronounced at the earliest time points and in some OP-1 (BMP-7) treated samples, the repair was so complete that it was difficult to find the defect site. Histologically, in the best OP-1 (BMP-7) treated sites, maturing cartilage was present which appeared similar to the intact articular cartilage, had thickness similar to the surrounding intact cartilage, was minimally degraded at the defect interface and was generally continuous with the repair cartilage. The control defects were filled primarily with fibrous tissue and/or what appeared to be fibrocartilage and a moderate degeneration of the cartilage at the defect interfaces was noted with large clusters of chondrocytes observed at the interface and fissures separating the intact cartilage from the repair tissue. The addition of a periosteal membrane or immobilization with the purpose of improv-

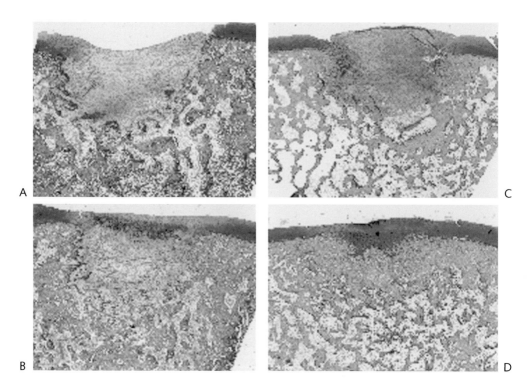

Figure 4
Canine osteochondral defect model
Histological results at six weeks (A, control and B, treated with OP-1(BMP-7) and at 12
weeks (C, control and D, treated with OP-1(BMP-7).

ing containment of the implant in the defect site did not result in any significant improvement in healing. In long-term animals, there appeared to be thinning of the repair cartilage in all groups. However, significant long-term studies with a large number of animals have not been done and thus this remains an important issue to be evaluated in future studies.

In a similar study with OP-1 (BMP-7), healing of 9 mm diameter osteochondral defects in goat knee chondyles was evaluated [49]. The defects were filled with OP-1 (BMP-7) mixed with a blood clot and contained at the site with a periosteal membrane. The OP-1 (BMP-7) was present with or without bone-derived type I collagen particles. Some defects had particles of ear perichondrium present in order to evaluate that tissue as a source of cells with chondrogenic potential and healing was evaluated at 1, 2 and 4 months after implantation. Contrary to the dog studies, there was no obvious difference between OP-1 (BMP-7) treated defects and controls at one and two months. However, at four months only one of three controls showed beginning signs of cartilage formation while all four OP-1 (BMP-7) treated defects were completely or partially filled with cartilage. The addition of ear perichondrium did not appear to improve the result and similar results were obtained with and without the collagen carrier added to the blood clot.

One large animal study has also been reported using BMP-2 [50]. The purpose of this study was to test the ability of an antiangiogenesis factor to inhibit the upgrowth of blood vessels from the marrow into the cartilage compartment in shallow osteochondral defects. The model utilized narrow defects (1.2 mm in width, 7–8 mm in length and 0.8 mm in depth) in the femoral patellar groove of minipigs. The lower (bony) portion of the defects were filled with a fibrin-gelatin matrix containing either free TGF-β or IGF-1 and liposome-encapsulated BMP-2. The upper (cartilage) portion was filled with the same matrix and factors, but in addition contained the antiangiogenic factor, suramin, in free or liposome-encapsulated form. At eight week sacrifice, the time-released suramin was highly efficacious in suppressing bone tissue upgrowth into the cartilage compartment and allowed tissue transformation into cartilage within the cartilage compartment at 50–80%. Without the antiangiogenic factor a significant amount of bone upgrowth occurred into the cartilage compartment, suppressing the amount of cartilage tissue present. It was suggested that such a factor is required in osteochondral defects to block the deleterious effects of blood/marrow on cartilage tissue formation and that preferentially stem cells for the cartilage repair should be derived from the synovium.

A pilot study has been conducted in sheep to ascertain whether a recombinant BMP could be used to augment the mosaicplasty (osteochondral autograft) procedure [51]. Specifically, the aim of this investigation was to determine if the addition of OP-1 (BMP-7) delivered with a combination of bone-derived type I collagen particles and carboxymethylcellulose could improve the histological outcome of: 1) the interface between transplanted and host cartilage; 2) interface between the transplanted and host bone; 3) donor site healing. In this study, two osteochondral defects were created in each femoral condyle. One of the defects on each condyle served as the graft donor site, (5.5 mm diameter, 10 mm deep) and the other as the graft recipient site (5.4 mm diameter, 10 mm deep). A further circumferential 1mm partial thickness defect (chondral only) was created around the periphery of the recipient sites to simulate the chondral gaps present between the multiple plugs used in the human clinical procedure. Into each recipient site one donor plug was transplanted with a press fit of 0.1 mm. In one knee, the OP-1 (BMP-7) was press fitted into the donor site as well as the recipient site prior to the plug implantation. The plug was then press fitted into the recipient site and additional OP-1 (BMP-7) was placed into the circumferential defect (Fig. 5). The contralateral knees served as controls without OP-1 (BMP-7) and all the animals were evaluated at 3, 6 and 12 weeks. The data showed little or no healing at the

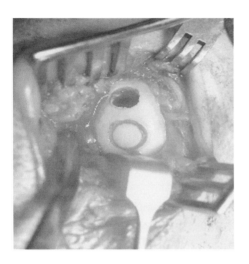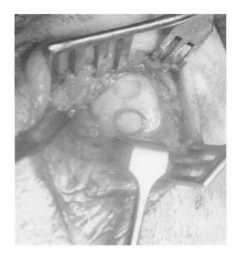

Figure 5
Sheep osteochondral defect model: mosaicplasty augmentation
Photographs of the empty, control defects on the left and the defects treated with OP-1
(BMP-7) collagen particles on the right. The donor holes are on the top and the recipient
holes with autograft plugs are on the bottom.

early time points in either the OP-1 (BMP-7)-treated or controls, but marked differences in repair were evident at 12 weeks in both the recipient and donor site groups. Histologically, the donor sites treated with OP-1/BMP-7 showed significant healing in both the cartilage and subchondral bone tissues (Fig. 6). The surface of the defect was covered by a contiguous layer of synovial membrane with few inflammatory cells. The bridging cartilage contained a noticeable amount of hyaline cartilage, and the cartilaginous formation at the margin of the defect was active and fused well with the existing adjacent cartilage. The control donor sites demonstrated disorganized healing in both the cartilage and subchondral bone, although all samples showed bridging of the defect. Histologically, the differences between OP-1 (BMP-7) treated and control graft healing were significant. The sites treated with OP-1 (BMP-7) appeared to have smoother articular surfaces, less severe chondrocyte clustering, closer to normal adjacent tissues and, most importantly, seamless subchondral bone integration without marked sclerotic transformation (Fig. 7A). The control mosaicplasty sites demonstrated much poorer cartilage integration and subchondral bone that was sclerotic and did not integrate with the existing bone (Fig. 7B). It was concluded from the data that mosaicplasty augmentation was a promising clinical application for BMPs and additional animal studies were warranted.

Figure 6
Sheep osteochondral defect model: mosaicplasty augmentation showing donor holes OP-1(BMP-7) treated on the left and untreated control on the right at 12 weeks.

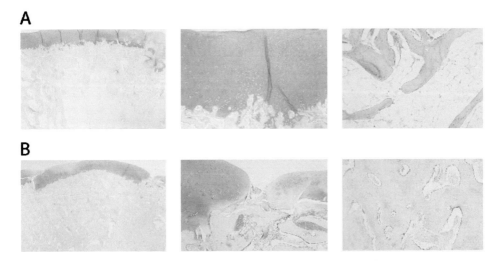

Figure 7
Sheep osteochondral defect model: mosaicplasty augmentation showing recipient holes with autograft plug
7A shows the OP-1(BMP-7) treated plug and 7B shows the control plug. The left photographs show the autograft with cartilage and bone compartments. The center photographs show a high magnification of one side of each autograft plug in order to demonstrate the cartilage integration. The right photographs show a magnification of the bone integration.

123

Chondral defects repair studies

Chondral defect repair studies have more clearly demonstrated that certain BMPs have the ability to be cartilage anabolic factors *in vivo*. These studies have shown that hyaline-like tissue repair can be induced in a non-repairing model where the defects are not exposed to stem cells and factors from the blood/marrow. However, the models are technically more difficult than the osteochondral defect models. Creation of defects that do not penetrate the subchondral bone is challenging and containment of these BMPs in the defects sites is even more challenging. Thus, these studies have required large animals where the cartilage is much thicker than in the rabbit and in fact one study used the horse which has cartilage thickness similar to that of humans. In regard to delivery of BMPs, more novel techniques have had to be developed such as the use of a miniosmotic pump for slow delivery the BMP into the joint.

The most intriguing study reported thus far evaluating a BMP for repairing chondral defects has been achieved with recombinant OP-1 (BMP-7) in the sheep knee [52]. In this study, large 10 mm defects were created such that the subchondral bone was not damaged and OP-1/BMP-7 dissolved in buffer was delivered to the synovial fluid *via* an extraarticularly positioned miniosmotic pump connected to the joint by a polyethylene tubing. Two defects were created in each knee; one on the medial condyle and the other on the trochlea of the femur, and infused over a two-week period with either (55 μg or 170 μg) OP-1 (BMP-7) or by an acetate buffer control. At three months following surgery, defects treated with OP-1 (BMP-7) were partially filled with newly formed cartilage, precartilagineous tissue and connective tissue at the top of the defect. The cartilage formation initially took place at the bottom, progressing towards the surface of the defect. In control knees, there was no sign of cell ingrowth into the defect area. In OP-1 (BMP-7) treated knees, condylar defects showed a 40–62% fill and the trochlear defects showed a 56–81% fill. At six months, defects treated with OP-1 (BMP-7) showed increased new cartilage over that seen at three months and this cartilage was well fused to the old cartilage and stained positive for type II collagen (Fig. 8). All of these defect sites showed significant filling in both chondyle (57–71%) and trochlea (74–92%) locations, but none of the control defects showed healing. In this regard, it is interesting that the repair process continued to progress between 3–6 months, even though OP-1 (BMP-7) was delivered for only the first two weeks and without a scaffolding material. It was hypothesized that the continuous presence of OP-1 (BMP-7) throughout the initial weeks following surgery may have attracted sufficient mesenchymal-like cells originating from the synovium into the defect area. Subsequently, OP-1 (BMP-7) could stimulate these cells to differentiate into chondrocytes which would then produce the appropriate extracellular matrix for filling the defect site. The actual filling of the defect site is a slow process and continues long after the BMP has been delivered.

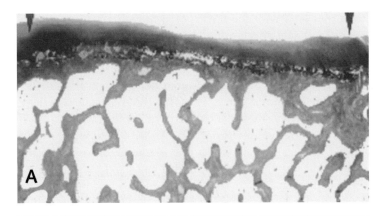

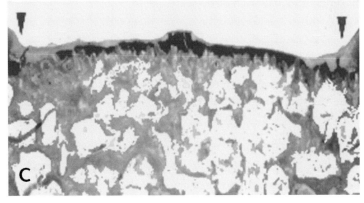

Figure 8
Sheep chondral defect model
Histological results at six months. The top photograph is an OP-1 (BMP-7) treated defect and the bottom photograph is a buffer control. The OP-1 (BMP-7) was slowly delivered by a miniosmotic pump into the joint over a two week period of time.

A chondral repair study has also been reported comparing recombinant BMP-2 and BMP-13 (also called GDF-6 or CDMP-2) in a much smaller defect in the minipigs knee [53]. In this study, long, narrow defects (0.5 mm wide, 5–6 mm long and 0.5 mm deep) were created and the BMPs, encapsulated in liposomes, were contained at the defect site by a fibrin clot. Both BMPs were efficacious in inducing cartilage-like tissue fill in approximately 90% of the defect site six weeks after surgery in this model. It should be noted that this study also included TGF-β, IGF-1, EGF, TGF-α and Tenascin-C for comparison. Of these only TGF-β showed the ability to induce cartilage tissue in this model. However, the amount of tissue was significantly

125

less with the TGF-β than that seen with the BMPs, even though a greater concentration was used and significant side effects were observed.

In addition to the studies using recombinant BMPs, one study has been reported using a BMP gene to repair a chondral defect [54]. The study evaluated the repair of large 15 mm diameter defects extending down to, but not through, the calcified cartilage layer in the horse petellofemoral joints. The OP-1 (BMP-7) gene was delivered to the defect site *via* transfected allogenic chondrocytes embedded in a fibrin clot. In comparison with the control cells without OP-1 (BMP-7), the BMP treated defects showed accelerated healing at four weeks and markedly more hyalin-like morphology. However, by eight months both the control and OP-1 (BMP-7) treated defects had similarly healed with cartilage repair tissue and it was concluded that the advantage of BMPs appear to be limited to an acceleration of cartilage healing in this model using modified allogenic chondrocytes.

Discussion

The purpose of this chapter was to review the current knowledge on BMPs in articular cartilage repair from the standpoint of both *in vitro* culture studies and animal repair studies. The data clearly show that BMPs have an important role in cartilage, both in normal homeostasis and in repair. The animal study results predict a bright future for the use of BMPs in the engineering of cartilage.

In cell culture studies there have been numerous demonstrations that BMPs are endogenously expressed in cartilage and that they act as anabolic factors for chondrocytes in culture. Recent studies have shown that many BMPs are endogenously expressed in cartilage, including BMP-2, -3, -4, -5, -6, -7 and CDMP-1 and -2. OP-1 (BMP-7) has also been localized to synovial fluid, synovium, ligament, tendon and meniscus. In regard to the anabolic activity, the role of certain BMPs in stimulating chondrocyte differentiation, extracellular matrix production and maintenance of the adult chondrocytic phenotype is well documented. However, few direct comparisons of BMPs have been reported and the only extensively studied BMP has been OP-1 (BMP-7). OP-1 (BMP-7) has been shown to stimulate the synthesis of all the major cartilage extracellular matrix proteins and to counteract the degenerative effect of numerous catabolic mediators. Furthermore, recent antisense studies have demonstrated that down-regulation of OP-1 (BMP-7) mRNA induced a significant decrease in proteoglycan synthesis in articular chondrocytes in culture. Thus, the data from *in vitro* studies have clearly demonstrated that at least one BMP, OP-1 (BMP-7), is very important in articular cartilage homeostasis. Certainly, a detailed analysis of the importance of other BMPs in this field needs to be done.

Data from both osteochondral and chondral repair studies in animals show that BMPs clearly have a therapeutic potential for cartilage repair. In a large chondral defect study in sheep, OP-1 (BMP-7) was shown to induce significant repair in a model where no repair took place in the controls. In addition, the repair tissue was hyaline-like and well bonded with the surrounding cartilage. In several large osteo-chondral defect studies, both OP-1 (BMP-7) and BMP-2 were observed to induce a significant improvement in repair of both the cartilage and bone compartment over that observed in control defects; the BMP-treated sites exhibited less fibrocartilage and more hyaline-like cartilage. Although the goal of most investigations is the restoration of perfect cartilage, the clinical demand is so great for new cartilage repair procedures that simply an improvement over the repair currently achieved can be an acceptable interim goal. Furthermore, the relevance of animal repair models to the repair of human cartilage is a serious issue and thus the question is asked as to how much more testing should be done in animals before clinical studies should begin.

In regard to using BMPs in clinical studies, the animal studies have clearly demonstrated that the BMPs can be safely administered to the joint. Recombinant OP-1 (BMP-7) and BMP-2 have been evaluated in various formulations, concentrations, frequencies of dosing and delivery routes for repairing both chondral and osteochondral defects in rabbits, dogs, goats, sheep and minipigs. In these studies there have been no reports of side effects, such as bone formation on the synovial surface or free floating objects in the synovial fluid. Furthermore there have not been reports of obvious deterioration/degeneration of the cartilage during the study periods up to 52 weeks nor have any inflammatory side effects such as synovitis, pannus formation or joint effusion been reported. The only conflicting data have come from a study evaluating BMP-2 injected into a normal mouse joint [55] where it was reported that after multiple injections of BMP-2 induced chondrogenesis (termed chondrophytes) in the region where the growth plates meet the joint space and at later times these chondrophytes developed into osteophytes. However, given the numerous reported repair studies done with BMP-2 in multiple animal species where no such side effects were observed, this result appears to be specific to the mouse.

In conclusion, a BMP based therapy for damaged cartilage would appear to have significant clinical potential. The BMP could be delivered locally to a focal defect site on an appropriate scaffold material or possibly slowly delivered to the joint without a scaffold as was demonstrated in the minipump delivery study. Specifically, a BMP therapy could be appropriate for augmentation of current procedures like mosaicplasty and microfracture, as well as a replacement for cell-based therapies, which involve removal of autologous cells derived from marrow or from cartilage, followed by expansion in culture and then by a second operation for implantation into the defect. Furthermore, arthroscopic administration of a BMP product should increase the attractiveness of the therapy enormously. Finally, the goal of injectable formulations of a BMP in a slow release material would seem to be the ideal route of administration in order to extend the therapeutic potential dramatically, particularly in the area of treatment and, ultimately, prevention of osteoarthritis.

Acknowledgements

Part of this work is supported by the NIH/NIAMS grants 2-AP-39239 and AR 47654 and grant from Stryker Biotech SC-001.

References

1   Chen P, Vukicevic S, Sampath TK, Luyten FP (1993) Bovine articular chondrocytes do not undergo hypertrophy when cultured in the presence of serum and osteogenic protein-1. *Biochem Biophy Res Commun* 197(3): 1253–1259

2   Nishida Y, Knudson CB, Kuettner KE, Knudson W (2000) Osteogenic protein-1 promotes the synthesis and retention of extracellular matrix within bovine articular cartilage and chondrocyte cultures. *Osteoarthritis and Cartilage* 8: 127–136

3   Lietman S, Yanagishita M, Sampath TK, Reddi AH (1997) Stimulation of proteoglycan synthesis in explants of porcine articular cartilage by recombinant osteogenic protein-1 (bone morphogenetic protein-7). *J Bone J Surg* 79–A: 1132–1137

4   Chen P, Vukicevic S, Sampath TK, Luyten FP (1995) Osteogenic protein-1 promotes growth and maturation of chick sternal chondrocytes in serum-free cultures. *J Cell Sci* 108 (Pt 1): 105–114

5   Flechtenmacher J, Huch K, Thonar EJ-MA, Mollenhauer JA, Davies SR, Schmid TM, Puhl W, Sampath TK, Adelotte MB, Kuettner KE (1996) Recombinant human osteogenic protein 1 is a potent stimulator of the synthesis of cartilage proteoglycans and collagens by human articular chondrocytes. *Arthritis Rheum* 39: 1896–1904

6   Huch K, Wilbrink B, Flechtenmacher J, Koepp HE, Aydelotte MB, Sampath TK, Kuettner KE, Mollenhauer JA, Thonar EJ-MA (1997) Effects of recombinant human osteogenic protein 1 on the production of proteoglycan, prostaglandin E2, and interleukin-1 receptor antagonist by human articular chondrocytes cultured in the presence of interleukin-1β. *Arthritis Rheum* 40: 2157–2161

7   Loeser RF, Pacione CA, Chubinskaya S (2003) The combination of insulin-like growth factor 1 and osteogenic protein 1 promotes increased survival of and matrix synthesis by normal and osteoarthritic human articular chondrocytes. *Arthritis Rheum* 48(8): 2188–2196

8   Chubinskaya S, Smith A, Hakimiyan A, Pacione C, Rueger DC, Loeser RF (2002) Synergistic effect of IGF-1 and OP-1 on matrix formation by normal and OA chondrocytes cultured in alginate beads. Abstract, 4th ICRS Symposium, Toronto, June 15–18

9   Chubinskaya S, Rueger DC, Berger RA, Kuettner KE (2002) Osteogenic protein-1 and its receptors in human articular cartilage. In: V Hascall, KE Kuettner (eds): *The many faces of osteoarthritis*. Birkhäuser Verlag, Basel, Switzerland, 81–89

10   Nishida Y, Knudson CB, Eger W, Kuettner KE, Knudson W (2000) Osteogenic protein-1 stimulates cell-associated matrix assembly by normal human articular chondrocytes: upregulation of hyaluronan synthase, CD 44 and aggrecan. *Arthritis Rheum* 43: 206–214

11   Vukicevic S, Luyten FP, Reddi AH (1989) Stimulation of the expression of osteogenic and chondrogenic phenotypes *in vitro* by osteogenin. *Proc Natl Acad Sci USA* 86: 8793–8797

12   Luyten FP, Yu YM, Yanagishita M, Vukicevic S, Hammonds RG, Reddi AH (1992) Natural bovine osteogenin and recombinant human bone morphogenetic protein-2B are equipotent in the maintenance of proteoglycans in bovine articular cartilage explant culture. *J Biol Chem* 267: 3691–3695

13   Luyten FP, Chen P, Paralkar V, Reddi AH (1994) Recombinant bone morphogenetic protein-4, transforming growth factor-β, and activin A enhance the cartilage phenotype of articular chondrocytes *in vitro*. *Exp Cell Res* 210: 224–229

14   Morris E (1996) Differential effects of TGF-beta superfamily members on articular cartilage metabolism: stimulation by rhBMP-9 and rhBMP-2 and inhibition by TGF-beta. *Trans Ortho Res Soc* 42: 175

15   van Susante JLC, Buma P, van Beuningen HM, van den Berg WB, Veth RPH (2000) Responsiveness of bovine chondrocytes to growth factors in medium with different serum concentrations. *J Ortho Res* 18: 68–77

16   Sailor LZ, Hewick RM, Morris EA (1996) Recombinant human bone morphogenetic protein-2 maintains the articular chondrocyte phenotype in long-term culture. *J Ortho Res* 14: 937–945

17   Stewart MC, Saunders KM, Burton-Wurster N, Macleod JN (2000) Phenotypic stability of articular chondrocytes *in vitro*: the effect of culture models, bone morphogenetic protein 2, and serum supplementation. *J Bone Miner Res* 15: 166–174

18   Vinall RL, Lo SH, Reddi AH (2002) Regulation of articular chondrocyte phenotype by bone morphogenetic protein 7, interleukin 1, and cellular context is dependent on the cytoskeleton. *Exp Cell Res* 272: 32–44

19   Bobacz K, Gruber R, Soleiman A, Erlacher L, Smolen JS, Graninger WB (2003) Expression of Bone morphogenetic protein 6 in healthy and osteoarthritic human articular chondrocytes and stimulation of matrix synthesis *in vitro*. *Arthritis Rheum* 48(9): 2501–1508

20   Yao J, Cole AA, Huch K, Kuettner KE (1996) The effect of OP-1 on IL-1beta induced gene expressions of matrix metalloproteinases and TIMP in human articular cartilage (abstract). *Trans Orthop Res Soc* 42: 305

21   Fan Z, Chubinskaya S, Rueger DC, Bau B, Haag J, Aigner T (2004) Regulation of anabolic and catabolic gene expression in normal and osteoarthritic adult human articular chondrocytes by osteogenic protein-1. *Clin Exp Rheumatol* 22(1): 103–106

22   Koepp HE, Sampath KT, Kuettner KE, Homandberg GA (1997) Osteogenic protein-1 (OP-1) blocks cartilage damage caused by fibronectin fragments and promotes repair by enhancing proteoglycan synthesis. *Inflamm Res* 47: 1–6

23   Im HJ, Pacione C, Chubinskaya S, Van Wijnen AJ, Sun Y, Loeser RF (2003) Inhibitory effects of insulin-like growth factor-1 and osteogenic protein-1 on fibronectin fragment- and interleukin-1beta-stimulated matrix metalloproteinase-13 expression in human chondrocytes. *J Biol Chem* 278(28): 25386–25394

24 Glansbeek HL, van Beuningen HM, Vitters EL, Morris EA, van der Kraan PM, van den Berg WB (1997) Bone morphogenetic protein-2 stimulates articular cartilage proteoglycan synthesis *in vivo* but does not counteract interleukin-1 alpha effects on proteoglycan synthesis and content. *Arthritis Rheum* 40(6): 1020–1028

25 Blanco FJ, Geng Y, Lotz M (1995) Differentiation-dependent effects of IL-1 and TGF-beta on human articular chondrocyte proliferation are related to inducible nitric oxide synthase expression. *J Immunol* 154(8): 4018–4026

26 van der Kraan PM, Vitters EL, van Beuningen HM, van de Loo FAJ, van den Berg WB (2000) Role of nitric oxide in the inhibition of BMP-2-mediated stimulation of proteoglycan synthesis in articular cartilage. *Osteoarthritis Cart* 8: 82–86

27 Fukui N, Zhu Y, Maloney WJ, Clohisy J, Sandell L (2003) Stimulation of BMP-2 expression by pro-inflammatory cytokines IL-1 and TNF-alpha in normal and osteoarthritic chondrocytes. *J Bone Joint Surg Am* 85-A (Suppl 3): 59–66

28 Chubinskaya S, Merrihew C, Cs-Szabo G, Mollenhauer JA, McCartney J, Rueger DC, Kuettner KE (2000) Human articular chondrocytes express osteogenic protein-1. *J Histochem Cytochem* 48(2): 239–250

29 Erlacher L, Ng C-K, Ullrich R, Krieger S, Luyten FP (1998) Presence of cartilage-derived morphogenetic proteins in articular cartilage and enhancement of matrix replacement *in vitro*. *Arthritis Rheum* 41: 263–273

30 Chubinskaya S, Kumar B, Merrihew C, Heretis K, Rueger DC, Kuettner KE (2002) Age-related changes in cartilage endogenous OP-1: New ELISA method. *Biochim Biophys Acta – Molecular Basis of Disease* 1588 (2): 126–134

31 Merrihew C, Kumar B, Heretis K, Rueger DC, Kuettner KE, Chubinskaya S (2003) Alterations in endogenous osteogenic protein-1 with degeneration of human articular cartilage. *J Orthop Res* 21(5): 899–907

32 Chubinskaya S, Kuettner KE (2003) Regulation of osteogenic proteins by chondrocytes. *Int J Biochem Cell Biol* 35(9): 1323–1340

33 Chubinskaya S, Frank B, Kumar B, Merrihew C, Hakimiyan A, Lenz ME, Rueger DC, Block JA, Thonar EJ-MA (2004) Characterization of endogenous osteogenic protein-1 in synovial fluid from normal human donors and patients with osteoarthritis or rheumatoid arthritis. *Trans Ortho Res Soc* 50: 956

34 Merrihew C, Soeder S, Rueger DC, Kuettner KE, Chubinskaya S (2003) Modulation of endogenous osteogenic protein-1 (OP-1) by interleukin-1 in adult human articular cartilage. *J Bone Joint Surg Am* 85-A (Suppl 3): 67–74

35 Muehleman C, Kuettner KE, Rueger DC, ten Dijke P, Chubinskaya S (2002) Immunohistochemical localization of osteogenic protein (OP-1) and its receptors in rabbit articular cartilage. *J Histochem Cytochem* 50: 1341–1350

36 Patwari P, Chubinskaya S, Hakimiyan A, Kumar B, Cole AA, Kuettner KE, Rueger DC, Grodzinsky AJ (2003) Injurious compression of adult human donor cartilage explants: investigation of anabolic and catabolic processes. *Trans Ortho Res Soc* 49: 695

37 Chubinskaya S, Oakes B, Shimmin A, Flynn J, Rueger D, Kildey R (2004) Anabolic

response in the articular joint induced by OP-1 in the goat model of osteochondral defects. 5th ICRS Meeting, Gent, Belgium, May 26–29

38   Anderson HC, Hodges PT, Aguilera XM, Missana L, Moylan PE (2000) Bone morphogenetic protein (BMP) localization in developing human and rat growth plate, metaphysis, epiphysis, and articular cartilage. *J Histochem Cytochem* 48(11): 1493–1502

39   Fahlgren A, Chubinskaya S, Aspenberg P (2004) A capsular incision leads to elevated levels of osteogenic protein-1 in rabbit knee joint cartilage. 5th ICRS meeting, Gent, Belgium, May 26–29

40   Soeder S, Hakimiyan A, Smith A, Rueger DC, Kuettner KE, Chubinskaya S (2003) Antisense inhibition of osteogenic protein-1 modulates articular cartilage integrity. *Trans Ortho Res Soc* 28: 598

41   Sellers RS, Peluso D, Morris EA (1997) The effect of recombinant human bone morphogenetic protein-2 (rhBMP-2) on the healing of full-thickness defects of articular cartilage. *J Bone Joint Surg Am* 79-A(10): 1452–1463

42   Sellers RC, Zhang R, Glasson SS, Kim HD, Peluso D, D'Augusta DA, Beckwith K, Morris EA (2000) Repair of articular cartilage defects one year after treatment with recombinant human bone morphogenetic protein-2 (rhBMP-2). *J Bone Joint Surg Am* 82-A(2): 151–160

43   Frenkel SR, Saadeh PB, Mehrar BJ, Chin GS, Steinbrech DS, Brent B, Gittes JK, Longaker MT (2000) Transforming growth factor beta superfamily members: role in cartilage modeling. *Plast Recon Surg* 105(3): 980–990

44   Grgic M, Jelic M, Basic V, Basic N, Pecina M, Vukicevic S (1997) Regeneration of articular cartilage defects in rabbits by osteogenic protein-1 (bone morphogenetic protein-7). *Acta Med Croatia* 51(1): 23–27

45   Cook SD, Rueger DC (1996) Osteogenic protein-1: biology and applications. *Clin Ortho Rel Res* (324): 29–38

46   Mattioli-Belmonte M, Gigante A, Muzzarelli RAA, Politano R, De Benedittis A, Specchia N, Buffa A, Biagini G, Greco F (1999) N,N-dicarboxymethyl chitosan as delivery agent for bone morphogenetic protein in the repair of articular cartilage. *Med Biol Eng Comp* 37(1): 130–134

47   Mason JM, Breibart AS, Barcia M, Porti D, Pergolizzi RG, Grande DA (2000) Cartilage and bone regeneration using gene-enhanced tissue engineering. *Clin Ortho Rel Res* (379S): S171–S178

48   Cook SD, Patron LP, Salkeld SL, Rueger DC (2003) Repair of articular cartilage defects with osteogenic protein-1 (BMP-7) in dogs. *J Bone Joint Surg* 85-A (Suppl 3): 116–123

49   Louwerse RT, Iheyligers IC, Klein-Nulend J, Sugiihara S, van Kampen GPJ, Semeins CM, Goei SW, de Koning MHMT, Wuisman PIJM, Burger EH (2000) Use of recombinant human osteogenic protein-1 for the repair of subchondral defects in articular cartilage in goats. *J Biomed Mater Res* 49(4): 506–516

50   Hunziker EB, Driesang MK (2003) Functional barrier principle for growth-factor-based articular cartilage repair. *Osteoarthritis Cartilage* 11: 320–327

51   Shimmin A, Young D, O'Leary S, Shih MS, Rueger DC Walsh WR (2003) Growth fac-

tor augmentation of an ovine mosaicplasty model. *Trans ICRS 4th Symposium*, no. 16, Toronto

52 Jelic M, Pecina M, Haspl M, Kos J, Taylor K, Maticic D, McCartney J, Yin S, Rueger D, Vukicevic S (2001) Regeneration of articular cartilage chondral defects by osteogenic protein-1 (bone morphogenetic protein-7) in sheep. *Growth Fact* 19: 101–113

53 Hunziker EB, Dreisang IMK, Morris EA (2001) Chondrogenesis in cartilage repair is induced by members of the transforming growth factor-beta superfamily. *Clin Ortho Rel Res* (391S): S171–S181

54 Hidaka C, Goodrich LR, Chen CT, Warren RF, Crystal RG, Nixon AJ (2003) Acceleration of cartilage repair by genetically modified chondrocytes over expressing bone morphogenetic protein-7. *J Ortho Res* 21: 573–583

55 van Beuningen HM, Glansbeek HL, van der Kraan PM, van den Berg W (1998) Differential effects of local application of BMP-2 or TGF-β1 on both articular cartilage composition and osteophyte formation. *Osteoarthritis Cartilage* 6: 306–317

Craniofacial reconstruction with bone morphogenetic proteins

Hendrik Terheyden[1] and Søren Jepsen[2]

[1]Department of Oral and Maxillofacial Surgery, University of Kiel, Arnold Heller Str. 16, 24105 Kiel, Germany; [2]Department of Restorative Dentistry and Periodontology, University of Bonn, 53111 Bonn, Germany

Introduction

Reconstructive surgery of the craniofacial skeleton comprises a large variety of indications which range from dental alveolar surgery to interdisciplinary cranial base interventions, from congenital malformations to acquired traumatic or tumor-related defects. In most applications the autogenous bone graft is the clinical gold standard. These grafts range from small intraorally harvested bone particles to large composite vascularized bone flaps. In the face, reconstruction always has functional and esthetic aspects. In both aspects, the shape of the reconstructed bone segment is very important. Regarding dental occlusion, a correct intermaxillary relation has to be achieved, especially if prosthetic rehabilitation with dental implants is intended. Due to the thin skin coverage, shape irregularities will end with a bad esthetic result. Furthermore, the regenerated mandibular segment has to resist an occlusal load up to 600 N on a single molar tooth.

From a biomechanical point of view, it is useful to distinguish between filling of bone gaps and augmentations above the existing anatomical bone level. As long as the osteoinductive components are used to fill pre-existing defects like bone cysts or some kinds of small mandibular continuity defects, the stability and space keeping effect of the carrier material is not so important. This changes in all kinds of augmentations or in large defects where the bone inducing implant has to resist soft tissue pressure which occurs during mastication, during movements of the tongue or the mimic muscles. These facts are especially important in alveolar ridge augmentation.

This review will focus on mandibular reconstruction and augmentations in dental implant surgery. These indications are standard situations of craniofacial reconstruction, which frequently occur in clinical routine.

Bone Morphogenetic Proteins: Regeneration of Bone and Beyond, edited by Slobodan Vukicevic and Kuber T. Sampath
© 2004 Birkhäuser Verlag Basel/Switzerland

Mandibular reconstruction

Direct application of BMP or BMP genes in mandibular continuity defects including distraction osteogenesis

The key study on mandibular reconstruction with bone morphogenetic proteins was performed by Toriumi and co-workers [1]. In dogs, a predictable and load bearing bridging of the defect occurred using rhBMP-2 and a collagen sponge carrier. However, some narrowing and reduction of height of the regenerated bone was noticed due to soft tissue pressure on the soft carrier material. A subsequent study with similar long-term results was later reported by the same group using rhBMP-2 and a biodegradable particular polylactide carrier [2]. Complete bridging as well as osseointegration of dental implants was observed in a monkey study by Boyne and co-workers using rhBMP-2 on a collagen sponge carrier [3]. In the latter study a wound dehiscence problem and impairment of bone formation occurred with the intraoral approach that is also typical of clinical work. The problem of monitoring bone regeneration in a mandibular defect was addressed in a sheep study [4]. A high level of preclinical research was achieved by a series of studies by a Japanese group [5–7] who applied rhBMP to successfully bridge large segmental mandibular continuity defects in a non-human primate model and restored occlusal function with the use of dental implants and prosthodontic bridgework. These authors noted excess bone formation at the beginning but remodeling and spontaneous anatomical shaping of the bone occurred, once functional load was applied to the regenerated bone segments.

Although the authors of the previous studies tried, it is practically impossible to strip all periosteum, especially in the alveolar parts in this kind of defect. Thus, the studies resemble clinically more a subperiosteal resection of a benign tumor. The prerequisites of bone healing in such defects are good because of the presence of periosteal bone cells. In a study in minipigs, our group [8] tried to remove all periosteal cells from a defect with another technique of mandibular resection, which is more common in malignant tumor surgery. In minipigs, the mandibular segment was removed with a layer of attached soft tissues including the periosteum. These defects were treated with rhOP-1 device (including bone collagen) which was stabilized with carboxymethylcellulosis (CMC). With this material, anatomical correct volume and shape of the regenerated bone was achieved and the regenerated bone had a mechanical stiffness of 75.3% of the stiffness of the contralateral hemimandible after three months (Figs 1–5).

In rodents, BMP was successfully applied to mandibular defects by a gene therapy approach using either an adenoviral vector directly [9] or *in vitro* transfected autologous cells [10].

RhOP-1 was successfully used to facilitate and accelerate callus maturation in mandibular distraction osteogenesis in a rat model. OP-1 was injected into the cal-

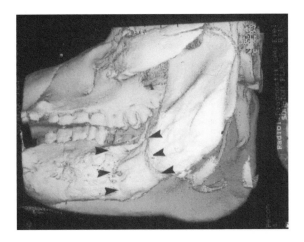

Figure 1
3D-reformatted CT Scan of a minipig. A critical size mandibular defect was filled with rhOP-1 collagen device. After 3 months there is a continuous bridging with functional bone. However, there is some narrowing of the contour probably because of the soft consistency of the collagen device.

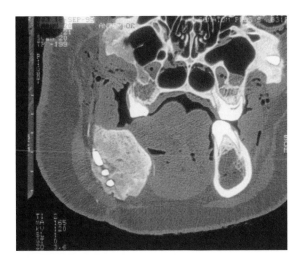

Figure 2
Vertical CT Scan through a critical size defect in the mandible of a minipig. The defect was filled with OP-1 putty device (3000 µg rhOP-1, 2 g collagen stabilized with 1 g carboxy-methylcellulose). In comparison to Figure 1 the stabilized OP-1 device produced a more bulky reconstructed bone which is likely to remodel during further function to normal shape. The bone formation took 3 months.

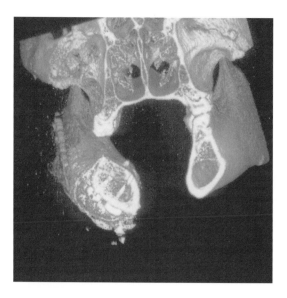

Figure 3
3D-reformatted CT scan of the animal of Figure 2 showing the large reconstructed volume and complete bridging of the defect after use of stabilized rhOP-1 device.

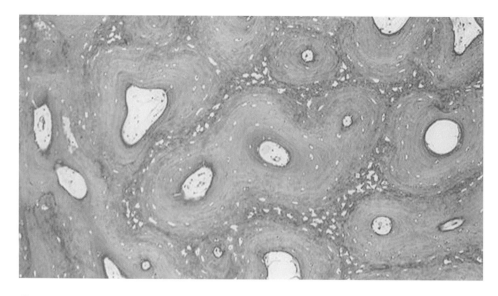

Fig. 4
Histological section of the regenerated bone of Figure 2 after three months of healing. Note the formation of compact bone with Haversian-like systems (toluidine blue, × 30).

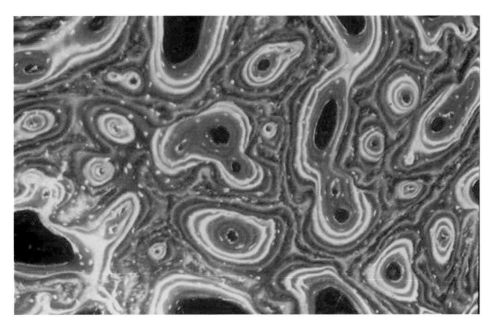

Figure 5
Histological section of the regenerated bone of Figure 2 after three months of healing. Fluorochrome labelling shows that bone formation in the defect started with orange (week 4 + 5) with irregular woven bone. The interspaces were then filled in the green phase (week 6 + 7) and in the red phase (week 8 + 9) until Haversian-like structures appeared (fluorescence microscopy, × 30).

lus after termination of the distraction phase. During three weeks of callus maturation, a continuous bone bridge with significantly increased mechanical stiffness formed in the OP-1 sites compared with the control sites [11] (Figs 6–8).

In conclusion, treatment of critical size defects with rhBMP leads to a predictable bone bridging in animal research without the need for autologous bone grafts. This has not been achieved with other types of bone substitutes or growth factors before. Improvements regarding the shape of the reconstructed segment were achieved with the use of CMC stabilized device. Furthermore, it was demonstrated that surface irregularities tend to remodel away after functional load was applied. The intraoral approach is a clinical standard for autologous bone grafting. In this case, bacterial contamination of BMP devices through saliva is an additional complicating factor. For application of the recombinant osteoinductive technology supplementation of the carrier material with antibiotic drugs may be a future field of research and development.

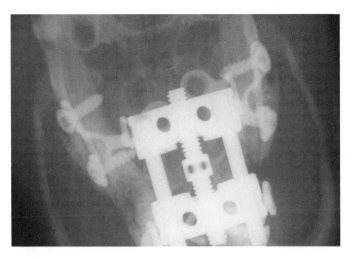

Figure 6
Bilateral distraction osteogenesis in a rat mandible. The total lengthening was 5 mm bilaterally obtained during 1 week. On the left mandible continuous bone bridging occurred 3 weeks after 2 injections of 50 µg rhOP-1 in aqueous solution into the distracted callus. On the control side after placebo injection the callus has not matured to complete mineralization within 3 weeks. There is still a fibrous gap between the segments.

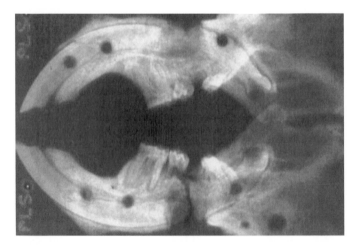

Figure 7
The microradiography shows the distracted hemimandibles. After 3 weeks of consolidation time complete bony bridging occurred on the test side (OP-1 injection, upper hemimandible) and not on the control side (placebo injection, lower hemimandible). Thus, rhOP-1 injection accelerated callus maturation, which also proved significantly in mechanical testing.

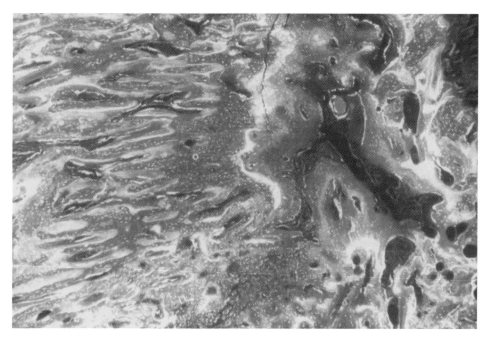

Figure 8
Histological section of the distracted bone of Figures 6, 7. During distraction osteogenesis woven bone in a parallel column-like fashion occurred (left part of this figure). After termination of the distraction in the consolidation the formation of woven bone was stimulated by rhOP-1 injection in the fibrous callus on top of the columnar bone which filled the distraction gap in the right half of the picture (orange 1 day after end of distraction, green 1 week after distraction, red 2 weeks after end of distraction, fluorescence microscopy, × 30).

Prefabrication of vascularized bone grafts

Clinically most mandibular defects occur after ablative surgery for malignant tumors. In this case, the usually combined intraoral/extraoral approaches are used. Microbiological contamination, extended operative time and extensive scar formation may decrease the success of primary reconstructive procedures. In most cases, additional radiotherapy will result in a bad recipient bed for bone grafts and BMP [12]. Clinically, in these cases a revascularized autogenous bone graft is applied. Usually, a fibular or iliac bone graft is harvested with a vascular pedicle (and sometimes with additional soft tissue flaps). These vessels are microsurgically connected with facial vessels and blood perfusion of the graft restored. The disadvantage of such technique is that harvesting of a vascularized bone graft is an operative burden

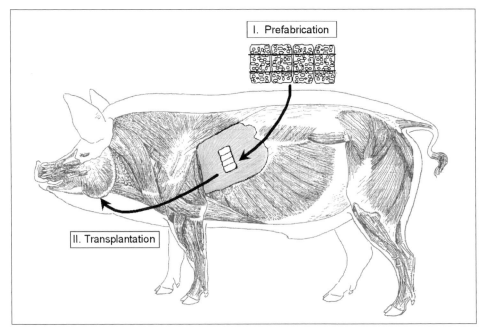

I. Prefabrication

II. Transplantation

Figure 9
Prefabrication of a bone graft: an osteoconductive scaffold (xenogenic bone blocks) in the shape of a segment of the mandibular angle was loaded with 600 µg rhOP-1 and placed in the latissimus dorsi muscle of minipigs. After 6 weeks the bone was harvested with a vascular pedicle and grafted to the mandibular defect. Perfusion was restored by microsurgical anastomosis with facial blood vessels. The prefabrication technique clinically is suitable for patients with mandibular defects after tumor resection who have been irradiated.

and therefore not suitable for every patient. Secondly, problems may occur with donor site morbidity, anatomical limitations of the donor sites and the shape of naturally occurring grafts.

The prefabrication technique allows the creation of a bone graft in an easily accessible soft tissue area which can be custom-shaped according to the requirements of the individual defect. Khouri and co-workers were the first who used BMP for custom prefabrication of a small artificial femur head in a rat [13]. Several authors followed with BMP-prefabricated bone flaps in small animals without using them as a graft for reconstruction [14–19]. A prefabrication and transplantation in a large animal model was performed by our group in minipigs [20] (Fig. 9). In 10 minipigs, an osteoconductive scaffold was placed in a soft tissue pouch inside the latissimus dorsi muscle [21]. The scaffold consisted of single blocks of xenogenic

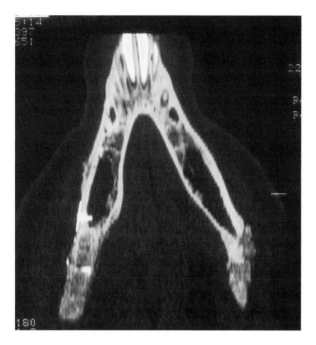

Figure 10
Prefabricated vascularized bone graft: Computed tomography 3 months after transplantation
surgery and microsurgical anastomosis of the vessels. The reformatted CT scan of the
mandibular arch shows that the prefabricated reconstructed segment perfectly fits into the
angle of the mandibular arch (left posterior hemimandible). On the contralateral side (right
posterior hemimandible) rhOP-1 had been directly applied without the prefabrication step
with less favourable reconstructive result.

bone (BioOss®, Geistlich, Wolhusen, Switzerland) which were connected with
resorbable threads forming an implant of $4.5 \times 2 \times 1$ cm size. Prior to surgery, 600 µg
of rhOP-1 in 1.2 ml acetate-mannitol buffer solution was poured over the scaffold
and soaked by the material. Bone growth in the blocks was studied by computed
tomography (Figs 10, 11) and histology. It was found that six weeks of prefabrication time are sufficient. Vascularization in the grafts was studied by bone scintigraphy (Fig. 12) and macro- and microangiography (Fig. 13) indicating a good vascularization on the microscopic level in areas of bone growth. In a subsequent study
in minipigs, a dose dependency of the parameters blood vessel density and bone density was observed. The best values were obtained with the dosage of 1,000 µg rhOP-
1 in a gram of carrier (xenogenic bone particles) [22]. In a subsequent study such
prefabricated grafts of $4.5 \times 2 \times 1$ cm in size were used to treat mandibular defects in

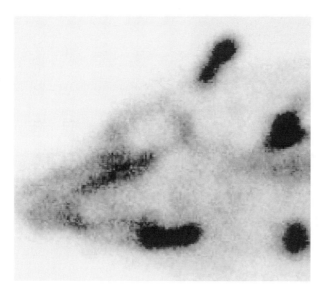

Figure 11
Prefabricated vascularized bone graft: Planar Tc99m - bone scintigraphy 7 days after transplantation demonstrates vitality and perfusion of the graft in the angle of the mandible. The remaining spots of tracer accumulation are clockwise the contralateral side after direct rhOP-1 application, the ear vein with the place of injection of the tracer and both lobes of the thyroid gland.

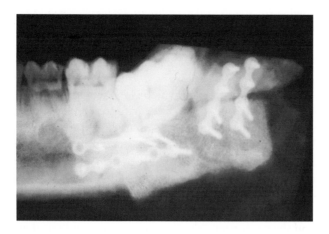

Figure 12
Prefabricated vascularized bone graft: Plain radiograph of the prefabricated bone flap fixed with titanium miniplates in the defect.

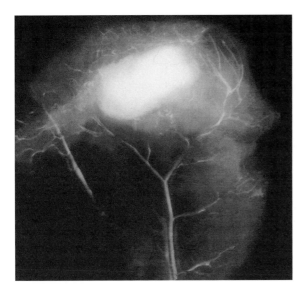

Figure 13
Prefabricated vascularized bone graft: Angiography of the latissimus dorsi flap containing the prefabricated bone graft. The thoracodorsal artery and vein (a, v) continuously branch in to the graft.

minipigs [23]. The grafts were harvested and grafted to a mandibular defect at the angle of the mandible in Göttingen miniature pigs. The defect was created in the mandibular angle using an epiperiosteal preparation and resection of the periosteum. The newly formed bone was stable enough to be fixed in the defects with conventional titanium miniplates and screws. Graft perfusion was restored by anastomosis with the facial vessels using a microsurgical technique. An identical defect of the contralateral side served as a control group and was treated by directly applied xenogenic bone scaffold and 600 µg rhOP-1. The first result of the study was that grafted prefabricated vascularized bone stayed viable. The continuous viability of large parts of the bone marrow was demonstrated by tracer uptake in bone scintigraphy and secondly shown in histology. Bone apposition in several areas was not interrupted by the transplantation process as proved by continuous polychromatic fluorescent labeling. As a second result, it was possible to restore the mandible with a prefabricated bone graft which was designed to fit into a certain mandibular defect. Histologically, it was observed that the growth of the newly formed bone was controlled by the osteoconductive scaffold which was filled with viable bone. Bone overgrowth was noted in only 2.3% of the volume. In CT scans performed three months postoperatively, a good restoration of the mandibular contour was observed and the regenerated bone showed good volume constancy, suitable, for instance, for the insertion of dental implants. An independent rating of CT scans with a numerical score system revealed a significantly better reconstructive result than with the directly applied material on the contralateral side.

In conclusion, the prefabrication technique is likely to open new possibilities in reconstructive surgery. The technology seems ready for clinical use once recombinant BMPs are approved for the clinical use. Further studies have to focus on technologies for custom shaping of individual parts for skeletal reconstruction.

Implantology

Aims for the use of BMP in implantology

It has been shown that long-term success of any dental implant under function depends on the achievement of direct bony anchorage [24]. Thus, the two basic aims of the use of growth factors and BMPs in implant dentistry are to increase bone implant contact (BIC) and to achieve a faster osseous integration of the dental implant, compared to standard clinical healing times of 3–6 months today. Furthermore, there is increasing evidence that in the near future BMPs will support or even replace autogenous bone grafting in augmentation of bone deficient implant sites. A future prospective for the use of BMPs may be to increase the quality of bone surrounding the implant in cases of osteoporotic bone and to reosseointegrate an implant after bone loss through peri-implant infection (perimplantitis).

Growth factors in implantology

A mixture of growth factors (PDGF/IGF-1) in a carboxymethylcellulose gel as a carrier was used in a few studies in implantology [25, 26] with some success. However, these studies have not been pursued recently. As a natural source of PDGF, platelet rich plasma (PRP) has been demonstrated to be useful to accelerate the maturation of particulate iliac bone grafts in reconstruction of the mandible [27]. The PRP method has been recognized by many dental clinicians. However, comparing directly PRP and BMPs, PRP in a rat model was not osteoinductive in contrast to rhOP-1 [28] and in typical relevant defect types in the maxillary area in a minipig model bone, regeneration by rhOP-1 was much more effective than PRP [29].

Enhancement and acceleration of BIC

Bone morphogenetic proteins (BMPs) have been reported to enhance osseous contact of dental implants. Some of these studies used naturally-sourced bovine BMP

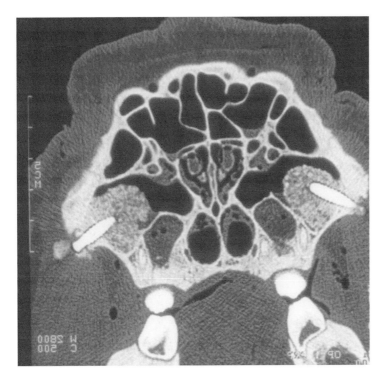

Figure 14
Sinus lift with rhOP-1: Axial CT-scan of miniature pig. Two implants are inserted from a laterocaudal direction into the augmented maxillary sinus area.

preparations in a canine mandibular site using a descriptive evaluation [30–32]. RhBMP-2 was used in an *in vitro* assay demonstrating a stimulation of osteoblastic cells on a titanium surface [33]. In a canine study, rhOP-1 induced new bone and enhanced osseous contact of HA-coated implants (BIC 80%) in combination with bone derived type I collagen in fresh extraction sites in the mandible [34]. Similar findings of enhanced BIC were reported in a canine study using rhBMP-2 [35, 36]. Our group observed 80% BIC with rhOP-1 and BioOss® compared to 32% with BioOss® alone in regenerated bone in a sinus augmentation study [37] (Figs 14, 15). 80% BIC is a noticeable value since a 60% BIC in mandibular bone is a representative value for a titanium implant [38]. Attempts have also been made to increase BIC by modifying the surface structure of the implants [39–41] or using HA coatings [42]. Although the studies are not easily compared due to different animal models, experimental periods and surface characteristics, it has to be emphasized that none of the studies achieved an osseous integration as high as the 80%. It can be concluded that rhOP-1 enhanced BIC.

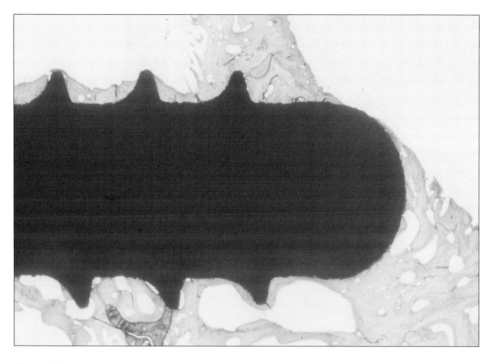

Figure 15
Sinus lift with rhOP-1: Newly formed bone covers the implant surface in the augmented area (rhOP-1 group) (toluidine blue, × 4).

BMPs have also been reported to accelerate bone formation around the implant in naturally derived [43] and recombinant form. In a minipig sinus augmentation study of our group, deposition of calcified material occurred on the implant surface after 2–3 weeks on the rhOP-1 side and after 8–9 weeks in the controls as monitored by polychromatic labeling [37]. In conclusion, BMPs can accelerate BIC formation. Clinical studies will have to elucidate whether clinically this may lead to earlier loading of implants and reduced recommendations for healing time which is actually six months in regenerated bone. Further research and development studies are required on biological improvement of dental implants, especially on BMP coating.

Sinus augmentation

From a biomechanical point of view it is useful to distinguish between inlay and onlay augmentations. Maxillary sinus augmentation is an inlay type of augmentation where the augmentation material is put into a relatively protected cavity with excellent contact to residual bone. The procedure is required when implants are planned in the edentulous parts of the lateral upper jaw where protrusion of the maxillary sinus led to an internal reduction of the height of alveolar bone. Sinus augmentation is a clinically very frequently used procedure.

BMP have been applied successfully in preclinical studies on sinus augmentation. In a study, utilizing rhBMP-2 and collagen sponges for a maxillary sinus floor augmentation in goats [44], bone growth was observed in the sinus floors. In a primate study, rhOP-1 on collagen carrier induced bone, but augmentation with BioOss® resulted in a better augmentative effect [45, 46]. Implants were not installed in those studies. In a sinus augmentation study [37] in miniature pigs (Figs 16–18) using 420 μg rhOP-1 in 1 ml acetate-mannitol buffer solution with 3 ml xenogenic bone mineral (BioOss®) as a carrier with simultaneous insertion of dental implants, our group reported a successful augmentation over the top of the simultaneously installed implants on the rhOP-1 side and on the control side after six months. In a subsequent study of our group in the same animal model, less BIC and augmentation height were observed with collagen carrier, compared to xenogenic bone or beta-tricalciumphosphate (Cerasorb®, Curasan, Kleinostheim, Germany) [47].

The results confirmed the findings of Margolin and co-workers [45] and support the view that for augmentation in the sinus the osteoconductive carrier alone was better than the soft collagen carrier and rhOP-1. However, osteoconduction takes time (six months or more) and the role of the BMP in this situation can be the acceleration and predictability of ossification. This was confirmed in our study by polychromatic labeling which revealed ossification on the implant surface as early as three weeks after implantation in contrast to the osteoconductive control where ossification on the implant occurred after nine weeks. As mentioned above, a predictable and significant increase in BIC was observed with simultaneous installation of the dental implant. The fact that implants are to be placed simultaneously with the osteoinductive proteins should be emphasized. In a site containing bone morphogenetic proteins, the implant is placed into the osteoinductive environment of the developing osteoprogenitor cells. Those cells interact with extracellular matrix and surfaces in their environment [48] and it is well known that the structure of the newly formed bone is influenced by the geometry of the environment [49]. Thus, it may be hypothesized that in implants placed secondarily to bone augmentation with BMP, the BIC rates would not be enhanced. In fact, in a second stage implantation study using rhBMP-2 on a collagen carrier in sinus augmentation in primates, the bone to implant contact was 41.4% not enhanced compared to the controls [50].

Human studies, as far as they are available, show inconsistent results. A small series of three human patient cases of sinus augmentation with rhOP-1 is reported in the literature [51–53]. The results range from good bone growth in one patient to absence of bone and persistent swelling in another patient. It was discussed that these inconsistent results may be attributed to the type of carrier used in the study (bone collagen). In a larger series [54] using rhBMP-2 and collagen sponge carrier, grossly good augmentative results but not always predictable augmentation height was reported.

In conclusion from animal and human studies for augmentative sinus procedures, a mineral osteoconductive carrier seems to be more suitable than soft collagen products and the role of BMP seems to be improving the predictability and speed of ossification and to enhance BIC in cases of primary implant installation.

Ridge augmentation

Alveolar ridge augmentation in implant dentistry is indicated when an edentulous part of the alveolar ridge has partially lost height and/or width due to ridge atrophy following tooth extraction. Ridge augmentation is an onlay type augmentation where the augmentation material is placed on top of the bone surface or into very shallow defects, where it has only limited contact to the residual bone. In this situation mechanical load (occlusal load and soft tissue pressure) acts towards the graft and the suture line. Furthermore, the augmentation is situated just below the suture line and is in higher risk of bacterial contamination and wound healing problems. This is more pronounced in vertical than in horizontal ridge augmentation.

Several studies dealt with BMP in ridge augmentation. In a basic study of our group using seven different carrier materials in mandibular augmentation in the rat, it was confirmed that mineralized calciumphosphate carriers for rhOP-1 result in a more predictable bone augmentation than collagens, and that the different osteoinductivity of carrier materials influences structure of the newly formed bone [49, 55]. In a canine study comparing peri-implant defects in the mandible treated with and without rhBMP-2 on collagen sponge carrier significant differences to the controls were noted after 12 weeks, but not after four weeks by radiographic evaluation [56]. In another canine study with rhBMP-2 and collagen sponge, a bone augmentation was observed. However, a low BIC of only 29.1% was reported after 16 weeks in regenerated bone [57]. Mixing collagen with HA particles resulted in a better augmentative effect but a compromised bone quality [58]. When a space maintaining membrane system was applied additionally with a collagen sponge carrier in the same model, the results improved [59]. Other carrier types than collagen sponge were successfully investigated in the canine ridge augmentation model such as calcium phosphate cement (alphaBSM) [60], polyglycolic acid/gelatin sponge complex [61], hyaluronan sponge [62] and allogenic demineralized bone matrix [63]. Ridge augmentation by rhBMP-2 in non-human primates was demonstrated with polymethylmethacrylate and BioOss® [64] and polylactic-co-glycolic acid-coated gelatin sponge (PGS) [65] as carrier materials. Even combinations of several growth factors with rhBMP-2 in a calcium phosphate bone cement were applied [66].

Clinical studies on ridge augmentation are of preliminary character [67]. A more substantial clinical study investigated in 11 patients after six months healing time the quality and maturation of regenerated bone in alveolar ridge defects after treatment with rhBMP-2 on a BioOss® carrier [68]. A significant difference was observed in comparison with the carrier material alone.

Reosseointegration and improvement of bone quality

An investigation on the use of a growth or differentiation factors for improvement of the local bone quality, for example in type IV bone, has not been reported yet. Reosseointegration after infection was observed with rhBMP-2 in a primate study [69]. This field remains to be an open question, although hypothetically, this seems to be a reasonable field of research.

Other fields of craniofacial reconstruction

There are plenty of indications for bone grafting in the craniomaxillofacial field. Cranial defects were successfully restored with Osteogenin as it is required in pediatric and adult craniofacial surgery [70]. RhBMP-2 was successfully applied with a collagen sponge carrier in a cleft palate defect in a monkey study [71] and with a polylactide beads carrier in a dog study [72].

Conclusive remarks

The question of carrier materials for rhBMP may be more important in craniofacial surgery than in other fields of reconstructive surgery. Volume and shape of the regenerated bone is important either in continuity reconstruction or in augmentations. A proven way to control the osteoinductive process is to use an osteoconductive scaffold for the induced bone cells [73]. The induced osteoprogenitors will adhere along the surface of this substratum and start matrix production in a controlled fashion. This theoretical principle has been proven in many of the reviewed studies. Porous hydroxyapatite as well as porous beta-tricalciumphosphate has been demonstrated to be suitable as a delivery agent, as space keeping material as well as osteoconductive scaffold for the bone cells. Further studies are required in the field of delivery materials.

As far as preclinical animal studies can predict clinical conditions, recombinant BMP has a strong potential to replace autogenous bone in many craniofacial applications.

References

1 Toriumi DM, Kotler HS, Luxemberg DP, Holtrop ME, Wang EA (1991) Mandibular reconstruction with an recombinant bone-inducing factor. *Arch Otolaryngol Head Neck Surg* 117: 1101–1112

2 Toriumi DM, O'Grady K, Horlbeck DM, Desai D, Turek TJ, Wozney J (1999) Mandibular reconstruction using bone morphogenetic protein 2: long-term follow-up in a canine model. *Laryngoscope* 109: 1481–1489

3 Boyne PJ (1996) Animal studies of application of rhBMP-2 in maxillofacial reconstruction. *Bone* 19 (Suppl 1): 83–92

4 Abu-Serriah M, Ayoub A, Boyd J, Paterson C, Wray D (2003) The role of ultrasound in monitoring reconstruction of mandibular continuity defects using osteogenic protein-1 (rhOP-1). *Int J Oral Maxillofac Surg* 32: 619–627

5 Marukawa E, Asahina I, Oda M, Seto I, Alam MI, Enomoto S (2001) Bone regeneration using recombinant human bone morphogenetic protein-2 (rhBMP-2) in alveolar defects of primate mandibles. *Br J Oral Maxillofac Surg* 39: 452–459

6 Marukawa E, Asahina I, Oda M, Seto I, Alam M, Enomoto S (2002) Functional reconstruction of the non-human primate mandible using recombinant human bone morphogenetic protein-2. *Int J Oral Maxillofac Surg* 31: 287–295

7 Seto I, Tachikawa N, Mori M, Hoshino S, Marukawa E, Asahina I, Enomoto S (2002) Restoration of occlusal function using osseointegrated implants in the canine mandible reconstructed by rhBMP-2. *Clin Oral Implants Res* 13: 536–541

8 Wang H, Springer IN, Schildberg H, Acil Y, Ludwig K, Rueger DR, Terheyden H (2004) Carboxymethylcellulose-stabilized collagenous rhOP-1 device-a novel carrier biomaterial for the repair of mandibular continuity defects. *J Biomed Mater Res* 68A: 219–226

9 Alden TD, Beres EJ, Laurent JS, Engh JA, Das S, London SD, Jane JA Jr, Hudson SB, Helm GA (2000) The use of bone morphogenetic protein gene therapy in craniofacial bone repair. *J Craniofac Surg* 11: 24–30

10 Park J, Ries J, Gelse K, Kloss F, von der Mark K, Wiltfang J, Neukam FW, Schneider H (2003) Bone regeneration in critical size defects by cell-mediated BMP-2 gene transfer: a comparison of adenoviral vectors and liposomes. *Gene Ther* 10: 1089–1098

11 Terheyden H, Wang H, Patrick H. Warnke P, Springer I, Erxleben A, Ludwig K, Rueger DC (2003) Acceleration of callus maturation using rhOP-1 in mandibular distraction osteogenesis in a rat model. *Int J Oral Maxillofac Surg* 32: 528–533

12 Khouri RK, Brown DM, Koudsi B, Deune EG, Gilula LA, Cooley BC, Reddi AH (1996) Repair of calvarial defects with flap tissue: role of bone morphogenetic proteins and competent responding tissues. *Plast Reconstr Surg* 98: 103–-109

13 Khouri RK, Koudsi B, Reddi H (1991) Tissue transformation into bone *in vivo*. A potential practical application. *JAMA* 266: 1953–1955

14 Viljanen VV, Gao TJ, Lindholm TS (1997) Producing vascularized bone by heterotopic bone induction and guided tissue regeneration: a silicone membrane-isolated latissimus dorsi island flap in a rat model. *Reconstr Microsurg* 13: 207–214

15 Mizumoto S, Inada Y, Weiland AJ (1993) Fabrication of vascularized bone grafts using ceramic chambers. *J Reconstr Microsurg* 1993: 441–449

16 Cavadas PC, Bonanad E, Baena-Montilla P, Vera-Sempere FJ (1996) Prefabrication of a free flap for tracheal reconstruction: an experimental study. Preliminary report. *Plast Reconstr Surg* 98: 1052–1062

17 Casabona F, Martin I, Muraglia A, Berrino P, Santi P, Cancedda R, Quarto R (1998) Prefabricated engineered bone flaps: an experimental model of tissue reconstruction in plastic surgery. *Plast Reconstr Surg* 101: 577–581

18 Levine J P, Bradley J, Turk AE, Ricci JL, Benedict JJ, Steiner G, Longaker MT, McCarthy JG (1997) Bone morphogenetic protein promotes vascularization and osteoinduction in preformed hydroxyapatite in the rabbit. *Ann Plast Surg* 39: 158–168

19 Kusumoto K, Bessho K, Fujimura K, Akioka J, Ogawa Y, Iizuka T (1998) Prefabricated muscle flap including bone induced by recombinant human bone morphogenetic protein-2: an experimental study of ectopic osteoinduction in a rat latissimus dorsi muscle flap. *Br J Plast Surg* 51: 275–280

20 Terheyden H, Jepsen S, Rueger D (1999) Mandibular reconstruction with prefabricated vascularized bone grafts using recombinant human osteogenic protein-1 – a preliminary study. Int J Oral Maxillofac Surg 28: 461–463

21 Terheyden H, Knak Ch, Jepsen S, Palmie S, Rueger D (2001) Prefabrication of vascularized bone grafts using recombinant human osteogenic protein-1. Pt. 1 Prefabrication. *Int J Oral Maxillofac Surg* 30: 373–379

22 Terheyden H, Menzel C, Wang H, Açil Y, Springer ING, Rueger D (2004) Prefabrication of vascularized bone grafts using recombinant human osteogenic protein-1 – Part. 3: dosage of rhOP-1, the use of external and internal scaffolds. *Int J Oral Maxillofac Surg* 33: 164–172

23 Terheyden H, Warncke P, Jepsen S, Dunsche A, Brenner W, Toth C, Rueger D (2001) Prefabrication of vascularized bone grafts using recombinant human Osteogenic Protein-1. Pt. 2 Transplantation. *Int J Oral Maxillofac Surg* 30: 469–478

24 Brånemark PI (1983) Osseointegration and its experimental background. *J Prosth Dent* 50: 399–410

25 Lynch SE, Buser D, Hernandez RA, Weber HP, Stich H, Fox CH, Williams RC (1991) Effects of the platelet-derived growth factor/insulin-like growth factor-I combination on bone regeneration around titanium dental implants. Results of a pilot study in beagle dogs. *J Periodontol* 62: 710–716

26 Becker W, Lynch SE, Lekholm U, Becker BE, Caffesse R, Donath K, Sanchez R (1992) A comparison of ePTFE membranes alone or in combination with platelet-derived growth factors and insulin-like growth factor-I or demineralized freeze-dried bone in promoting bone formation around immediate extraction socket implants. *J Periodontol* 63: 929–940

27 Marx RE, Carlson ER, Eichstaedt RM, Schimmele AR, Strauss JE, Georgeff KR (1998) Platelet-rich plasma: Growth factor enhancement for bone grafts. *Oral Surg Oral Med Oral Pathol Oral Radiol Endod* 85: 638–646

28 Roldan JC, Jepsen S, Washington G, Freitag S, Açil Y, Terheyden H (2004) Bone formation in the presence of platelet-rich plasma *versus* bone morphogenetic protein-7. *Bone* 34: 80–90

29 Roldan JC, Jepsen S, Schmidt C, Knüppel H, Rueger D, Acil Y, Terheyden H (2004) Sinus floor augmentation with simultaneous placement of dental implants in presence of platelet-rich plasma *versus* human bone morphogenetic protein-7. *Clin Oral Implan Res; in press*

30 Yan J, Xiang W, Baolin L, White FH (1994) Early histologic response to titanium implants complexed with bovine bone morphogenetic protein. *J Prosth Dent* 71: 289–294

31 Wang X, Jin Y, Liu B, Zhou S, Yang L, Yang X, White FH (1994) Tissue reactions to titanium implants containing bovine bone morphogenetic protein: a scanning electron microscopic investigation. *Int J Oral Maxillofac Surg* 23: 115–199

32 Wang X, Liu B, Jin Y, Yang X (1993) The effect of bone morphogenetic protein on osseointegration of titanium implants. *J Oral Maxillofac Surg* 51: 647–651

33 Ong JL, Cardenas HL, Cavin R, Carnes DL Jr (1997) Osteoblast responses to BMP-2-treated titanium *in vitro*. *Int J Oral Maxillofac Implants* 12: 649–654

34 Cook SD, Salkeld SL, Rueger DC (1995) Evaluation of recombinant human osteogenic protein-1 (rhOP-1) placed with dental implants in fresh extraction sites. *J Oral Implantol* 21: 281–289

35 Sykaras N, Triplett RG, Nunn ME, Iacopino AM, Opperman LA (2001) Effect of recombinant human bone morphogenetic protein-2 on bone regeneration and osseointegration of dental implants. *Clin Oral Implants Res* 12: 339–349

36 Fiorellini JP, Buser D, Riley E, Howell TH (2001) Effect on bone healing of bone morphogenetic protein placed in combination with endosseous implants: a pilot study in beagle dogs. *Int J Periodontics Restorative Dent* 21: 41–47

37 Terheyden H, Jepsen S, Möller B, Tucker MM, Rueger DC (1999) Sinus floor augmentation with simultaneous placement of dental implants using a combination of deproteinized bone xenografts and recombinant human osteogenic protein-1. A histometric study in miniature pigs. *Clin Oral Impl Res* 10: 510–521

38 Arvidson K, Bystedt H, Ericsson I (1990) Histometric and ultrastructural studies of tissues surrounding Astra dental implants in dogs. *Int J Oral Maxillofac Impl* 5: 127–134

39 Buser D, Schenk RK, Steinemann S, Fiorellini JP, Fox CH, Stich H (1991) Influence of surface characteristics on bone integration of titanium implants. A histomorphometric study in miniature pigs. *J Biomed Mater Res* 25: 889–902

40 Ericsson I, Johansson CB, Bystedt H, Norton MR (1994) A histomorphometric evaluation of bone-to-implant contact on machine-prepared and roughened titanium dental implants. A pilot study in the dog. *Clin Oral Implants Res* 5: 202–206

41 Gotfredsen K, Wennerberg A, Johansson C, Skovgaard LT, Hjørting-Hansen E (1995) Anchorage of TiO_2-blasted, HA-coated, and machined implants: an experimental study with rabbits. *J Biomed Mater Res* 29: 1223–1231

42 Weinlaender M, Kenney EB, Lekovic V, Beumer J3d, Moy PK, Lewis S (1992) Histo-

morphometry of bone apposition around three types of endosseous dental implants. Int *J Oral Maxillofac Implants* 7: 491–496

43 Rutherford RB, Sampath TK, Rueger DC, Taylor TD (1992) Use of bovine osteogenic protein to promote rapid osseointegration of endosseous dental implants. *Int J Oral Maxillofac Implants* 7: 297–301

44 Kirker-Head CA, Nevins M, Palmer R, Nevins ML, Schelling SH (1997) A new animal model for maxillary sinus floor augmentation: evaluation parameters. *Int J Oral Maxillofac Implants* 12: 403–411

45 Margolin MD, Cogan AG, Taylor M, Buck D, McAllister TN, Toth C, McAllister B (1998) Maxillary sinus augmenation in the non human primate. A comparative radiographic and histologic study between recombinant human osteogenic protein-1 and natural bone mineral. *J Periodontol* 69: 911–919

46 McAllister BS, Margolin MD, Cogan AG, Taylor M, Wollins J (1998) Residual lateral wall defects following sinus grafting with recombinant human osteogenic protein-1 or Bio-Oss in the chimpanzee. *Int J Periodontics Restorative Dent* 18: 227–239

47 Terheyden H, Mueller H, Schulz-Walz JE, Jepsen S, Rueger D (2000) Comparison of three carrier materials for rhOP-1 in sinus augmentation. *J Dent Res* 79: 512

48 Ripamonti U, Reddi AH (1994) Periodontal regeneration: potential role of bone morphogenetic proteins. *J Periodont Res* 29: 225–235

49 Terheyden H, Jepsen S, Vogler S, Tucker MM, Rueger DC (1997) Recombinant human osteogenic protein-1 (rhBMP-7) in the rat mandibular augmentation model: differences in bone morphology are dependent on the type of carrier. *Mund Kiefer Gesichtschir* 1: 272–275

50 Hanisch O, Tatakis DN, Rohrer MD, Wöhrle PS, Wozney JM, Wikesjö UME (1997) Bone formation and osseointegration stimulated by rhBMP-2 following subantral augmentation procedures in nonhuman primates. *Int J Oral Maxillofac Implants* 12: 785–792

51 Groeneveld EH, van-den-Bergh JP, Holzmann P, ten-Bruggenkate CM, Tuinzing DB, Burger EH (1999) Histomorphometrical analysis of bone formed in human maxillary sinus floor elevations grafted with rhOP-1 device, demineralized bone matrix or autogenous bone. Comparison with non-grafted sites in a series of case reports. *Clin Oral Implants Res* 10: 499–509

52 Groenveld HH, van-den-Bergh JP, Holzmann P, ten-Bruggenkate CM, Tuinzing-DB, Burger-EH (1999) Histological observations of a bilateral maxillary sinus floor elevation 6 and 12 months after grafting with osteogenic protein-1 device. *J Clin Periodontol* 26: 841–846

53 van den Bergh JPA, ten Bruggenkate CM, Groeneveld EHJ, Burger EH, Tuinzing DB (2000) Recombinant human bone morphogenetic protein-7 in maxillary sinus floor elevation surgery in 3 patients compared to autogenous bone grafts. A clinical pilot study. *J Clin Periodontol* 27: 627–636

54 Boyne P, Marx RE, Nevins M, Triplett G, Lazaro E, Lilly LC, Adler M, Nummikowski

P (1997) A feasibility study evaluating rhBMP-2/absorbable collagen sponge for maxillary sinus floor augmentation. *Int J Periodont Rest Dent* 17: 11–25

55   Terheyden H, Jepsen S, Vogler S, Tucker M, Rueger DC (1996) Recombinant human osteogenic protein 1 (rhBMP-7) in the rat mandibular augmentation model using different carrier materials. *J Cran Maxillofac Surg* (Suppl 1) 24: 114

56   Cochran D, Nummikoski PV, Jones AA, Makins SR, Turek TJ, Buser D (1997) Radiographic analysis of regenerated bone around endosseous implants in the canine using recombinant human bone morphogenetic protein-2. *Int J Oral Maxillofac Implants* 12: 739–748

57   Sigurdsson TJ, Fu E, Takakis D, Rohrer M, Wikesjö UME (1997) Bone morphogenetic protein-2 for peri-implant bone regeneration and osseointegration. *Clin Oral Impl Res* 8: 375–385

58   Barboza EP, Duarte ME, Geolas L, Sorensen RG, Riedel GE, Wikesjo UM (2000) Ridge augmentation following implantation of recombinant human bone morphogenetic protein-2 in the dog. *J Periodontol* 71: 488–496

59   Wikesjo UM, Qahash M, Thomson RC, Cook AD, Rohrer MD, Wozney JM, Hardwick WR (2003) Space-providing expanded polytetrafluoroethylene devices define alveolar augmentation at dental implants induced by recombinant human bone morphogenetic protein 2 in an absorbable collagen sponge carrier. *Clin Implant Dent Relat Res* 5: 112–123

60   Wikesjo UM, Sorensen RG, Kinoshita A, Wozney JM (2002) RhBMP-2/alphaBSM induces significant vertical alveolar ridge augmentation and dental implant Osseointegration. *Clin Implant Dent Relat Res* 4: 174–182

61   Nagao H, Tachikawa N, Miki T, Oda M, Mori M, Takahashi K, Enomoto S (2002) Effect of recombinant human bone morphogenetic protein-2 on bone formation in alveolar ridge defects in dogs. *Int J Oral Maxillofac Surg* 31: 66–72

62   Hunt DR, Jovanovic SA, Wikesjo UM, Wozney JM, Bernard GW (2001) Hyaluronan supports recombinant human bone morphogenetic protein-2 induced bone reconstruction of advanced alveolar ridge defects in dogs. A pilot study. *J Periodontol* 72: 651–658

63   Sigurdsson TJ, Nguyen S, Wikesjo UM (2001) Alveolar ridge augmentation with rhBMP-2 and bone-to-implant contact in induced bone. *Int J Periodontics Restorative Dent* 21: 461–473

64   Boyne PJ, Shabahang S (2001) An evaluation of bone induction delivery materials in conjunction with root-form implant placement. *Int J Periodontics Restorative Dent* 21: 333–343

65   Marukawa E, Asahina I, Oda M, Seto I, Alam MI, Enomoto S (2001) Bone regeneration using recombinant human bone morphogenetic protein-2 (rhBMP-2) in alveolar defects of primate mandibles. *Br J Oral Maxillofac Surg* 39: 452–459

66   Meraw SJ, Reeve CM, Lohse CM, Sioussat TM (2000) Treatment of peri-implant defects with combination growth factor cement. *J Periodontol* 71: 8–13

67   Howell TH, Fiorellini J, Jones A, Alder M, Nummikoski P, Lazaro M, Lilly L, Cochran D (1997) A feasibility study evaluating rhBMP-2/absorbable collagen sponge device for

local alveolar ridge preservation or augmentation. *Int J Periodontics Restorative Dent* 17: 124–139

68   Jung RE, Glauser R, Scharer P, Hammerle CH, Sailer HF, Weber FE (2003) Effect of rhBMP-2 on guided bone regeneration in humans. *Clin Oral Implants Res* 14: 556–568

69   Hanisch O, Tatakis DN, Boskovic MM, Rohrer MD, Wikesjö UME (1997a) Bone formation and reosseointegration in peri-implantitis defects following surgical implantation of rhBMP-2. *Int J Oral Maxillofac Implants* 12: 604–610

70   Ripamonti U, Ma SS, Cunningham NS, Yeates L, Reddi AH (1993) Reconstruction of the bone -bone marrow organ by osteogenin, a bone morphogenetic protein, and demineralized bone matrix in calvarial defects of adult primates. *Plast Reconstr Surg* 91: 27–36

71   Boyne PJ, Nath R, Nakamura A (1998) Human recombinant BMP-2 in osseous reconstruction of simulated cleft palate defects. *Br J Oral Maxillofac Surg* 36: 84–90

72   Mayer M, Hollinger J, Ron E, Wozney J (1996) Maxillary alveolar cleft repair in dogs using recombinant human bone morphogenetic protein-2 and a polymer carrier. *Plast Reconstr Surg* 98: 247–59

73   Ripamonti U, Ma S, Reddi AH (1992) The critical role of geometry of porous hydroxyapatite delivery system in induction of bone by osteogenin, a bone morphogenetic protein. *Matrix* 12: 202–212

Clinical experience of osteogenic protein-1 (OP-1) in the repair of bone defects and fractures of long bones

Gary E. Friedlaender

Department of Orthopaedics and Rehabilitation, Yale University School of Medicine, P.O. Box 208071, New Haven, CT 06520-8071, USA

Introduction

Approximately six million fractures occur each year in the United States, of which 1.5 million involve long bones. These injuries produce considerable morbidity and impairment in individuals of both genders, all age groups and regardless of socioeconomic circumstances. On an annual basis, fractures account for over 36 million lost days from work, more than seven million days missed from school, approximately 6.4 million days of hospitalization, nearly 9.4 million visits to healthcare providers, and cost society an estimated US$ 21 billion [1]. Among the most problematic and burdensome fractures are those 5–10% that demonstrate delayed healing or nonunion [2].

Standard treatment of long bone fractures involves both biomechanical and biological considerations, often some form of internal fixation and bone graft [3]. Autogenous bone graft has substantial biological potential without the risks of disease transfer from donor to recipient; nor does it, by definition, pose any histocompatibility issues. Autografts do, however, come in limited supply, shapes and sizes. More so, they are associated with significant donor site morbidity. Persistent pain at the donor site occurs in approximately 30% of patients for six months or longer following the removal of autogenous bone [4].

Collectively, the limitations, morbidities and complications associated with the use of bone autografts have provided the incentive to search for alternative materials with sufficient biological potential and avoidance of the need for painful donor sites. Bone morphogenetic proteins, including osteogenic protein-1 (BMP-7), have considerable clinical potential for addressing these needs [5]. BMPs are cytokines central to the bone remodeling and repair cycle, specifically responsible for the osteoinductive nature of bone and bone grafts [6, 7]. Building upon the original observations of Urist, Sampath and Reddi [8] and Wozney and co-workers [9] isolated, characterized and reproduced these TGF-β superfamily molecules by recombinant techniques. Both OP-1 and BMP-2 are now commercially available and each approved by regulatory authorities, including the Food and Drug Administration, for specific clinical applications. OP-1 is approved in the US, Canada, Europe and Australia for the treatment of long bone nonunions and spine fusions, based upon a wide range of preclinical studies that demonstrated both safety and efficacy as well as a growing clinical experience that supports these same conclusions [10]. More than 10,000 OP-1 implants have been used worldwide, providing a large and growing experience with this important molecule.

Preclinical experience

Among preclinical studies are demonstrations of the ability of the OP-1 Implant™ (3.5 mg of rhOP-1 combined with 1 g of bovine bone derived type I collagen) to cause the healing of critical sized long bone defects in rabbit, canine and nonhuman primate models [11–14].

Clinical experience

Tibial nonunion study

In 1992, the FDA approved OP-1 for investigational use in a prospective, randomized multicenter clinical trial designed to evaluate this implant for the treatment of tibial nonunions, and accrual to the study was completed in 1996 [15]. 122 patients with 124 established tibial nonunions were randomly assigned to receive treatment of this difficult injury with either an intramedullary rod and fresh autogenous bone graft (61 patients with 61 nonunions) or the same internal fixation with OP-1 (61 patients with 63 nonunions). These groups were comparable in their duration of nonunion (median 17 months in each group) with statistically insignificant differences in their degree of comminution, frequency of open injury and grade, prior treatment approaches, age, gender and weight. The OP-1-treated group had significantly more atrophic nonunions (41% *versus* 25%) and a strong trend towards more smokers (74% *versus* 57%, p = 0.057).

Assessment criteria included the severity of pain at the fracture site, ability to walk while full weight-bearing, the need for additional operative intervention directed at the fracture, radiographic evaluation of bony union and the physician's satisfaction with the outcome. Patients were closely followed, including 3, 6, 9, 12 and 24 months following their surgical procedure, and the nine-month postoperative visit was used as the primary endpoint of this study.

Clinical success was demonstrated in the OP-1-treated group in 81% of patients at nine months following their operative procedure and a comparable 85% of the autograft-treated group. The OP-1-treated group also had significantly less blood loss (254 cc *versus* 345 cc), a shorter hospital stay (3.7 days *versus* 4.1 days) and a trend towards a reduction in the operative time (169 min *versus* 178 min) compared with the autograft-treated group. As anticipated, all autograft donor sites were associated with discomfort, 20% of patients noting pain at six months. No serious adverse events were associated with the use of OP-1.

Radiographic evidence of healing at this same endpoint was 75% in the OP-1-treated group and 84% in those patients receiving supplemental autogenous bone graft, a statistically insignificant difference (p = 0.218). These radiographic results persisted throughout the two-year follow-up of this study, and anecdotally well beyond this time frame).

Physician satisfaction with outcome was also comparable in both groups; 86% in the OP-1-treated group and 90% in the autograft treated patients. Based upon these results, the safety and efficacy of the OP-1 Implant™ was demonstrated and regulatory approval was granted in the US, Canada, Europe and Australia as described above.

Other clinical experiences

A number of other clinical experiences have confirmed and expanded upon the demonstrated usefulness and safety profile of OP-1.

In the first reported study of recombinant human BMP in clinical practice, Geesink et al. [16] showed that five of six patients implanted with rhOP-1 into a critical-sized fibular defect healed this gap.

McKee et al. [17] recently reported their experience treating recalcitrant long bone nonunions with OP-1. The study included 31 patients with six tibial, nine clavicular, 10 humeral, two ulnar and four femoral nonunions. All patients underwent standard internal fixation and the addition of rhOP-1 Putty (3.4 mg OP-1, 1 g collagen type I bovine collagen and 230 mg carboxymethylcellulose) to the nonunion defect site. All 31 patients exhibited abundant new bone formation and all were considered healed at a mean of 13 weeks. No adverse clinical events related to the OP-1 were noted.

A prospective, randomized clinical trial involving seven Canadian institutions has enrolled 124 patients with fresh open tibial shaft fractures who underwent intramedullary rod fixation [18]. Equal numbers of patients were treated with and without the addition of OP-1 at the time of initial surgical intervention. McKee and his collaborators have previously reported their preliminary findings, which included a reduced number of secondary interventions for failure to heal in the OP-1 treated group (17 *versus* 8, p = 0.02) and a trend towards improved functional outcome in the OP-1-treated group (reflecting more patients with no pain with activity, 8% *versus* 56%, p = 0.04, and full weight bearing, 95% *versus* 84%, p = 0.11). Again, no OP-1-related adverse events occurred. The longer term results of this study are anticipated in the near future.

McQueen et al. [19] have prospectively evaluated the use of OP-1 Putty compared with autograft in the healing of corrective metaphyseal osteotomies of the distal radius for clinically symptomatic malunions. 20 patients were randomly assigned to one of the treatment groups (10 patients per group), and all patients were internally fixed with a dorsal plate. Evaluation was accomplished using radiographs, functional testing and clinical review. The defect was considered healed radiographically if filled at least 75% by new bone in two standard views at 90 degrees to each other.

At 12 weeks, all 10 patients treated with autograft and nine of 10 patients receiving OP-1 were healed by radiographic evaluation. Five of 10 autograft-treated patients developed a clinically significant donor site hematoma. Otherwise complications were similar in both groups and there were no adverse events related to the OP-1.

Susarla et al. [20] recently presented the results of a retrospective study to assess time-to-healing associated with the use of OP-1 Implant™ in humeral nonunions. 12 patients were evaluated for an average of 323 days following their fracture, five of whom underwent prior internal fixation and five were smokers. All patients were treated with internal fixation and OP-1. 11 of the 12 patients achieved radiographic union at an average of 162 days and a median of 122 days, significantly less time than the authors ascribe to the historical use of autogenous bone grafts. Clinical results paralleled the radiographic analysis and no infections or OP-1-related adverse events were apparent.

OP-1 Putty has also been extensively evaluated in an Individual Patient Usage Australian study approved by the Therapeutic Goods Administration and reported by Shimmin [21]. A wide variety of skeletal disorders were included in this experience, including 113 nonunions of which most were long bones and the remainder involved the scaphoid, navicular and pelvis. OP-1 was used in each case and most patients were also treated with other adjunctive therapies, primarily autografts, allografts and osteoconductive fillers; 70% of these very challenging cases resolved their nonunions by clinical criteria and 65% by radiographic evaluation.

Conclusions

There has now been considerable clinical experience with rhOP-1. Over 10,000 patients have been treated worldwide. Several studies have focused on the treatment of fractures, especially long bone nonunions, and in each circumstance the efficacy and safety of OP-1 has been demonstrated. Clinical outcome measures and radiographic evidence of healing have been comparable in OP-1-treated patients compared with autogenous bone graft, but without the substantial morbidity associated with graft donor sites. Other musculoskeletal applications of OP-1 are being evaluated, especially related to the spine and joint reconstruction, and this experience parallels the success described for long bone nonunions. This profile of efficacy and safety has been the basis for regulatory approval of OP-1 in numerous countries, with an expanding list of applications as study results become available.

This experience represents the beginning of a new approach to the repair and regeneration of musculoskeletal tissues based upon the clinical application of bone morphogenetic proteins and other cytokines intrinsic to the bone remodeling cycle. More will be learned in the coming years regarding dose, timing, carriers, routes of administration and indications for these exciting molecules.

References

1   Praemer A, Furner S, Rice DP (1999) *Musculoskeletal conditions in the United States.* American Academy of Orthopaedic Surgeons, Rosemont, Illinois

2   Connolly JF (1991) *Tibial nonunion: diagnosis and treatment.* American Academy of Orthopaedic Surgeons, Park Ridge, Illinois

3   Wiss DA, Stetson WB (1996) Tibial nonunion: treatment alternatives. *J Amer Acad Orthop Surgeons* 4: 249–257

4   Younger EM, Chapman MW (1989) Morbidity at bone graft donor sites. *J Orthop Trauma* 3: 192–195

5   Schmitt JM, Hwong K, Winn SR, Hollinger JO (1999) Bone morphogenetic proteins: an update on basic biology and clinical relevance. *J Orthop Res* 17: 269–278

6   Urist MR (1965) Bone formation by autoinduction. *Science* 150: 893–899

7   Urist MR, Strates BS (1971) Bone morphogenetic protein. *J Dent Res* 50: 1392–1406

8   Sampath TK, Reddi RH (1981) Disassociative extraction and reconstitution of extracellular matrix components involved in local bone differentiation. *Proc Natl Acad Sci USA* 78: 7599–7603

9   Wozney JM, Rosen V, Celeste AJ, Mitsock LM, Whitters MJ, Kriz RW, Hewick RM, Wang EA (1988) Novel regulators of bone formation: molecular clones and activities. *Science* 242: 1528–1534

10 Vukicevic S, Sampath KT (eds) (2002) *Bone morphogenetic proteins: from laboratory to clinical practice.* Birkhäuser Verlag, Basel

11 Cook SD, Baffes GC, Wolfe MW, Sampath TK, Rueger DC (1994) Recombinant human bone morphogenetic protein-7 induces healing in a canine long-bone segmental defect model. *Clin Orthop* 301: 302–312

12 Cook SD, Wolfe MW, Salkeld SL, Rueger DC (1995) Effect of recombinant human osteogenic protein-1 on healing of segmental defects in non-human primates. *J Bone Joint Surg Am* 77: 734–750

13 Cunningham BW, Kanayama M, Parker LM, Weis JC, Sefter JC, Fedder IL, McAfee PC

(1999) Osteogenic protein *versus* autologous interbody arthodesis in sheep thoracic spine. *Spine* 24: 509–518

14 Grauer JN, Patel TCh, Erukar JS, Troiano NW, Panjabi MM, Friedlaender GE (2001) Evaluation of OP-1 as a graft substitute for posterolateral lumbar fusion. *Spine* 26: 127–133

15 Friedlaender GE, Perry CR, Cole JD, Cook SD, Cierney G, Muschler GF, Zych GA, Calhoun JH, LaForte AJ, Yin S (2001) Osteogenic protein-1 (bone morphogenetic protein-7) in the treatment of tibial nonunions. *J Bone Joint Surg Am* 83 (Suppl 1): S151–S158

16 Geesink RGT, Hoefnagels NHM, Bulstra SK (1999) Osteogenic activity of OP-1 bone morphogenetic protein (BMP-7) in a human fibular defect. *J Bone and Joint Surg Br* 81: 710–718

17 McKee MD, Schemitsch EH, Waddell JP, Wild L (2004) The treatment of long bone nonunion with rhBMP: results of a prospective pilot study. Poster Presentation (#P242) at the 71st Annual American Academy of Orthopaedic Surgeons Meeting, 10–14 March, San Francisco, CA

18 McKee MD, Schemitsch EH, Wild L, Waddell JP (2002) The treatment of complex, recalcitrant long-bone nonunion with human recombinant bone morphogenic protein: results of a prospective pilot study. Presented at the 18th Annual Meeting of the Orthopaedic Trauma Association, 11 October, Toronto, Canada

19 McQueen MM, Hajducka C, Court-Brown CM (2003) A comparison of rhBMP-7 (Ossigraft) and autogenous graft for treatment of metaphyseal defects after osteotomy of the distal radius. Presented at the 19th Annual Orthopaedic Trauma Association Meeting, 10 October, Salt Lake City, Utah

20 Susarla A, Liporace F, Tejwani NC, Koval KJ, Egol KA (2004) OP-1 Implant as an adjunct to mechanical fixation in humeral nonunions (Poster P241) Presented at the 71st Annual Meeting of the American Academy of Orthopaedic Surgeons, 10–14 March, San Francisco, California

21 Giltaij LR, Shimmin A, Friedlaender GE (2002) Osteogenic protein-1 (OP-1) in the repair of bone defects and fractures of long bones: clinical experience. In: Vukicevic S, Sampath KT (eds): *Bone morphogenetic proteins: from laboratory to clinical practice*. Basel, Birkhauser, 193–205

Development of the first commercially available recombinant human bone morphogenetic protein (rhBMP-2) as an autograft replacement for spinal fusion and ongoing R&D direction

Bill McKay

Medtronic Sofamor Danek, 1800 Pyramid Place, Memphis, TN 38132, USA

Introduction

After several decades of promising research on bone morphogenetic proteins (BMPs), a product that surgeons can use clinically is now commercially available. More than two decades after Urist's first observation of demineralized bone osteoinductivity in 1965, a single bone morphogenetic protein was isolated from bone and produced recombinantly (rhBMP-2) [1, 2]. The recombinant production of BMP-2 opened the door for the development of a commercial BMP product, since the extraction of these proteins from bone was not a viable commercial option.

Despite the fact that the literature was full of publications showing the potency of rhBMP-2 in animals and its superiority over autogenous bone, it wasn't for another decade and a half that rhBMP-2 was FDA approved and finally commercially available (July 2002) [3]. There were several reasons for this length of time between having a source of recombinant protein and its commercial approval. One of the biggest hurdles was learning how to deliver this molecule in a fashion to obtain consistent bone formation. It was determined that rhBMP-2 fluid is rapidly cleared from the surgical site, thus requiring a carrier to both maintain the protein at the surgical site and act as a scaffold for new osteoid deposition. The pharmacokinetics of rhBMP-2 with various delivery carriers was investigated to determine the best method of delivery at the optimal concentration and dose. In addition, the safety profile of rhBMP-2 had to be assessed and well understood before moving into the clinic.

Early clinical experience with rhBMP-2 in a fresh fracture clinical trial was disappointing because of a suboptimal carrier and a concentration of rhBMP-2 that was too low (0.43 mg/ml). Bone formation was observed, but not as consistently as had been seen in the animal studies. This trial was subsequently followed by an interbody spinal fusion cage study, with a collagen sponge carrier and higher rhBMP-2 concentration (1.5 mg/ml), which demonstrated equivalence to use of autogenous iliac crest bone graft. This clinical trial ultimately led to the first commercial approval of rhBMP-2 (INFUSE® Bone Graft, Medtronic Sofamor Danek on

Bone Morphogenetic Proteins: Regeneration of Bone and Beyond, edited by Slobodan Vukicevic and Kuber T. Sampath
© 2004 Birkhäuser Verlag Basel/Switzerland

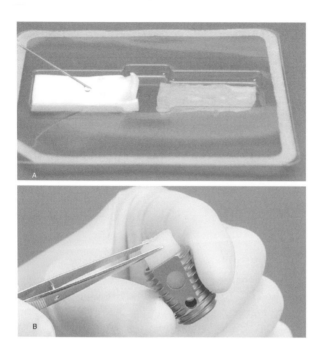

Figure 1
Collagen sponge carrier is soaked with rhBMP-2 solution at the time of surgery (A), rolled and then placed into an interbody fusion cage (B).

July 2002) [3]. RhBMP-2 was also found to have a good safety profile. Only desired local bone formation was observed, and no systemic adverse effects have been seen. More recently rhBMP-2 delivered on the same collagen sponge carrier was approved in Europe and the United States for tibial fresh fractures (InductOs™).

Research and Development continues in the search for additional delivery vehicles and expanded clinical indications such as posterolateral fusion. A lot has been learned in the development of the first BMP product that can be applied in the development of future BMP products. Some day a complete portfolio of rhBMP-2 products will be available to the surgeon.

Description of the commercial rhBMP-2 product (INFUSE® Bone Graft)

The first commercially available rhBMP-2 (INFUSE® Bone Graft, Medtronic Sofamor Danek, Memphis, TN) for spinal fusion was approved in the United States in July 2002 [3]. The product consists of freeze-dried rhBMP-2 at a concentration of

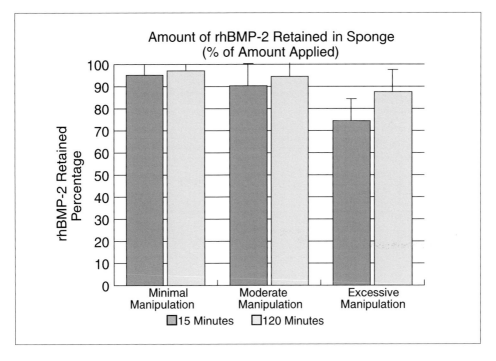

Figure 2
In vitro *study showing percent rhBMP-2 remaining within collagen sponge carrier after minimal, moderate, and excessive manipulation after 15, and 120 min soak times*
Under typical minimal manipulation over 95% of rhBMP-2 is retained in the collagen sponge.

1.5 mg/ml which is reconstituted with sterile water at the time of surgery. The reconstituted rhBMP-2 is evenly applied to a type I collagen sponge at the time of surgery and allowed to soak for 15–120 min (Fig. 1).

This soaking period allows for the rhBMP-2 to bind to the sponge prior to implantation. Significant binding occurs so that during surgical implantation less than 5% of the BMP is lost during minimal manipulation of the sponge after only 15 min of soaking, and less than 2.5% if allowed to soak for 120 min [4] (Fig. 2).

INFUSE® was approved for use with the LT-CAGE™ Lumbar Tapered Fusion Device in anterior lumbar interbody fusions (Fig. 1). The rhBMP-2 hydrated sponge is placed inside the LT-CAGE™ and inserted in an intervertebral disc space to facilitate bone growth and fusion of adjacent vertebrae. INFUSE® Bone Graft was found, in a multicenter prospective randomized clinical trial, to result in a fusion rate equivalent to autogenous bone graft from the iliac crest [5].

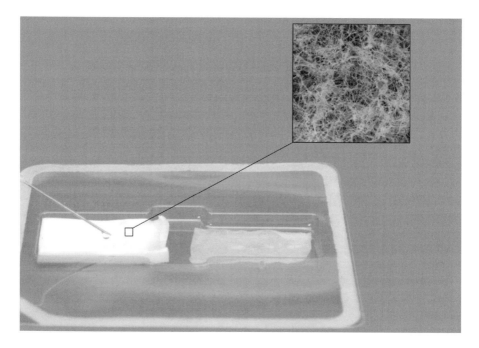

Figure 3

Collagen sponge carrier shrinks to the volume of rhBMP-2 solution applied Scanning electron microscopy (SEM) image of the collagen sponge showing its highly porous fibular matrix structure.

The rhBMP-2 molecule is a homodimer consisting of two chains of 114 and 131 amino acids. The collagen sponge carrier is derived from type I bovine collagen from the Achilles tendon. The Achilles tendon contains a high concentration of type I collagen, making it good source. The collagen sponge is highly porous (99%) fibular collagen that is hydrated to 30% of its total "soak load" hydration capacity (Fig. 3). It was determined that a 30% soak load gave optimal handling characteristics without fluid loss during minimal manipulation. Because the sponge is so porous the sponge ultimately assumes the same volume of rhBMP-2 solution applied to it. For example, 8 cc of rhBMP-2 solution placed on the collagen sponge results in a hydrated sponge of 8 cc, or sufficient graft to fill an 8cc bone defect. INFUSE® is available in three kit sizes, 2.8 cc, 5.6 cc, and 8 cc, all delivering a concentration of 1.5 mg/ml.

The hydrated collagen sponge is soft and pliable, and easily folded or rolled to place inside an interbody fusion cage. The collagen sponge proved to be an ideal delivery vehicle for the rhBMP-2 when used in a protected environment, such as an interbody fusion cage. Several important lessons have been learned in the use of

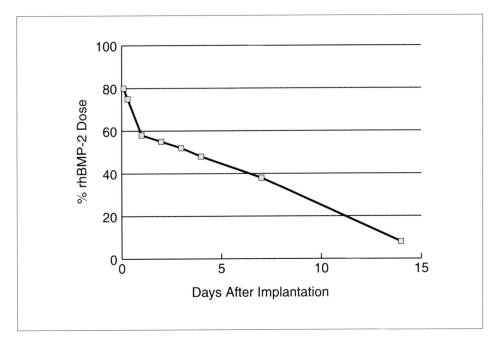

Figure 4
In vivo *residence of implanted rhBMP-2 from the collagen sponge at an orthotopic site*
The half-life is approximately 2.5 days and cleared from the site in approximately 3 weeks.

rhBMP-2 on the collagen sponge. It is important to completely fill the area where new bone is desired with the collagen sponge. Gaps may not necessarily fill in with bone without the presence of the collagen matrix delivering BMP-2 to that area and acting as a scaffold for new bone formation. It is also important not to overly compress or compact the sponge, which may restrict cell infiltration. Other precautions are not to use irrigation or place a drain directly on the rhBMP-2 soaked sponge after implantation to avoid washing away the protein from the surgical site or diluting the effective concentration of rhBMP-2.

Once implanted, the collagen sponge is degraded over a 2-4 week period of time. *In vivo*, the half-life of rhBMP-2 has been found to be approximately 2.5 days, and it is detectable at the surgical site for approximately 3 weeks (Fig. 4).

The ideal release kinetics for any BMP has yet to be determined. The optimal timing of release of the BMP from its carrier is still unknown. It is unknown whether it is the initial release off the carrier 2–3 days after implantation that is more important, or the tail end of the release kinetics curve when sufficient quantities of responding cells may have migrated into the area. Designing an experiment to

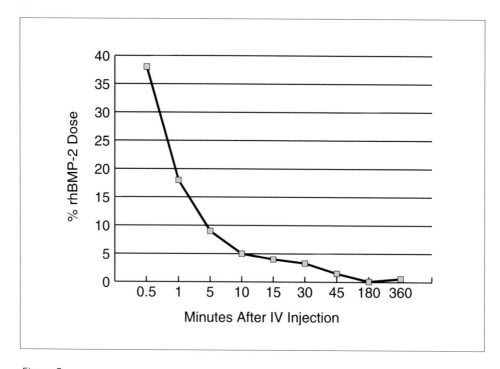

Figure 5
In vivo *residence of IV injected rhBMP-2*
The half-life is approximately 30 seconds and cleared from the blood in approximately 1 h.

answer this question would be very difficult, so optimal delivery of BMP for different clinical applications is an active area of ongoing research, which requires empirical *in vivo* experimentation.

The dose and concentration of rhBMP-2 in INFUSE® Bone Graft has been the subject of many discussions. Large super-physiologic doses of rhBMP-2 have been required to obtain the desired new bone formation response. The reason for this is still unknown, but no adverse systemic or toxicity responses attributable to rhBMP-2 have been observed in the hundreds of preclinical (animal) studies or the many clinical studies conducted to date [7, 8]. Preclinical toxicity studies, conducted with doses up to about 1000 × those being used clinically, did not result in any adverse effects. The preclinical toxicity studies involved both the direct injection of large doses of rhBMP-2 (5.3 mg/kg) into the blood stream and implantation at orthotopic sites (1.6 mg/kg). One of the most likely reasons for the extremely clean safety profile is that rhBMP-2 is rapidly cleared from the blood stream. Direct IV injection studies found that rhBMP-2 has a half-life of less than 1 min (30 s) and it was essentially cleared in about 1 h (Fig. 5).

From the orthotopic implantation study the maximum amount of rhBMP-2 detectable in the bloodstream was only 0.1% of the amount implanted. This maximum level occurred immediately after implantation, and thereafter rhBMP-2 was undetectable. So, not only does an insignificant amount of rhBMP-2 get into the blood stream after implantation, it is rapidly cleared from the blood stream once in it. This combination of properties is most likely the reason no systemic adverse effects have been observed with rhBMP-2. In addition, because of these inherent properties of rhBMP-2, a systemic treatment for osteoporosis is probably not possible; it would have to be applied locally. Therefore, use of super-physiologic doses of rhBMP-2 results in the desired bone formation without any adverse systemic or toxicity effects.

INFUSE® Bone Graft is the first of many rhBMP-2 products being developed. Additional products will be approved for additional clinical applications using new carriers and concentrations of rhBMP-2. Determination of the safety and efficacy of new carrier-rhBMP-2 concentration combinations requires both preclinical and clinical investigations. Caution is advised against attempting to utilize other carriers with the commercially available rhBMP-2 in the absence of these preclinical and clinical studies. For example, preclinical research has found that many carriers with fast degradation rates such as calcium sulfate, tricalcium phosphate, and demineralized bone are poor carriers for rhBMP-2 and should be avoided.

Clinical safety and efficacy of rhBMP-2 established in spinal interbody fusion applications

INFUSE® Bone Graft was the first rhBMP successfully evaluated in a multicenter prospective randomized clinical trial for spinal fusion [5]. The trial involved the use of titanium interbody fusion cages (LT-CAGE™ Lumbar Tapered Fusion Device, Medtronic Sofamor Danek, Memphis, TN) in a single level anterior lumbar interbody fusion (ALIF) procedure in patients diagnosed with degenerative disc disease. The study design was a prospective randomized trial in which all the patients received the LT-CAGE™ filled with either autogenous iliac crest bone or INFUSE® Bone Graft. The study was designed to show equivalence to autogenous bone graft. Therefore, it was the first clinical trial to evaluate the potential to eliminate the need to harvest iliac crest bone graft and prove that rhBMP-2 could act as a bone graft replacement in spinal fusion.

Two interbody cages were used in the disc space. The rhBMP-2 soaked collagen sponges were only placed inside the hollow cages, and no graft was placed around the outside of the cages within the disc space (Fig. 6). The internal volume of the cages varied depending on the cage size used in each particular patient. Therefore, the volume of INFUSE® used depended on the cage size selected, and varied from 2.8–8.0 cc (for two cages). At an rhBMP-2 concentration of 1.5 mg/ml, a total dose of 4–12 mg of rhBMP-2 was used per level.

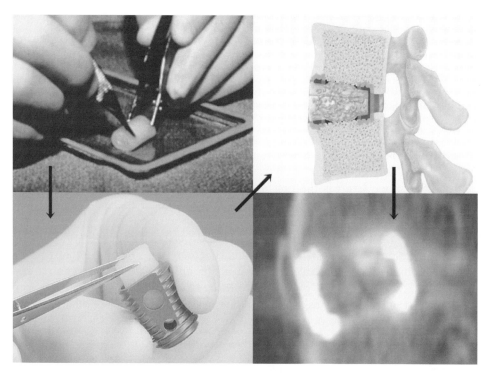

Figure 6
The rhBMP-2 soaked collagen sponge is rolled up and placed into the spinal fusion cage.
The cage has opening on the top and bottom surface allowing new bone growth through the
device which is observable on a CT scan several months postoperatively.

A total of 279 patients were enrolled in the study, 143 rhBMP-2 patients and 136 autograft patients. All patients were followed for two years. Demographically, both treatment arms were statistically the same, verifying the validity of the randomization procedure (Fig. 7).

Since the INFUSE® treated patients did not undergo a second surgical incision to harvest iliac crest autogenous bone graft, both the overall surgical procedure time and blood loss was statistically reduced. At two years post-surgery, 32% of the autograft treated control patients reported some hip pain, which is avoided with INFUSE®. Clinical outcome was evaluated using validated Oswestry and SF-36 outcome questionnaires. No statistical differences were found in clinical outcomes between the two groups (Figs 8 and 9).

Both plain radiographs and CT scans were obtained at 6, 12 and 24 months. Since the main function of the rhBMP-2 was to induce new bone formation and

Demographic Information*		
	Autograft	INFUSE®
Age (yrs)	42.3	43.3
Weight (lbs)	181.1	179.1
Sex (male/female)	68/68	78/65
Worker's Compensation (%)	34.6	32.9
Spinal Litigation (%)	16.2	12.6
Tobacco Use (%)	32.9	36.0
History of previous surgeries	40.4	37.8

*No statistical difference between groups

Figure 7
Demographics of patients randomized into the prospective randomized multicenter clinical study comparing INFUSE® Bone Graft to iliac crest autograft

fusion through the cages, it was important to utilize 1 mm thin slice CT scans, the most discriminating assessment of new bone formation. Two independent radiologists read the films, and if there was disagreement in their fusion assessment, a third independent radiologist became the tiebreaker. Fusion results at 24 months were 100% for INFUSE® and 99.3% for autograft [9]. This difference was not statistically significant. Figures 10 and 11 show typical serial CT scans of an autograft treated and an INFUSE® treated patient. Two important observations to point out are: 1) INFUSE® induced bone formation will continue to remodel well out to 12 months, and; 2) even though INFUSE® was only placed inside the cages, new bone formation was often seen adjacent to the outside of the cages in the disc space. One possibility for this observation is that blood clot surrounding the outsides of the cage may act as a sink for any BMP diffusing out of the collagen sponge, leading to its ossification.

Safety of rhBMP-2 was also demonstrated in this clinical trial with no rhBMP-2 related systemic adverse events reported.

This clinical study demonstrated that INFUSE® Bone graft was equivalent to iliac crest autogenous bone graft and led to its FDA approval in July 2002.

Two additional ALIF clinical trials were conducted involving the LT-CAGE™ Lumbar Tapered Fusion Device that allowed for an integrated analysis of a larger patient population to compare the efficacy of rhBMP-2 to autogenous bone graft [10]. A total 679 patients were available for analysis.

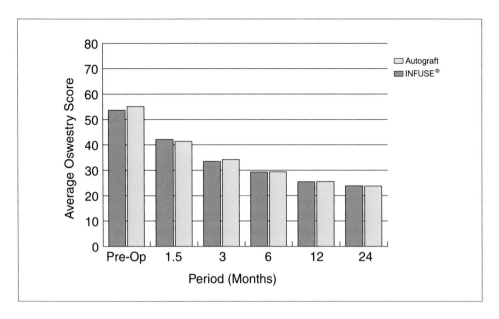

Figure 8
Oswestry pain scores of INFUSE® Bone Graft and autograft treated patients over two year follow-up

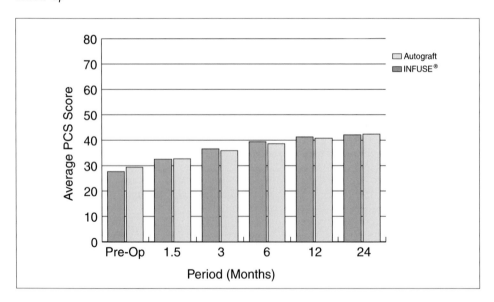

Figure 9
SF-36 pain scores of INFUSE® Bone Graft and autograft treated patients over two year follow-up

277 INFUSE®/LT-CAGE™
402 autograft/LT-CAGE™
679 total patient population in integrated analysis

The demographics of this patient population pool were not statistically different. Analysis showed that the fusion rates of the INFUSE® treated patients was statistically higher than the autograft treated patients (94.4% and 89.4% respectively, p = 0.022). This confirms what has been seen consistently in published animal studies, that rhBMP-2 results in statistically higher fusion rates than autogenous bone graft [11–21].

More recently, INFUSE® Bone Graft obtained expanded approval for use with two other interbody fusion cages, the Interfix™ and Interfix RP™ (Medtronic Sofamor Danek, Memphis, TN) [22]. In addition, rhBMP-2 was approved for use in tibial fresh fractures in both the United States and Europe. It is commercially available under the tradename, InductOs™, in Europe.

Important lessons learned from clinical trials

Embarking on the first clinical trial with rhBMP-2 for spinal fusion has taught many important lessons. The first lesson learned was, while animal studies are helpful in screening potential new biologic agents, the true safety and efficacy of these biological agents must be established in prospective randomized clinical trials. Animals heal bone defects much easier and faster than humans, even with lower doses of BMP. This first became apparent when the first rhBMP-2 clinical trial in fresh fractures failed, using a 0.43 mg/ml concentration on a non-cohesive bioerodable particle carrier, despite successful results in animal studies. This phenomena was further confirmed when BMP concentrations had to be increased when moving from lower order animals to higher order animals and then humans in the spinal fusion applications. Caution must be exercised using untested biologic technologies in humans based on just lower order animal data.

Secondly, new bone formation was observed adjacent to the outside of the interbody fusion cages when the rhBMP-2 soaked sponge was placed only inside them. This may suggest that blood clot forming around the cages may act as a sink for BMP diffusing out of the sponge. In this particular surgical procedure, the additional bone formation was a positive finding since it resulted in a larger fusion mass within the disc space. In some surgical procedures, this additional bone may not be desired, so it is advised to control hemostasis in these situations.

Finally, antibodies to rhBMP-2 were found in 0.7% of both treatment groups in the Pivotal INFUSE® Bone Graft/LT-CAGE™ Lumbar Tapered Fusion Device clinical trial [5]. The fact that it was the same in both groups was surprising, but may be due to the sensitivity of the assay. The important lesson learned was that the

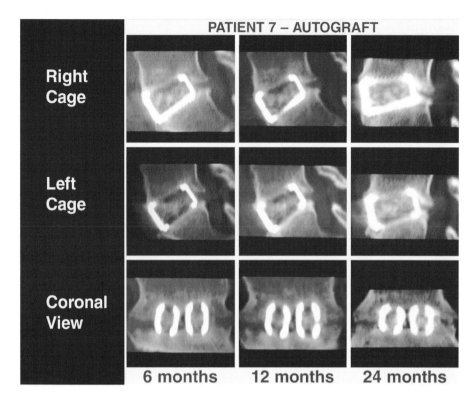

Figure 10
Serial CT scans of an autograft treated patient with iliac crest autograft in the LT-CAGE™ at 6, 12, and 24 months
Note the presence of autograft chips at six months still and their remodeling and consolidation by 24 months.

patients with elevated antibodies to rhBMP-2 in the rhBMP-2 group did not have any adverse effects and went on to fuse as well as the other patients. In addition, the patients with elevated antibodies only had a transient elevation at three months and returned to normal by six months.

Identification of the optimal local delivery vehicle for rhBMP-2 in other spinal fusion applications

One of the overriding reasons development of rhBMP-2 took so long to finally become commercially available was determining how to best deliver the protein to the surgical site and generate new bone formation in a desired location and space.

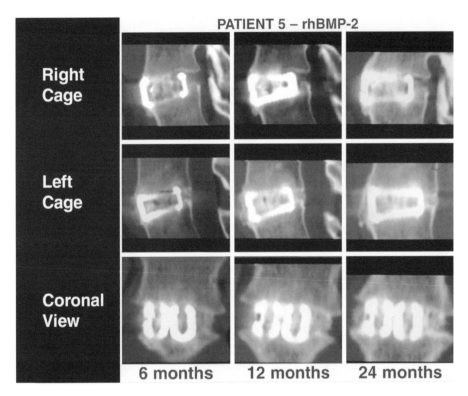

Figure 11
Serial CT scans of an INFUSE® treated patient in the LT-CAGE™ at 6, 12, and 24 months
Note the increasing radiopacity, or new bone formation, within the cage from six months to
24 months.

The use of a collagen sponge was found to work well inside bone defects and pro-tected environments such as an interbody fusion cage. Animal studies consistently showed that rhBMP-2 is superior to autogenous iliac crest bone graft [11–21] (Fig. 12). In more challenging environments, such as sites with less protection from overlying soft tissue compression or with less bleeding bone surfaces, alternate car-riers and/or rhBMP-2 concentrations had to be developed, or alternate ways to uti-lize the collagen sponge had to be found [23]. All three product development routes are being pursued.

To expand upon the use of the currently available collagen sponge, simple alter-native techniques have been investigated with some success. Simply adding "bulk-ing agents" to the collagen sponge to give it some compression resistance to overly-ing soft tissue, has shown some efficacy in more challenging environments such as

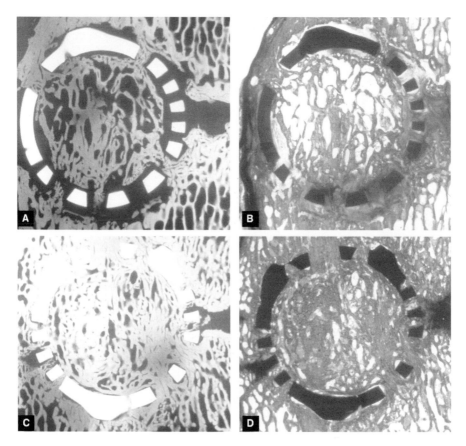

Figure 12
Microradiographs (A,C) and corresponding histological sections (B,D) of interbody cages filled with iliac crest autograft (A,B) or rhBMP-2 on a collagen sponge (C,D) in sheep at six months [11]
Fibrous tissue (pink) was present around perimeter of autograft filled cages more than rhBMP-2 filled cages.

posterolateral fusion [24]. Bulking agent materials such as autograft, allograft, and slow resorbing ceramics are most effective. Fast degrading materials such as calcium sulfate, tricalcium phosphate and demineralized bone are not good bulking agents for rhBMP-2. They degrade too fast, resulting in loss of space maintenance and lack of the desired volume of new bone formation. It was observed in the pre-clinical studies, that the presence of rhBMP-2 can cause the accelerated resorption of various biomaterials, and therefore the fast resorbing materials mentioned above will resorb even faster in the presence of rhBMP-2.

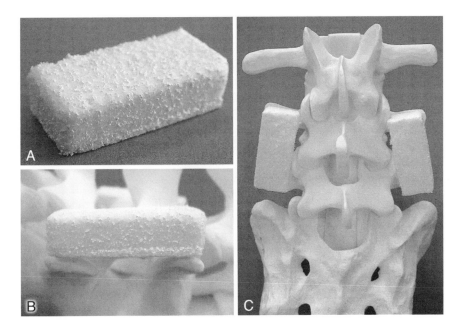

Figure 13
Photograph of the compression resistant matrix (CRM) consisting of a collagen sponge containing resorbable calcium phosphate granules (A)
A lateral view shows the CRM sponge bridging adjacent transverse processes and how it is designed to maintain a space for new bone formation (B). A single sponge is placed across the transverse spineous processes on each side of the spine (C).

Figure 14
Photograph of the biphasic calcium phosphate (BCP) granular ceramic
The rhBMP-2 solution is applied and absorbed by the 80% porous ceramic granules.

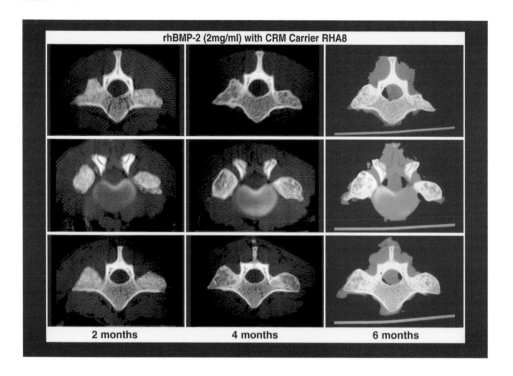

Figure 15
Serial CT scans of a monkey spine at 2, 4, and 6 months, treated with the CRM carrier at 2 mg/ml rhBMP-2 concentration
Note the resorption of the carrier and replacement by new bone formation apparent the by increase in radiopacity and formation of outer cortical shell.

More "compression resistant" carriers for rhBMP-2 are being developed that do not require the addition of bulking agents. These new carriers contain slow resorbing calcium phosphates to provide the compression resistance. Two such carriers are: 1) "compression resistant matrix" (CRM) consisting of a composite collagen-ceramic sponge, and; 2) biphasic calcium phosphate (BCP) ceramic granules (Figs 13 and 14). CRM is a type I collagen sponge containing a suspension of 1 mm diameter biphasic calcium phosphate granules (15% hydroxyapatite/85% tricalcium phosphate). BCP is a 3 mm diameter biphasic calcium phosphate granule composition of 60% hydroxyapatite/40% tricalcium phosphate. Both of these carriers have been shown to result in 100% fusions in the challenging non-human primate posterolateral fusion model [25, 26]. These carriers provided good space maintenance evident by the large volume of new bone formed in the fusion masses (Figs 15 and 16). The calcium phosphate contents of CRM and BCP degrade over a peri-

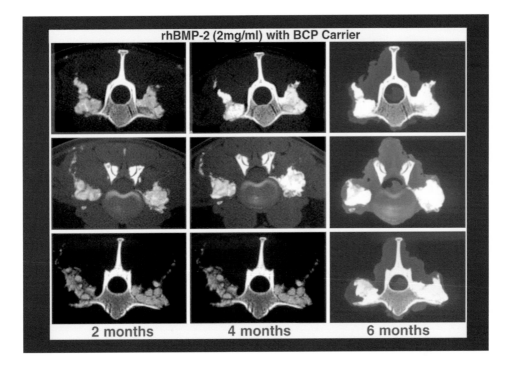

rhBMP-2 (2mg/ml) with BCP Carrier

2 months 4 months 6 months

Figure 16
Serial CT scans of a monkey spine at 2, 4, and 6 months, treated with the BCP carrier at
2 mg/ml rhBMP-2 concentration
The presence of new bone formation is apparent by the increase in radiopacity.

od of several months as opposed to just weeks for collagen. Their slower resorption provides a longer lasting scaffold for new bone deposition. Calcium phosphate's highly osteoconductive inherent nature also may facilitate early osteoid deposition.

Both of these carriers are currently being evaluated in prospective randomized clinical trials. These trials are studying rhBMP-2 at a 2 mg/ml concentration in single level instrumented posterolateral fusions in patients diagnosed with degenerative disc disease. The rhBMP-2 concentration was increased from the 1.5 mg/ml concentration in the commercially available INFUSE® kit, because the posterolateral fusion site is a more challenging environment. In this application, there is minimal bleeding bone present and a much larger gap in which to form new bone compared to an interbody fusion location. The posterolateral fusion technique is attempting to form bridging bone from one transverse process to the next (a 3–4 cm gap sur-

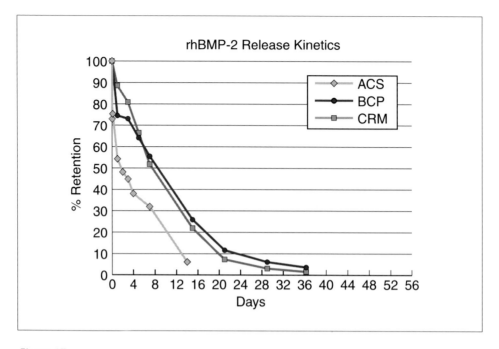

Figure 17
In vivo *residence of implanted rhBMP-2 from the CRM and BCP carriers at an orthotopic site (posterolateral fusion model)*
The release kinetics are similar for the two carriers, with a half-life of approximately eight days and clearance from the site in approximately six weeks.

rounded by soft tissue), as opposed to an interbody fusion technique that attempts to form bridging bone from one vertebral body endplate to the next (a 1–2 cm gap surrounded by bleeding bone).

The rhBMP-2 solution (2 mg/ml) is applied directly to these carriers at the time of surgery and allowed to soak for 5–90 minutes. The release kinetics of the CRM and BCP carriers is much different than that of the collagen sponge in the INFUSE® kit. The half-life of rhBMP-2 *in vivo* from the CRM and BCP carriers is approximately four times longer (8–9 days) than the collagen sponge (2–3 days) [6] (Fig. 17). The rhBMP-2 is also present at the site of implantation approximately twice as long, six weeks as opposed to three weeks for the collagen sponge. The longer presence of rhBMP-2 at the surgical site may explain why these carriers were found to be more effective in challenging environments. It is interesting to note that even though the CRM and BCP carriers have different resorption properties (BCP degrades much slower), the rhBMP-2 release kinetics are essentially identical.

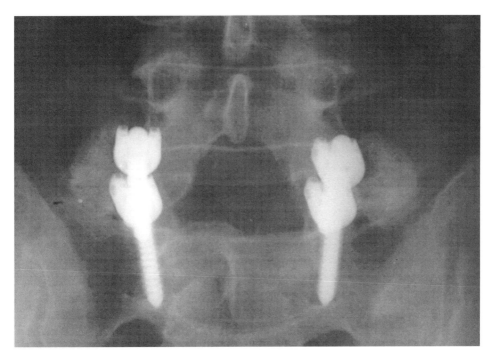

Figure 18
Radiograph of an rhBMP-2 treated patient with the CRM carrier at six months, showing bridging bone from one transverse process to the next

Clinical evaluation of rhBMP-2 in spinal posterolateral fusion applications

Two separate clinical trials are ongoing utilizing rhBMP-2 (2 mg/ml) in posterolateral fusion applications with both the CRM and BCP carriers. They are prospective randomized clinical studies in patients diagnosed with degenerative disc disease requiring a single level instrumented posterolateral fusion. Enrollment has been completed in both studies and follow-up is continuing, so data are not available at this time. Early radiographic findings appear promising (Figs 18–21).

A pilot study with the BCP carrier was completed prior to initiating these pivotal studies with rhBMP-2 (2 mg/ml). This pilot study was a prospective randomized study in patients diagnosed with single level degenerative disc disease, which were randomized into one of three treatment arms: 1) autograft with instrumentation; 2) rhBMP-2/BCP with instrumentation, and; 3) rhBMP-2/BCP without instrumentation. Published results of the two-year follow-up data reported a 100% fusion rate with rhBMP-2/BCP and only 40% for the autograft treated patients [27].

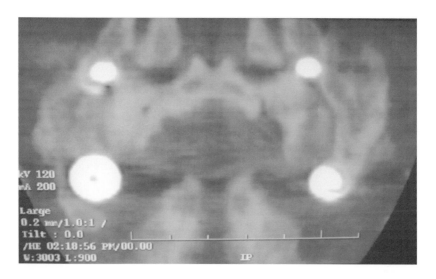

Figure 19
CT scan of same rhBMP-2 treated patient with the CRM carrier in Figure 18, showing bridging bone from one transverse process to the next

rhBMP-2/BCP Patient 3

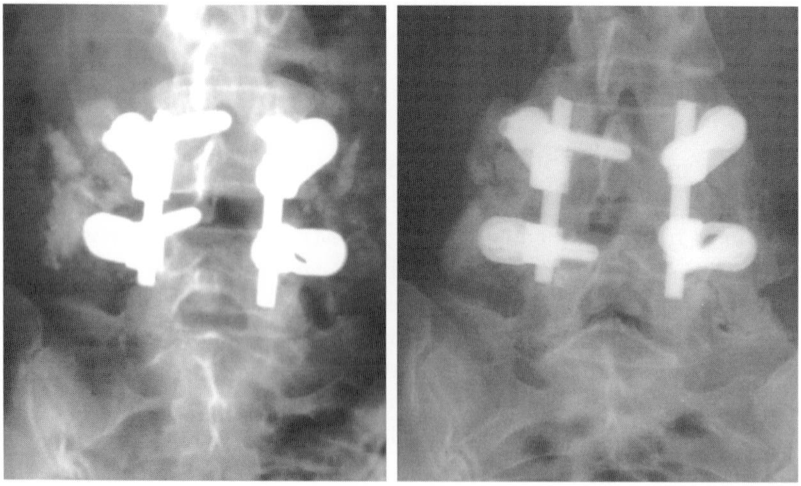

Immediate post-op	5 years

Figure 20
Radiograph of an rhBMP-2 treated patient with the BCP carrier at 0 and 5 years, showing bridging bone from one transverse process to the next

rhBMP-2/BCP Patient 3

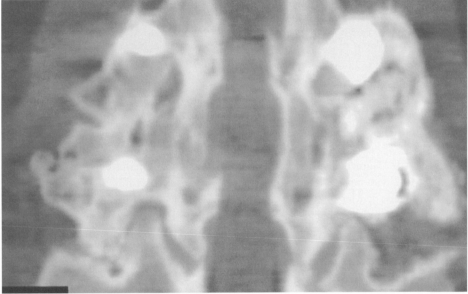

5 years

Figure 21
CT scan of same rhBMP-2 treated patient with the BCP carrier in Figure 20, showing bridg-
ing bone from one transverse process to the next

Future directions of rhBMP-2 research and development

Research and development on rhBMP-2 is ongoing in both the identification of new delivery carriers as well as seeking approval in expanded clinical indications. In the years to come, a portfolio of rhBMP-2 products should become available to the surgeon.

Carrier development is focused on providing "compression resistance" to soft tissue compression, to provide more uniform bone formation in the volume and shape of the carrier implanted. Development is also occurring to identify a flowable or putty type carrier that is optimal for packing into more complicated bone defects. Finally, as more is learned about the activity of rhBMP-2 during the different stages of its release kinetics profile, carriers can be designed to further optimize its release.

References

1   Urist MR (1965) Bone: Formation by autoinduction. *Science* 150: 893–899
2   Wozney JM, Rosen V, Celeste AJ, Mitsock LM, Whitters MJ, Kriz RW, Hewick RM, Wang EA (1988) Novel regulators of bone formation: molecular clones and activities. *Science* 242: 1528–1534
3   Medtronic press release on July 2, 2002
4   Spector M, Zanella JM, Peckham JM, Sandhu HS (2004) Release of recombinant human bone morphogenetic 2 from absorbable collagen during the mechanical compression associated with its use for spinal fusion. *Orthopaedic Research Society* 1140: 7–10
5   Burkus JK, Gornet MF, Dickman C, ZdeblickTA (2002) Anterior interbody fusion using rhBMP-2 with tapered interbody cages. *J Spinal Disorders* 15(5): 337–349
6   Louis-Ugbo J, Kim HS, Boden SD, Mayr MT, Li RC, Seerherman H, D'Augusta D, Blake C, Jiao A, Peckham S (2002) Retention of 125I-labeled recombinant human bone morphogenetic protein-2 by biphasic calcium phosphate or a composite sponge in a rabbit posterolateral spine arthrodesis model. *J Orthopaedic Research* 20(5): 1050–1059
7   McKay B, Sandhu HS (2002) Use of recombinant human bone morphogenetic protein-2 in spinal fusion applications. *Spine* 27(S16): 66–85
8   Ashley R, Lane JM (2002) Safety profile for the clinical use of bone morphogenetic proteins in the spine. *Spine* 27(S16): 40–48
9   Burkus JM, Dorchak DD, Sanders DL (2003) Radiographic assessment of interbody fusion using recombinant human bone morphogenetic protein type 2. *Spine* 28(4): 372–377
10   Burkus JM, Heim SE, Gornet MF, Zdeblick TA (2003) Is INFUSE Bone Graft superior to autograft bone? An integrated analysis of clinical trials using the LT-Cage lumbar tapered fusion device. *J Spinal Disorders* 16(2): 113–122
11   Sandhu HS, Toth JM, Diwan AD, Seim HB, Kanim LEA, Kabo JM, Turner AS (2002) Histological evaluation of the efficacy of rhBMP-2 compared with autograft bone in sheep spinal anterior interbody fusion. *Spine* 27(6): 567–575
12   Boden S, Martin G, Horton WC, TrussT, Sandhu H (1998) Laparoscopic anterior spinal arthrodesis with rhBMP-2 in a titanium interbody threaded cage. *J Spinal Disorders* 11(2): 95–101
13   Zdeblick TA, Ghanayem AJ, Rapoff AJ, Swain C, Bassett T, Cooke ME (1998) Cervical interbody fusion cage – an animal model with and without bone morphogenetic protein. *Spine* 23(7): 758–766
14   Hecht BP, Fischgrund JS, Herkowitz HN, Penman L, Toth J, Shirkhoda A (1999) The use of recombinant human bone morphogenetic protein-2 (rhBMP-2) to promote spinal fusion in a non-human primate anterior interbody fusion model. *Spine* 24(7): 629–636
15   David SM, Gruber HE, Meyer RA, Murakami T, Tabor OB, Howard BA, Wozney JM, Hanley EN (1999) Lumbar spinal fusion using recombinant human bone morphogenetic protein in the canine. *Spine* 24(19): 1973–1979

16 Hollinger EH, Trawick RH, Boden SD, Hutton WC (1996) Morphology of the lumbar intertransverse process fusion mass in the rabbit model: A comparison between two bone graft materials – rhBMP-2 and autograft. *J Spinal Disorders* 9(2): 125–128

17 Schimandle JH, Boden SD, Hutton WC (1995) Experimental spinal fusion with recombinant human bone morphogenetic protein-2. *Spine* 20(12): 1326–1337

18 Sandhu HS, Kanim LEA, Kabo JM, Toth JM, Zeegen EN, Liu D, Delamarter RB, Dawson EG (1996) Effective doses of recombinant human bone morphogenetic protein-2 in experimental spinal fusion. *Spine* 21(18): 2115–2122

19 Fischgrund JS, James SB, Chabot MC, Hankin R, Herkowitz HN, Wozney JM, Shirkhda A (1997) Augmentation of autograft using rhBMP-2 and different carrier media in the canine spinal fusion model. *J Spinal Disorders* 10(6): 467–472

20 Helm GA, Sheehan M, Sheehan JP, Jane JA, diPierro CG, Simmons NE, Sweeney TM (1997) Utilization of type I collagen gel, demineralized bone matrix, and bone morphogenetic protein-2 to enhance autologous bone lumbar spinal fusion. *J Neurosurgery* 86: 93–100

21 Sheehan JP, Kallmes DF, Sheehan JM, Jane JA, Fergus AH, diPierro CG, Simmons NE, Makel DD, Helm GA (1996) Molecular methods of enhancing lumbar spine fusion. *Neurosurgery* 39(3): 548–554

22 Medtronic press release December 1, 2004

23 Martin GJ, Boden SD, Morone MA, Moskovitz PA (1999) Posterolateral Intertransverse process spinal arthrodesis with rhBMP-2 in a non-human primate: Important lessons learned regarding dose, carrier, and safety. *J Spinal Disorders* 12(3): 179–186

24 Akamaru T, Suh D, Boden SD, Kim HK, Minamide A (2003) Simple carrier matrix modifications can enhance delivery of recombinant human bone morphogenetic protein-2 for posterolateral spine fusion. *Spine* 28(5): 429–434

25 Suh DY, Boden SD, Louis-Ugbo J, Mayr M, Murakami H, Kim HS, Minamide A, Hutton WC (2002) Delivery of recombinant human bone morphogenetic protein-2 using a compression-resistant matrix in posterolateral spine fusion in the rabbit and in the non-human primate. *Spine* 27(4): 353–360

26 Boden S, Martin G, Morone MA, Ugbo J, Moskoviz P (1999) Posterolateral lumbar intertransverse process spine arthrodesis with recombinant human bone morphogenetic protein-2/hydroxyapatite-tricalcium phosphate after laminectomy in the nonhuman primate. *Spine* 24(12): 1179–1185

27 Boden SD, Kang J, Sandhu HS, Heller JG (2002) Use of rhBMP-2 to achieve posterolateral lumbar spine fusion in humans: a prospective randomized clinical pilot trial. *Spine* 27(23): 2662–2673

Bone morphogenetic proteins and the synovial joints

Frank P. Luyten[1], Rik Lories[1], Dirk De Valck[1], Cosimo De Bari[2] and Francesco Dell'Accio[2]

[1]Laboratory of Skeletal Development and Joint Disorders, Department of Rheumatology, University Hospitals Leuven, Herestraat 49, B-3000 Leuven, Belgium; [2]GKT School of Medicine, King's College London, Department of Rheumatology, 5th Floor, Thomas Guy House, SE1 9RT London, UK

Bone morphogenetic protein signaling during joint development

Morphological events of joint formation

Bone morphogenetic proteins (BMPs) are involved in a broad array of morphogenetic processes. These span from the specification of the dorso–ventral body axis to patterning, organogenesis and differentiation of most tissues. Nevertheless, the initial discovery of BMPs as protein preparations that induced ectopically and *in vivo* a cascade of endochondral bone formation in rats, has strongly stimulated the study of their role in the development of the skeleton and in patterning of the synovial joints [1–3]. Additionally, with their remarkable cartilage and bone morphogenetic activity, BMPs represent an attractive therapeutic option for skeletal and joint disorders. Indeed, growing scientific evidence supports the concept that tissue repair and regeneration recapitulates to a certain extent the process of tissue formation during embryonic development. Recent advances in unraveling the molecular basis of developmental processes support BMP signaling as potential targets in reparative approaches in joint diseases.

Joint development has been extensively studied in a variety of animal species including human [4–12] chick [13–18], mouse [19, 20], and rat [21]. Since the molecular cascades driving organogenesis and tissue specification are highly conserved across species, with some limitations, one could integrate the available data into a common scheme.

The appendicular skeleton develops from a primitive, densely packed cellular mesenchyme derived from the lateral plate mesoderm [22–24]. Limb outgrowth is proceeding in a proximal–distal fashion. The condensation of mesenchymal cells leads to the formation of uninterrupted rod-like structures called *anlagen*. Subsequently, within the condensations, cells undergo chondrogenic differentiation to form cartilaginous templates surrounded by a sheath of spindle-shaped cells, the perichondrium. In the middle of each cartilaginous element, prechondrocytes undergo a series of maturation steps yielding hypertrophic chondrocytes, to be eventually replaced by bone and bone marrow in the process of endochondral ossification.

Synovial joints form through a process of segmentation of the cartilaginous anlagen but prior to the endochondral bone formation. In the region of the prospective joint, a few cell layers wide zone of mesenchymal cells does not undergo cartilage differentiation [19] but forms a cellular condensation called joint interzone. Morphologically [21], the joint interzone represents the first evidence of joint formation. The interzone, at 12.5 days *post coitum* (dpc), consists of a few layers of morphologically homogeneous elongated cells. By 15.5 dpc, the interzone differentiates into three distinguishable layers. Two chondrogenic, perichondrium-like dense layers covering the articulating surfaces of the cartilage elements contain flattened elongated cells at the articular side, and rounded chondrocyte-like cells at the epiphyseal side. One layer of a loose cellular tissue, with low cell density and enlarged intercellular spaces is present between these two layers. The dense zones further differentiate into articular cartilage at both ends of the future joint. Following a phase of vascular invasion, which selectively involves the peripheral part of the interzone, one giving rise to the capsulo-synovial apparatus, a process of the formation of a joint space takes place in the central loose layer of the interzone. This process, called joint cavitation, begins with the appearance of small clefts within the interzone, eventually forming the synovial cavity. In rats, for instance, joint cavitation is seen first in proximal joints at 16.5 dpc and is completed in distal joints by 20 dpc. Peri- and intra-articular joint-associated structures such as joint capsule, menisci, and ligaments differentiate from the mesenchymal cells surrounding the interzone and from the cells constituting the interzone, respectively [21]. In contrast to mammals, in the avian embryo some joint interzones form after the entire mesenchymal condensation underwent cartilage differentiation. This happens presumably by invasion of mesenchymal cells from the perichondrium or by de-differentiation of chondrocytes at the site of interzone formation [25, 26]. In addition, the avian joint is somewhat different from the mammalian joint, since the articular surface is covered by a perichondrium-like fibrocartilage layer, the articular cap, which is absent in mammals [14], with the exception of the temporomandibular joint.

Molecular signaling in joint formation

From the molecular point of view, joint development consists of two main phases: joint patterning, with the specification of the site where a joint will form within a mesenchymal condensation, and tissue differentiation. There is compelling evidence for a role of BMPs in both specification and tissue differentiation in joint development.

The identification of the signals responsible for the determination of the site where the joint will form is still posing a research challenge. Hints to resolve this point come from genetic studies, transgenic models, and natural mutations in which joint formation is disrupted in various species including humans. One of the best documented candidate genes to play a role in joint determination is cartilage-derived morphogenetic protein-1/growth-differentiation factor-5 (CDMP1/Gdf5) [27–30], a bone morphogenetic protein (BMP) family member. In developing mouse limbs, CDMP1/Gdf5 is expressed in the perichondrium and in every joint interzone 24–36 hours before its morphological appearance [28]. In naturally occurring loss-of-function mutations in the CDMP1/Gdf5 gene in mice (brachypodism) [29] and in humans (Hunter-Thompson chondrodysplasia) [31] the distal elements of the appendicular skeleton develop poorly and a specific subset of phalangeal joints does not form. Although CDMP1/Gdf5 is expressed in all joint interzones early in limb development, only a subset of joints is affected by CDMP1/Gdf5 null mutations, indicating that other molecules, possibly other BMPs, can compensate CDMP1/Gdf5 function. This hypothesis finds support in the phenotype of another spontaneous mutation of the CDMP1/Gdf5 gene in humans, Grebe chondrodysplasia (OMIM 200700) [32]. In contrast to the Hunter-Thompson variant, in which CDMP1/Gdf5 protein is presumably absent as the result of a frameshift mutation in the mature region [31], the Grebe chondrodysplasia [33] is associated with a point mutation in the CDMP1/Gdf-5 gene. This mutation results in a protein that is not secreted, is inactive *in vitro* and can form non-functional heterodimers with other BMP family members thereby probably preventing their secretion [33] and possibly acting as a dominant negative for other BMP family members. This in turn could explain why this phenotype is much more severe than the Hunter-Thompson Syndrome but interestingly, still in a proximo–distal fashion. These studies provide support to the intriguing hypothesis that the morphogenesis of different skeletal elements is regulated by different BMP family members, possibly as a result of gene duplications within the BMP family, followed by gain and loss of specific regulatory elements [1]. Additional support for this hypothesis has come from the recently described skeletal phenotypes of Gdf6$^{-/-}$ and Gdf5/Gdf6 double knockout (KO) mice [34]. Gdf6 and Gdf7, closely related to Gdf5, are expressed in different subsets of developing joints. Gdf6$^{-/-}$ mice show defects in joint, cartilage and ligament formation at sites different from the Gdf5 mutants. Gdf5$^{-/-}$/Gdf6$^{-/-}$ mice show additional defects not limited to skeletal fusions (or lack of joint formation) but also severe reduction and absence of skeletal elements in the limb [34].

Disruption of joint formation is obtained in a number of different experimental models. BMP-7/OP-1 is highly expressed in the differentiating perichondrium of chick limb cartilages at stages 29–34 HH (Hamburger/Hamilton) [35], with characteristic interruptions in the zones of future joint formation [36]. Implantation of BMP-7/OP-1 soaked beads at these stages in the joint region disrupts joint formation [36]. Thus, it has been suggested that BMP-7/OP-1 would act as an inhibitory factor for joint formation, preventing joints from forming in non-physiological sites, and that, at least in the chick model, the discontinuities in its expression in the perichondrium would have a permissive role [36]. In contrast to BMP-7/OP-1, BMP-2 transcripts exhibit linear domains of expression in the joint interzones over the same developmental stages [36]. Bmp-2 has been also detected with a similar pattern in mice as early as stage 13.5 dpc, and its expression becomes prominent at stage 15.5 dpc [37]. Overexpression of Bmp-2 and Bmp-4 also disrupts joint formation [38]. In summary, it becomes clear that the correct patterning of the appendicular skeleton and the joint formation process requires an interplay of different signaling molecules restricted in their spatio-temporal activity. One of the best candidate protein families to be involved in that process is the BMP-family.

BMP signaling is regulated in many ways: at the extracellular level by several secreted antagonists (e.g., Noggin, Chordin and DAN/Gremlins), at the receptor level by alternative expression of different receptors, at the intracellular level by both cytosolic proteins including *smads*, and nuclear proteins such as Smad-interacting proteins, and finally by several transcriptional regulators at the DNA level [39–42].

Noggin (Nog) is a secreted molecule that physically interacts with BMP family members and inhibits their activity [43]. It is expressed in developing murine limbs in the condensing mesenchyme and in immature chondrocytes [44]. Its expression pattern and *in vitro/in vivo* function suggest that its developmental role is to establish boundaries of BMP activity. In Noggin deficient mice the resulting excess of BMP activity leads to enlarged appendicular skeletal elements and joint fusions [44]. This skeletal phenotype closely resembles that of CDMP1/Gdf5 overexpression [45–47]. The absence of joints is likely due to failure of joint formation rather than joint fusion, since the CDMP1/Gdf5 expression domain is disrupted in Nog$^{-/-}$ mice, while the expression of other BMPs such as bmp-2, bmp-4, Bmp-5, and Bmp-6 is unaffected [44]. Autosomal dominant missense mutations in a highly conserved region of the NOG gene have been identified in five independent families that segregate proximal symphalangism (SYM1; OMIM 185800) and one dominant missense mutation in a family segregating multiple synostoses syndrome (SYNS1; OMIM 186500) [48]. Three different missense mutations have been described in carpal–tarsal coalition syndrome (OMIM 186570). Remarkably, two of these mutations are identical to those in SYM1 [49]. The principal feature of these syndromes is joint fusion. Truncating mutations in the NOG gene in two families with autosomal dominant stapes ankylosis with broad thumbs and toes, hyperopia, and skeletal anomalies (OMIM 184460) have also been described. Interestingly, these patients do not show symphalangism [50]. The mechanism by which these mutations alter Noggin function and cause the phenotypes is currently not known. Functional haplo-insufficiency is one potential mechanism, as has previously been suggested for CDMP1/Gdf5 mutations in Brachydactyly C families [51]. Alternatively, different mutations may impair the ability of the peptide to bind a subset of TGF-β family members, accounting for the differences in the two syndromes and between the families [48]. These data also suggest that the requirement of noggin for joint morphogenesis may vary between species.

At the receptor level, BMP signaling is regulated by the expression of different BMP receptors [39–42]. Activin-like Kinase-6(ALK6)/BMPR-IB type I BMP receptor is expressed early throughout the prechondrogenic mesenchymal condensations, and its expression pattern becomes later restricted to a narrow domain flanking both distally and proximally that of CDMP1/Gdf5 in the joint interzones (Fig. 1). Although *Alk6* is expressed in all the skeletal elements, *Alk6*-deficient mice display only mild skeletal deformities, lacking both the first and the second phalanges [52, 53]. This phenotype is overlapping, but not identical to that of CDMP1/Gdf5 deficient mice since, in contrast to CDMP1/Gdf5 [bpJ–/–] mice, the metacarpal elements are of normal length and articulate directly to a normal distal phalanx. The double homozygous Alk6[–/–] CDMP1/Gdf5 [bpJ–/–], however, resemble more closely the Gdf5 [bpJ–/–], again with subtle but discrete differences [52, 53]. These genetic data, taken together with the results obtained from *in vitro* studies [54, 55] seem to indicate that Gdf5 and Alk6 function within the same pathway, and that their absence can be compensated by other signaling pathways in most skeletal elements. The discrete differences between the phenotypes described indicate that CDMP1/Gdf5 signals prevalently, but not exclusively, through the Alk6 receptor, as well as Alk6 does not transduce only CDMP1/Gdf5 signaling.

Little is known about the molecules upstream of CDMP1/Gdf-5. Possible candidates are Hox genes, a family of transcription factors, which are thought to control the positional information of skeletal elements [56]. Indeed, mutations of Hoxa and Hoxd genes cause fusion of carpal joints [57–59]. Recently, the characterization of regulatory elements of Gdf5 has been described [60, 61]. The knowledge and availability of these elements should allow further analysis of signaling pathways critically involved in the joint formation process using genetic approaches.

The process of joint cavity formation

Various mechanisms have been proposed to be involved in the induction of cavitation in synovial joints. To date, the factors considered as being involved in this process are fetal movements [62–67], programmed cell death (PCD) [68–70], and selective secretion and turnover of extracellular matrix (ECM) components [14, 71].

The role of movement in joint cavitation is controversial. The observation that synovial joints fail to develop in immobilized chick embryos [62–65] has led to hypothesize that mechanical disruption of intercellular matrix could occur under forces generated by muscle activity. However, in myogenin-deficient mice, which do not develop contracting skeletal muscles, joint cavitation takes place normally [72].

During mammalian morphogenesis, PCD is an essential mechanism to eliminate selectively cell populations and accomplish histogenesis and organogenesis. In the rat embryo, PCD has been observed histologically within the interzone before cavitation [21]. It has been suggested that cells with chondrogenic potential would be

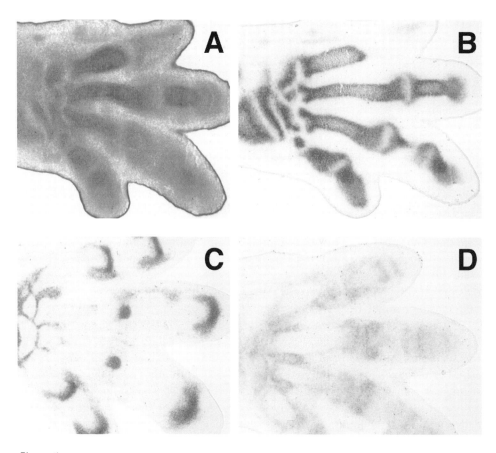

Figure 1

Gene expression pattern of selected BMP signal transduction components during joint morphogenesis

Cryosectioned 14.5 dpc mouse forelimbs were stained with toluidine blue (a) or analyzed by in situ *hybridization with digoxigenin-labeled cRNA probes for Col2a1 (b), Gdf5/Cdmp1 (c) and Bmpr-Ib/Alk6 (d).*

eliminated in this way from the interzone, thus preventing cartilage differentiation [68, 18].

Another mechanism envisages synthesis and deposition of large amounts of hyaluronic acid (HA) as a mechanical factor to separate the opposing joint surfaces [14, 71, 73]. This theory is corroborated by the histochemical localization of free HA at the chick metatarsophalangeal joint interzones concomitant with the first signs of cavitation at stage 37 HH [71], and confirmed by the local increased activity of uridine diphosphoglucose dehydrogenase and HA synthase, enzymes involved in HA synthesis [73]. The swelling pressure of the HA is assumed to physically separate the cells, thereby inducing joint cavitation, to increase and maintain the cavity, and prevent secondary fusion across the joint space [71, 73].

More recently, PCD has been described to occur within joint interzones of developing digits in mouse fetuses between 13.5 and 14.5 dpc, thus shortly before cavitation starts (14–15 dpc) [70]. These data have also been confirmed in the chick embryo at stages 33–35 HH [74]. CDMP1/Gdf5 and BMP-2, expressed in the joint interzone within the same time window, are good candidates in mediating this process, since BMPs have been shown to induce apoptosis in mesenchymal cells at certain sites and stages during development [35, 75, 76]. In Alk6$^{-/-}$ mice, as a secondary event, Gdf5 is overexpressed with an expanded expression domain [52]. This expression domain overlaps with an area of intense cell death [52]. These data seem to indicate that CDMP1/Gdf5 stimulates chondrogenesis and cartilage growth through the Alk6 receptor, while triggering apoptosis in the absence of Alk6, therefore through a different receptor. Since the Alk6 expression domains are flanking the narrow stripe of CDMP1/Gdf5 expression at the joint interzones, a role of CDMP1/Gdf5 in inducing apoptotic events associated with joint cavitation may be anticipated.

Finally, it is important to mention that the combined genetic and experimental evidence establish the existence of a signaling center in the joint interzone, directly or indirectly, orchestrating limb growth. For instance, loss of function of CDMP1/Gdf5 results in delayed chondrogenesis and shorter limbs. Overexpression of the same polypeptide modulates dose dependently the size of the limbs and epiphysis, both in the chick and mouse model [46, 47, 77].

Bone morphogenetic protein signaling in postnatal synovial joints

Bone morphogenetic protein signaling in articular cartilage

In the last decade, our understanding of the molecular events leading to joint formation has been rapidly expanding. However, the whole picture is still far from being drawn. The set of molecules known to be involved is not yet complete. In addition, information of how these molecules interact with each other orchestrating the processes of skeletal and joint morphogenesis and tissue differentiation is limited.

Even more limited is our knowledge and data about molecular signaling in postnatal joints. There is reasonable evidence that nature may utilize postnatally the same signaling pathways for comparable roles and functions during development. In other words, the molecular events that regulate tissue differentiation and organogenesis during development may also be involved postnatally in tissue homeostasis and repair. For example, BMPs and Hedgehog proteins, critically involved in the formation of cartilage and bone during embryogenesis, are also expressed in fracture healing and distraction osteogenesis [78, 79]. It is conceivable that at least some of the molecules herein discussed in the context of joint development have also a role in the maintenance of joint tissues, and in the processes of tissue repair and regeneration.

An example comes from CDMP1. This molecule, of which mRNA expression during development is strongly associated with the initiation of the joint interzone [80], is also present in normal human adult articular cartilage [81]. Its expression, as determined by immunohistochemistry, is mostly restricted to the superficial cartilage in normal joints, while in osteoarthritic cartilage its expression domain is extending to damaged areas [81]. These data suggest a possible role for CDMP1 in the homeostasis of normal cartilage, as well as in repair processes. Accordingly, recombinant CDMP1 increases proteoglycan biosynthetic activity in adult articular cartilage that has been partially matrix-depleted by mild trypsin treatment [81].

The effects of CDMP1 on articular chondrocytes may not be limited to a stimulation of matrix synthesis. A 30 min incubation of adult pig articular chondrocytes with recombinant CDMP1 at a final concentration of 100 ng/ml resulted not only in enhanced matrix deposition, but also in an increased cell number when injected as a cell suspension intramuscularly in nude mice. The wet weight of the implant of hyaline-like cartilage recovered after three weeks was 2- to 3-fold higher. In addition, the cartilage tissue stained more intensely with safranin O as compared with the untreated control (Fig. 2). CDMP1, therefore, may be implicated in the proliferation and metabolic activity of articular chondrocytes.

The presence of BMP-7/OP-1 was demonstrated in normal adult human articular cartilage, as determined by *in situ* hybridization, Western blotting, and immunohistochemistry [82]. BMP-7/OP-1 mRNA was found in the superficial and middle layers of the cartilage, whereas in the deep layer levels of expression were very low. The topographic distribution of the protein within the tissue was quite interesting as revealed by immunostaining performed using two different antibodies, one recognizing the active mature form, and the other reacting with the inactive pro-form. Mature BMP-7/OP-1 was found predominantly in the superficial and middle layers of the tissue, whereas pro-BMP-7/OP-1 was predominantly detected in the deep layer of the cartilage [82]. The distinct localization of pro- and mature forms of BMP7/OP-1 suggests that the processing of pro-BMP-7/OP-1 into mature BMP-7/OP-1 may occur primarily in the superficial chondrocytes. The detection of BMP receptors type IA and IB, and type II in normal human articular cartilage [83], further corroborates a possible autocrine/paracrine function for BMPs in the maintenance and repair of the articular surface.

Cartilage is critical for both bone and joint morphogenesis. Articular cartilage and growth plate cartilage are biologically distinct. In contrast to the articular chondrocytes, the transient chondrocytes in the growth plate determine the longitudinal and radial growth of the cartilage skeletal elements, which are replaced by bone through a process called endochondral ossification. BMP-2/4 and BMP-7/OP-1,

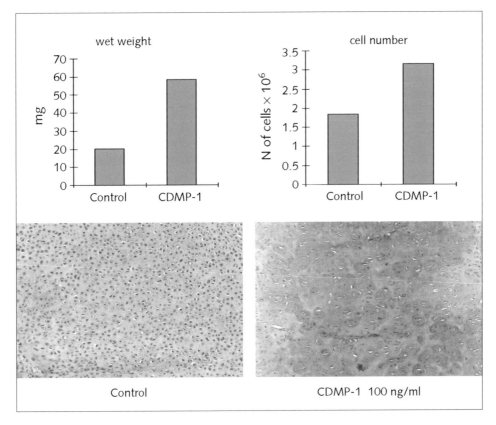

Figure 2
Treatment with GDF-5/CDMP-1 enhances the capacity of articular chondrocytes to organize cartilage tissue in vivo
Swine articular chondrocytes from metatarsal joints were treated with 100 ng/ml CDMP1 or with control medium, washed and injected intramuscularly into nude mice. Three weeks later, the samples were weighed, and either destined to histological analysis (safranin O staining) or digested in 0.2 crude collagenase in DMEM for cell count.

BMP receptors (BMPRIA, BMPRIB, and BMPRII), and their intracellular signaling transducers Smads have been detected immunohistochemically in the epiphyseal plate of growing animals [84, 85]. Their temporal and spatial expression patterns suggest a morphogenic role for BMPs in the multistep cascade of endochondral ossification in the epiphyseal growth plate.

In contrast to the transient cartilage template, articular cartilage is stable throughout life, being resistant to vascular invasion and endochondral ossification. Factors responsible for the maintenance of articular cartilage include TGF-β super-

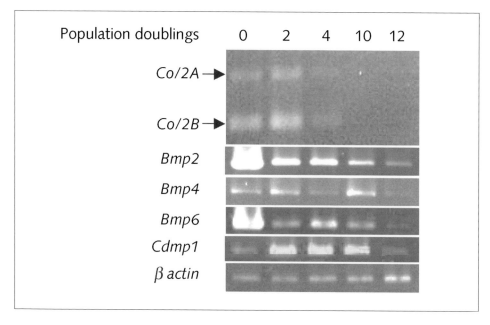

Population doublings   0   2   4   10   12

Co/2A →

Co/2B →

Bmp2

Bmp4

Bmp6

Cdmp1

β actin

Figure 3
Modulation of the expression of some BMPs during chondrocyte expansion in vitro
Human adult articular chondrocyte lose their phenotypic traits during in vitro *expansion.*
Bmp2 *and* Bmp6 *mRNA levels decrease throughout passaging, paralleling the downregulation of both collagen type IIA and type IIB mRNAs.*

family signaling molecules. The induction of osteoarthritis in adult mice with tissue specific overexpression of a dominant negative TGF-β type II receptor [86] would support this concept (see below).

We have determined by semi-quantitative reverse transcriptase polymerase chain reaction (RT-PCR) the expression of BMPs and related receptors by articular chondrocytes, isolated from normal adult human knee cartilage. BMP-2, 4, and 6, as well as CDMP1 were expressed by freshly isolated cell populations (Fig. 3). We have found a correlation between the BMP expression profile and the phenotype of chondrocytes during *in vitro* expansion. While passaging, chondrocytes are known to undergo a dedifferentiation or rearrangement of their phenotypic traits [87]. The expression levels of BMP-2 and 6 were downregulated during *in vitro* passaging in parallel with cartilage matrix protein collagen type II (Fig. 3). These findings underscore the potential role of BMPs in the maintenance of the chondrocyte phenotype. A variety of *in vitro* models have provided evidence that BMPs promote chondrogenesis [88–90], enhance cartilage matrix synthesis [91–93], and support re-expression of the cartilage phenotype [94, 95].

Bone morphogenetic protein signaling in postnatal joint associated tissues

The synovial joint is a complex organ that encompasses different tissues, i.e., cartilage, subchondral bone, menisci, and the capsulo-ligamentous apparatus. The synovial membrane lines the inner surface of the joint capsule and covers most intra-articular structures except for the cartilage.

Increasing evidence supports the hypothesis that multipotent stem cells are present postnatally in different organs and tissues. Although their *in vivo* role is not well understood, these cells appear to contribute to postnatal growth, and participate in tissue homeostasis by replacing differentiated cells lost to physiological turnover, injury, or senescence. A hypothetical role for BMPs in adult tissues can be the maintenance and recruitment of a pool of stem/progenitor cells for tissue homeostasis and regeneration. Fine balances of BMPs would be required for either the maintenance of this cell population in a quiescent phenotype, or their activation and commitment to a specific lineage.

Mesenchymal stem cells (MSCs) have the potential to differentiate into lineages of mesenchymal tissues, including cartilage, bone, fat, and muscle. Isolation and characterization of MSCs from bone marrow [96] and periosteum [97, 98] have been described. We have identified a population of multipotent MSCs derived from adult human synovial membrane [99]. These cells possess *in vitro* high self-renewal capacity with limited senescence. Under appropriate culture conditions, expanded synovial membrane-derived MSCs can be induced to differentiate *in vitro* toward chondrogenesis, osteogenesis, myogenesis, and adipogenesis. We have also demonstrated that these cells can contribute to *in vivo* muscle repair [100]. Remarkably, synovial MSC failed from ectopic, stable cartilage, resistant to vascular invasion, *in vivo* [101]. As determined by RT-PCR, synovial membrane-derived MSCs express all known BMP receptors and many BMPs (Tab. 1). Recent studies presented no BMP receptors detected in normal synovial membrane by immunostaining [83]. This indicates that the cell isolation technique and the subsequent expansion of this selected cell subpopulation under fetal bovine serum containing culture conditions results in an enrichment of BMP receptor expressing cells.

Repair processes require tissue regeneration as a recreation of destroyed cells and ECM, followed by the restoration of tissue architecture and appropriate relationships between different tissues. TGF-β superfamily members including BMPs/CDMPs are good candidates for the orchestration of these regenerative processes. As morphogens, they would be involved in the coordination of different events such as positional information, patterning, and they could participate in the regulation of the proliferation rate and the progress in the differentiation cascade and maturation process. GDF5, 6, or 7 appear to be able to induce neotendon/ligament formation when implanted at ectopic sites *in vivo* [102], suggesting that they can influence progenitor cells to differentiate along a tendon/ligament pathway. Implantation of GDF5 or 6 on collagen sponges has been reported to enhance tendon healing in rats

Table 1 - Expression of BMPs/CDMPs and receptors by human synovial membrane-derived mesenchymal stem cells, as determined by RT-PCR

Receptors		BMPs/CDMPs	
ALK1	+	BMP-2	+
ALK2	+	BMP-3	–
ALK3	+	BMP-4	+
ALK4	+	BMP-5	–
ALK5	+	BMP-6	+
ALK6	+	BMP-7/OP-1	–
BMPR2	+	GDF5/CDMP1	+
		GDF6/CDMP2	–
		TGF-β1	+
		TGF-β2	+
		TGF-β3	+

[103]. The elucidation of the functions of morphogens including the BMPs/CDMPs will lead to the identification of additional therapeutic targets and novel tissue engineering protocols to enhance and control repair processes in joint disorders, thereby possibly delaying or limiting major surgery.

Bone morphogenetic protein signaling in joint disease

Recently, new data have been reported on the potential role of BMPs in joint disease [104]. Given their well documented functions in bone and joint development, as well as their potential contribution to joint tissue homeostasis, it seems likely that these molecules also have a role in different diseases affecting the joint. They may influence the disease process itself, or be involved in eventual repair processes taking place as a response to injury.

As for every "organ", different types of disease can affect the synovial joint: degenerative disease, inflammatory and auto-immune disorders, infectious diseases, metabolic diseases as well as benign and malignant tumors.

Transforming growth factor-β/Bone morphogenetic protein signaling in degenerative joint diseases

Osteoarthritis (OA) is a common disorder, occurring mostly in middle and older aged persons, characterized by articular cartilage destruction and subchondral bone remodeling, leading to loss of joint function, and increasing disability. Although several risk factors have been recognized, such as obesity, familial history, skeletal malformations and trauma, the precise pathological events causing the disease and associated with disease progression are not yet clear. The key features appear to be subchondral bone sclerosis, potentially changing the weight-bearing properties and therefore the internal mechanics and dynamics of the joints, together with localized articular cartilage damage. However, the complete picture is likely to be far more complex. The whole joint organ is involved. The presence of new bone formation at the joint margins, so-called osteophytes, suggests repair efforts which are either insufficient or poorly coordinated, since they do not result in repair of the damaged tissue with preserved function. In OA models, several stages of the disease have been described [105, 106]. The early stage of the degenerative process is characterized by hypertrophy of the articular cartilage with a net increase in matrix synthesis and content. This phase, occurring before macroscopic cartilage damage can be demonstrated, is followed by a phase with net matrix loss by depletion of matrix components, resulting in focal damage and loss of function. In the late phase, it is suggested that the release of matrix components and particles from the cartilage lead to synovial activation and inflammation, including the secretion of inflammatory cytokines such as IL-1 and TNF-α. The resulting cytokine imbalance further enhances protease and matrix metalloproteinase (MMP) synthesis, stimulation of cyclo-oxygenase and further damage of joint tissues. The complex interactions between these signaling molecules, effector enzymes and the different cell populations are likely to be influenced by the presence of growth and differentiation factors such as BMPs, not only in the hypertrophic phase but also in the later stages.

Some evidence regarding the role of TGF-β superfamily signaling in skeletal and joint diseases has been obtained in mouse genetic models. Skeletal tissue specific overexpression of a truncated, kinase-deficient TGF-β type II receptor, acting as a dominant-negative effectively neutralizing TGF-β signaling, resulted in skeletal malformations. They include progressive skeletal degeneration after birth, leading to kyphoscoliosis, and stiff and torqued joints in heterozygous mice by the age of 4–8 months [107]. Strikingly, the histological changes resemble those seen in osteoarthritis. The first signs of joint degeneration in this mouse model are seen at four weeks of postnatal growth: patches of the articular surface appear denuded and an increase in hypertrophic chondrocytes is seen in the deeper layers of the articular cartilage. In six-month-old mice, articular cartilage is fibrillated and disorganized. Chondrocytes are organized in clusters; there is an increased number of hypertrophic chondrocytes and a disruption of the tidemark with excessive bone formation. Osteophytes can be recognized as outgrowths of chondroid tissue in the articular margins undergoing enchondral bone formation. Proteoglycan synthesis, as shown by safranin O staining, is decreased in the "osteoarthritic" transgenic mice. Type X collagen, molecular marker of non-proliferating hypertrophic chondrocytes, is expressed in the joints of older transgenic mice, localized to fibrillated articular cartilage, osteophytes and cartilage growing in the joint space as can also be seen in human osteoarthritis [108]. A similar phenotype is seen in mice deficient in Smad3, a TGF-β receptor smad [109]. Smad3$^{-/-}$ homozygotes (KO mice) display skeletal abnormalities, including inwardly turned paws, kyphosis of the spine, osteopetrosis and abnormal ossification of the joints. At six months many mutant mice develop an osteoarthritis-like disease, characterized by progressive loss of articular cartilage, surface fibrillation, formation of large osteophytes, upregulation of type X collagen and decreased proteoglycan synthesis. The presence of osteoarthritic changes in a model in which TGF-β signaling is impaired, suggests that TGF-β is important for the maintenance of tissue integrity, and that the balance between TGF-β and BMP signaling influences joint homeostasis.

Using intra-articular joint injections, Van Den Berg et al. have extensively studied the *in vivo* effects of TGF-βs and BMPs on cartilage metabolism, and potential interactions with IL-1. BMP-2 strongly enhances proteoglycan synthesis after injection into the knee joint of normal mice [110]. The effect, however, is short-termed as compared to the effect of TGF-β1 injection. Unlike BMP, following TGF-β1 injection, proteoglycan synthesis rises at lower rate and to lower levels but the response is maintained for 20 days. This is probably due to stimulation of endogenous TGF-β or BMP production and/or upregulation of receptors. Remarkably, TGF-β1 counteracts the IL-1 induced suppression of proteoglycan synthesis whereas BMP-2 does not [111, 112]. However the relative dose of TGF-β used in these experiments is higher than that of BMPs. The effect and the counterbalance of TGF-β and IL-1 are only seen in articular cartilage, but not at the joint margin where osteophytes are formed. TGF-β probably induces cartilage formation from the periosteum, as has been demonstrated in an *in vitro* model [98] and this process seems not to be influenced by IL-1 in the *in vivo* mouse model. Nonetheless BMP-2 and BMP-7/OP-1 mRNA and protein have been detected in the growing osteophyte [109, 113, 114]. CDMP1 and CDMP2 have been detected in osteoarthritic and normal cartilage, and have been shown to promote cartilage matrix recovery after enzymatic depletion *in vitro*, with restoration or maintenance of the normal phenotype thus pointing to a potentially important repair mechanism [115]. Data by Chubinskaya et al. show the presence of BMP-7/OP-1 in human articular cartilage as mentioned above [116].

Interestingly, recent work has shown that inhibition of TGF-β and BMP signaling in a mouse model of osteoarthritis using adenoviral overexpression of Smad6, Smad7 and the antagonist latency-associated protein results in diminished osteophyte formation [117].

Bone morphogenetic proteins and inflammatory joint disease

Although many systemic inflammatory disorders can also involve the synovial joints, most forms of chronic arthritis can be categorized into two distinct groups:

rheumatoid arthritis (RA) and the spondylarthropathies (SpA), the latter consisting of ankylosing spondylitis, psoriatic arthropathy, enteropathic SpA, reactive arthritis (such as Reiter syndrome) and undifferentiated SpA. It is remarkable that although most of the key inflammatory mediators such as TNF-α and IL-1 have been found within the synovium and the synovial fluid in both disease groups, and at least some of the destructive mechanisms appear to be driven by the same molecular players, the pathological endpoints are strikingly different. RA is mostly characterized by periarticular osteoporosis, extensive cartilage and bone destruction and no appreciable repair efforts. The SpAs have mostly no periarticular osteoporosis, often less destruction and remarkable "repair", frequently apparently "overdoing" it, leading to bony bridging of the joint cavity and ankylosis. Many of these presumed repair processes morphologically closely resemble bone and cartilage formation during development, and therefore a role of TGF-β family members can be expected. It is noteworthy that Braun et al. detected by *in situ* hybridization expression of TGF-β2 in biopsies from the sacroiliac joints of patients with ankylosing spondylitis [118]. More comprehensive and in depth research in this field is only beginning to acquire a momentum. Most data on joint pathology have come from samples obtained at joint replacement surgery, and therefore only represent severe and end-stage disease. However, the development of needle arthroscopy as a diagnostic tool in daily rheumatology practice, and the availability of biopsies at distinct stages of the disease, is rapidly increasing our knowledge of the pathology and the molecular players involved.

We have demonstrated that different BMPs, including BMP-2, BMP-6 and BMP-7 are expressed in synovial biopsies obtained from patients with chronic arthritis [104]. Protein levels of BMP-2 and BMP-6 are significantly higher in patients with RA and SpA as compared to non-inflammatory controls. BMP-2 and BMP-6 expression is found in both macrophages and fibroblast-like synoviocytes as demonstrated by immunohistochemistry [104]. BMP-2 and BMP-6 expression *in vitro* is upregulated by proinflammatory cytokines such as IL-1 and TNF-α, but not by Interferon-γ. We have also demonstrated that BMP-2 is associated with fibroblast-like synoviocyte apoptosis *in vitro* and *in vivo* [104]. These data give additional support to the hypothesis that BMPs are pleiotropic cytokines with different function in the complex signaling network in chronic arthritis. Upregulation of BMP expression by monokines has also been demonstrated in articular chondrocytes *in vitro* [119].

Bone morphogenetic proteins and infectious arthritis

Bacterial synovial joint infection is probably the most destructive and rapidly progressive pathological process within the organ. Septic arthritis is either caused by a contiguous process or by bacteremia in the subsynovial vessels from a distant focus. Some bacteria preferentially localize within the joint. Bacterial products such as endotoxin, cell fragments, immune complexes, and bacterial opsonization cause an extensive inflammatory reaction from the innate as well as from the acquired immune system including the production of TNF-α and IL-1, activation of proteolytic enzymes and MMPs, antibody production and generation of effector and memory T cells. Moreover, phagocytosis by neutrophils causes autolysis, thereby releasing lysosymic tissue-destructive enzymes within the joint cavity. Bacterial products are also capable of inducing chondrocyte proteinases that often subsist even after the bacteria have been cleared by the host immune system. Infection also leads to activation of the subsynovial endothelial cells, resulting in thrombosis and ischemia. We were able to detect by Western blot BMP-4, CDMP1 and CDMP2 in the synovial fluid of patients with septic arthritis (Fig. 4). It should therefore not be surprising that BMPs may be involved in either modulation of the reaction or in a failing attempt to repair the occurring damage. However, it is not yet clear which cells and tissues are responsible for the BMP release into the fluid. BMP release could be caused by upregulation of BMP-production as part of a repair effort, but it might also be explained by the release of BMPs previously trapped in the articular cartilage matrix. Nonetheless, these preliminary observations provide sufficient impetus to further investigate the potential role of BMPs in infectious joint disease.

Bone morphogenetic proteins and skeletal and joint tumors

Synovial joint tumors are rare disorders. In view of the role of BMPs in skeletal growth and differentiation, they may be involved in tumors containing bone and chondroid tissue.

Only a few research teams have looked at BMPs in these disorders to date, and therefore available data are rather limited in scope. Most research in this field has been done by Yoshikawa et al. [120–123]. Osteosarcomas, not necessarily joint-associated, were analyzed for ectopic bone formation activity, as a way to measure BMP activity, by implanting the lyophilized fraction of the tumor in a nude mice model. Not only did the BMP-activity containing tumors have some distinct radiological and pathological properties, they also showed a higher resistance to doxorubicin-methotrexate chemotherapy, and a higher tendency to metastasize [120–123]. Subsequently BMP-2 or BMP-4 were detected immunohistochemically in osteosarcomas, except in nine chondroblastic subtypes, in malignant fibrous histiocytomas (MFH) and in several sarcomas but not in synovial, rhabdomyo- and fibrosarcoma. However, the sensitivity of the technique can be questioned since no BMP was detected in any normal human tissue, or in a 16-week-old human fetus [122].

Guo et al. studied BMP expression in 36 osteosarcomas, 6 Ewing's sarcoma, 20 synovial sarcomas and 20 chondrosarcomas by RT-PCR [124]. BMP-2 and BMP-4 mRNAs were detected in almost all sarcomas, BMP-6 in 22 osteosarcomas and

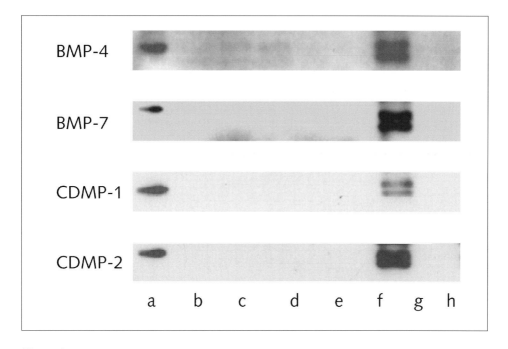

Figure 4
Western blot of BMP-4, BMP-7, CDMP-1 and CDMP-2 in synovial fluid
Growth factors in 1% hyaluronidase treated synovial fluid were concentrated by heparin
sepharose binding in 8M urea, 10 mM Tris, 150 mM NaCl (pH 7.3), washed with 8M urea
10 mM Tris 3M NaCl and precipitated with ice cold trichloroacetic acid 30% (w/v); the
resulting binding protein pellet was redissolved in 8M urea 0.05 M Tris and subsequently ran
on reducing SDS-PAGE gels. Western blots were performed with polyclonal anti-BMP-4,
anti-BMP-7, anti-CDMP-1 or anti-CDMP-2 antibodies [114], then incubated with biotiny-
lated secondary antibody and analyzed with peroxidase/luminol staining. Lanes (a) 10 ng of
human recombinant protein (b) patient with undifferentiated spondylarthropathy (c) patient
with chondrocalcinosis (d) patient with rheumatoid arthritis (e) patient with mono-arthritis
of unknown etiology (f) patient with chondrocalcinosis (g) patient with septic arthritis (h)
patient with rheumatoid arthritis.

seven chondrosarcomas. BMP type II receptor was found in 25 osteosarcomas, eight chondrosarcomas, four Ewing's sarcomas and 15 synovial sarcomas. The expression of the type II BMP receptor correlates with metastasis in osteosarcoma and synovial sarcomas. Recently, a new series of experiments was reported [125] in which nine out of 11 osteosarcomas displayed the presence of BMPs and BMP-receptors, detectable by immunohistochemistry. The two negative samples were osteosarcomas of the chondroblastic type. In another case, eight out of 10 malignant fibrous histiocytomas also expressed BMPs but not their receptors, thus suggesting that a lack of BMP responding cells is associated with non-ossification of malignant fibrous histiocytomas. Recently Sulzbacher et al. reported a series of 47 osteosarcomas in which BMP expression was studied by immunohistochemistry [126]. BMP-7 and BMP-8 are highly expressed in osteosarcoma. Moreover, high expression of BMP-6 correlates with a chondroid differentiation. In contrast to some conclusions derived from previously mentioned studies, these results indicated that the expression of BMPs does not contribute to speculations about the tumor type, tumorigenicity, metastasis or the outcome of patients.

References

1   Kingsley DM (1994) What do BMPs do in mammals? Clues from the mouse short-ear mutation. *Trends Genet* 10: 16–21
2   Luyten FP (1997) A scientific basis for the biologic regeneration of synovial joints. *Oral Surg Oral Med Oral Pathol Oral Radiol Endod* 83: 167–169
3   Dell'Accio F, De Bari C, Luyten FP (1999) Molecular basis of joint development. *Jpn J Rheumatol* 9: 17–29
4   Bernays A (1878) Die Entwicklungsgeschichte des Kniegelenkes des Menschen, mit Bemerkungen über die Gelenke im Allgemeinen. *Morphologiesches Jahrbuch* 4: 403–446
5   Andersen H, Bro-Rasmussen F (1961) Histochemical studies on the histogenesis of the joints in human fetuses with special references to the development of the joint cavities in the hand and foot. *Am J Anat* 108: 111–122
6   Andersen H (1962) Histochemical studies on the histogenesis of the human elbow joint. *Acta Anatomica* 51: 50–68
7   Andersen H (1962) Histochemical studies of the development of the human hip joint. *Acta Anatomica* 48: 258–292
8   Andersen H (1963) Histochemistry and development of the human shoulder and acromio-clavicular joint with particular reference to the early development of the clavicle. *Acta Anatomica* 55: 124–165
9   Gardner E, Gray DJ, O'Rahilly R (1959) The prenatal development of the skeleton and joints of the human foot. *J Bone Joint Surg Am* 41A: 847–876
10   Haines RW (1947) The development of joints. *J Anat* 81: 33–55
11   Andersen H (1961) Histochemical studies on the histogenesis of the knee joint and superior tibio-fibular joint in human foetuses. *J Anat* 46: 274–303
12   Merida Velasco JA, Sanchez Montesinos I, Espin Ferra J, Rodriguez Vazquez JF, Merida Velasco JR, Jimenez Collado J (1997) Development of the human knee joint. *Anat Rec* 248: 269–278
13   Morrison EH, Ferguson MW, Bayliss MT, Archer CW (1996) The development of articular cartilage: I. The spatial and temporal patterns of collagen types. *J Anat* 189: 9–22

14 Archer CW, Morrison H, Pitsillides AA (1994) Cellular aspects of the development of diarthrodial joints and articular cartilage. *J Anat* 184: 447–456

15 Gardner E, O'Rahilly R (1962) The development of the elbow joint of the chick and its correlation with the embryonic staging. *J Anat Entwicklungsgesch* 123: 174–179

16 Henrikson RC, Cohen AS (1965) Light and electron microscopic observation of the developing chick interphalangeal joints. *J Ultrastruct Res* 13: 129–162

17 O'Rahilly R, Gardner E (1956) The development of the knee joint of the chick and its correlation with embryonic staging. *J Morphol* 98: 49–88

18 Mitrovic DR (1977) Development of the metatarsophalangeal joint of the chick embryo: morphological, ultrastructural and histochemical studies. *Am J Anat* 150: 333–347

19 Ginsburg GT, Royster D, Kassabian G, Shuler CF, Dougherty WR, Sank AC (1995) Mesenchymal commitment to digital joint formation. *Ann Plast Surg* 35: 95–104

20 Takabatake K, Yamamoto T (1991) Morphology of the synovium during its differentiation and development in the mouse knee joint. A histochemical, SEM and TEM study. *Anat Embryol Berl* 183: 537–544

21 Mitrovic D (1978) Development of the diarthrodial joints in the rat embryo. *Am J Anat* 151: 475–485

22 Chevallier A, Kieny M, Mauger A (1977) Limb-somite relationship: origin of the limb musculature. *J Embryol Exp Morphol* 41: 245–258

23 Christ B, Jacob HJ, Jacob M (1977) Experimental analysis of the origin of the wing musculature in avian embryos. *Anat Embryol Berl* 150: 171–186

24 Kenny-Mobbs T (1985) Myogenic differentiation in early chick wing mesenchyme in the absence of the brachial somites. *J Embryol Exp Morphol* 90: 415–436

25 Craig FM, Bentley G, Archer CW (1987) The spatial and temporal pattern of collagens I and II and keratan sulphate in the developing chick metatarsophalangeal joint. *Development* 99: 383–391

26 Thorogood PV, Hinchliffe JR (1975) An analysis of the condensation process during chondrogenesis in the embryonic chick hind limb. *J Embryol Exp Morphol* 33: 581–606

27 Chang SC, Hoang B, Thomas JT, Vukicevic S, Luyten FP, Ryba NJ, Kozak CA, Reddi AH, Moos M, Jr (1994) Cartilage-derived morphogenetic proteins. New members of the transforming growth factor-beta superfamily predominantly expressed in long bones during human embryonic development. *J Biol Chem* 269: 28227–28234

28 Storm EE, Kingsley DM (1996) Joint patterning defects caused by single and double mutations in members of the bone morphogenetic protein (BMP) family. *Development* 122: 3969–3979

29 Storm EE, Huynh TV, Copeland NG, Jenkins NA, Kingsley DM, Lee SJ (1994) Limb alterations in brachypodism mice due to mutations in a new member of the TGF beta-superfamily. *Nature* 368: 639–643

30 Luyten FP (1997) Cartilage-derived morphogenetic protein-1. *Int J Biochem Cell Biol* 29: 1241–1244

31 Thomas JT, Lin K, Nandedkar M, Camargo M, Cervenka J, Luyten FP (1996) A human

chondrodysplasia due to a mutation in a TGF-beta superfamily member. *Nat Genet* 12: 315–317

32   Grebe H (1952) Die Achondrogenesis: ein einfach rezessives. *Erbmerkmal Folia Hered Path* 2: 23–28

33   Thomas JT, Kilpatrick MW, Lin K, Erlacher L, Lembessis P, Costa T, Tsipouras P, Luyten FP (1997) Disruption of human limb morphogenesis by a dominant negative mutation in CDMP1. *Nat Genet* 17: 58–64

34   Settle SH, Rountree RB, Sinha A, Thacker A, Higgins K, Kingsley DM (2003) Multiple joint and skeletal patterning defects caused by single and double mutations in the mouse *Gdf6* and *Gdf5* genes. *Dev Biology* 254(1): 116–130

35   Hamburger V, Hamilton HL (1951) A series of normal stages in the development of the chick embryo. *J Morphol* 88: 49–92

36   Macias D, Ganan Y, Sampath TK, Piedra ME, Ros MA, Hurle JM (1997) Role of BMP-2 and OP-1 (BMP-7) in programmed cell death and skeletogenesis during chick limb development. *Development* 124: 1109–1117

37   Rosen V, Thies RS, Lyons K (1996) Signaling pathways in skeletal formation: a role for BMP receptors. *Ann NY Acad Sci* 785: 59–69

38   Duprez D, Bell EJ, Richardson MK, Archer CW, Wolpert L, Brickell PM, Francis West PH (1996) Overexpression of BMP-2 and BMP-4 alters the size and shape of developing skeletal elements in the chick limb. *Mech Dev* 57: 145–157

39   Massague J, Chen YG (2000) Controlling TGF-beta signaling. *Genes Dev* 14: 627–644

40   Massague J, Wotton D (2000) Transcriptional control by the TGF-beta/Smad signaling system. *EMBO J* 19: 1745–1754

41   Merino R, Rodriguez-Leon J, Macias D, Ganan Y, Economides AN, Hurle JM (1999) The BMP antagonist Gremlin regulates outgrowth, chondrogenesis and programmed cell death in the developing limb. *Development* 126: 5515–5522

42   Hogan BL (1996) Bone morphogenetic proteins: multifunctional regulators of vertebrate development. *Genes Dev* 10: 1580–1594

43   Piccolo S, Sasai Y, Lu B, De Robertis EM (1996) Dorsoventral patterning in *Xenopus*: inhibition of ventral signals by direct binding of chordin to BMP-4. *Cell* 86: 589–598

44   Brunet LJ, McMahon JA, McMahon AP, Harland RM (1998) Noggin, cartilage morphogenesis, and joint formation in the mammalian skeleton. *Science* 280: 1455–1457

45   Francis West PH, Richardson MK, Bell E, Chen P, Luyten F, Adelfattah A, Barlow AJ, Brickell PM, Wolpert L, Archer CW (1996) The effect of overexpression of BMPs and GDF-5 on the development of chick limb skeletal elements. *Ann NY Acad Sci* 785: 254–255

46   Francis-West PH, Abdelfattah A, Chen P, Allen C, Parish J, Ladher R, Allen S, MacPherson S, Luyten FP, Archer CW (1999) Mechanisms of GDF-5 action during skeletal development. *Development* 126: 1305–1315

47   Tsumaki N, Tanaka K, Arikawa-Hirasawa E, Nakase T, Kimura T, Thomas JT, Ochi T, Luyten FP, Yamada Y (1999) Role of CDMP-1 in skeletal morphogenesis: promotion of mesenchymal cell recruitment and chondrocyte differentiation. *J Cell Biol* 144: 161–173

48 Gong Y, Krakow D, Marcelino C, Wilkin D, Chitayat D, Babul-Hirji R, Hudgins L, Cremers CW, Cremers FPM, Brunner HG et al (1999) Heterozygous mutations in the gene encoding noggin affect joint morphogenesis. *Nat Genet* 21: 302–330

49 Dixon ME, Armstrong P, Stevens DB, Bamshad M (2001) Identical mutations in NOG can cause either tarsal/carpal coalition syndrome or proximal symphalangism. *Genet Med* 3 (5): 349–353

50 Brown DJ, Kim TB, Petty EM, Downs CA, Martin DM, Strouse PJ, Moroi SE, Milunsky JM, Lesperance MM (2002) Dominant stapes ankylosis with broad thumbs and toes, hyperopia, and skeletal anomalies is caused by heterozygous nonsense and frameshift mutations in NOG, the gene encoding noggin. *Am J Hum Genet* 71: 618–624

51 Polinkovsky A, Robin NH, Thomas JT, Irons M, Lynn A, Goodman FR, Reardon W, Kant SG, Brunner HG, van der Burgt I et al (1997) Mutations in CDMP1 cause autosomal dominant brachydactyly type C. *Nat Genet* 17: 18–19

52 Baur ST, Mai JJ, Dymecki SM (2000) Combinatorial signaling through BMP receptor IB and GDF5: shaping of the distal mouse limb and the genetics of distal limb diversity. *Development* 127: 605–619

53 Yi SE, Daluiski A, Pederson R, Rosen V, Lyons KM (2000) The type I BMP receptor BMPRIB is required for chondrogenesis in the mouse limb. *Development* 127: 621–630

54 Nishitoh H, Ichijo H, Kimura M, Matsumoto T, Makishima F, Yamaguchi A, Yamashita H, Enomoto S, Miyazono K (1996) Identification of type I and type II serine/threonine kinase receptors for growth/differentiation factor-5. *J Biol Chem* 271: 21345–21352

55 Erlacher L, McCartney J, Piek E, Ten Dijke P, Yanagishita M, Oppermann H, Luyten FP (1998) Cartilage-derived morphogenetic proteins and osteogenic protein-1 differentially regulate osteogenesis. *J Bone Miner Res* 13: 383–392

56 Erlebacher A, Filvaroff EH, Gitelman SE, Derynck R (1995) Toward a molecular understanding of skeletal development. *Cell* 80: 371–378

57 Davis AP, Capecchi MR (1994) Axial homeosis and appendicular skeleton defects in mice with a targeted disruption of hoxd-11. *Development* 120: 2187–2198

58 Mortlock DP, Post LC, Innis JW (1996) The molecular basis of hypodactily (*Hd*): a deletion in *Hoxa* 13 leads to arrest of digital arch formation. *Nat Genet* 13: 284–289

59 Favier B, Rijli FM, Fromental-Romain C, Fraulob V, Chambon P, Pascal D (1996) Functional cooperation between the non-paralogous genes *Hoxa-10* and *Hoxd-11* in the developing forelimb and axial skeleton. *Development* 122: 449–460

60 Rountree R, Schoor M, Kingsley D (2000) Using Gdf5 control sequences to test the role of genes in joint development. *International Conference Bone Morphogenetic Proteins 2000* June 7–11, 2000, Granlibakken, Lake Tahoe, California

61 Sugiura T, Hötten G, Kawai S (1999) Minimal promoter components of the human growth/differentiation factor-5 gene. *Biochem Biophys Res Commun* 263: 707–713

62 Persson M (1983) The role of movements in the development of sutural and diarthrodial joints tested by long-term paralysis of chick embryos. *J Anat* 137: 591–599

63 Mitrovic D (1971) [Effect of pharmacological paralysis on the formation and evolution

of articular fissures of the digital joints in chick embryo fleet] Effet de la paralysie phar-macologique sur la formation et l'évolution des fentes articulaires des articulations dig-itales des pattes chez l'embryon de poulet. *CR Acad Sci Hebd Seances Acad Sci D* 273: 1748–1751

64   Mitrovic D (1982) Development of the articular cavity in paralysed chick embryos and in chick embryo limb buds cultured in chorioallantoic membranes. *Acta Anatomica* 113: 313–324

65   Drachman DB, Sokoloff L (1966) The role of movement in embryonic joint develop-ment. *Dev Biol* 14: 401–420

66   Lelkes G (1958) Experiments *in vitro* on the role of movement in the development of joints. *J Embryol Exp Morphol* 6: 183–186

67   Drachman DB, Sokoloff L (1966) The role of movement in embryonic joint develop-ment. *Dev Biol* 14: 401–420

68   Mitrovic D (1971) [Physiological necrosis in the articular mesenchyma of rat and chick embryos] La necrose physiologique dans le mesenchyme articulaire des embryons de rat et de poulet. *CR Acad Sci Hebd Seances Acad Sci D* 273: 642–645

69   Mitrovic D (1972) [Presence of degenerated cells in the developing articular cavity of the chick embryo] Presence de cellules degenerees dans la cavite articulaire en developpe-ment chez l'embryon de poulet. *CR Acad Sci Hebd Seances Acad Sci D* 275: 2941–2944

70   Mori C, Nakamura N, Kimura S, Irie H, Takigawa T, Shiota K (1995) Programmed cell death in the interdigital tissue of the fetal mouse limb is apoptosis with DNA fragmen-tation. *Anat Rec* 242: 103–110

71   Craig FM, Bayliss MT, Bentley G, Archer CW (1990) A role for hyaluronan in joint development. *J Anat* 171: 17–23

72   Hasty P, Bradley A, Morris JH, Edmonson DG, Venuti JM, Olson EN (1993) Muscle deficiency and neonatal death in mice with a targeted mutation in the myogenin gene. *Nature* 364: 501–506

73   Pitsillides AA, Archer CW, Prehm P, Bayliss MT, Edwards JC (1995) Alterations in hyaluronan synthesis during developing joint cavitation. *J Histochem Cytochem* 43: 263–273

74   Nalin AM, Greenlee TK Jr, Sandell LJ (1995) Collagen gene expression during develop-ment of avian synovial joints: transient expression of types II and XI collagen genes in the joint capsule. *Dev Dyn* 203: 352–362

75   Ganan Y, Macias D, Duterque Coquillaud M, Ros MA, Hurle JM (1996) Role of TGF beta s and BMPs as signals controlling the position of the digits and the areas of inter-digital cell death in the developing chick limb autopod. *Development* 122: 2349–2357

76   Zou H, Niswander L (1996) Requirement for BMP signaling in interdigital apoptosis and scale formation. *Science* 272: 738–741

77   Merino R, Macias D, Ganan Y, Economides AN, Wang X, Wu Q, Stahl N, Sampath KT, Varona P, Hurle JM (1999) Expression and function of Gdf-5 during digit skeletogene-sis in the embryonic chick leg bud. *Dev Biol* 206: 33–45

78   Vortkamp A, Pathi S, Peretti GM, Caruso EM, Zaleske DJ, Tabin CJ (1998) Recapitu-

lation of signals regulating embryonic bone formation during postnatal growth and in fracture repair. *Mech Dev* 71: 65–76

79   Liu Z, Luyten FP, Lammens J, Dequeker J (1999) Molecular signaling in bone fracture healing and distraction osteogenesis. *Histol Histopathol* 14: 587–595

80   Dell'Accio F, De Bari C, Luyten FP (1999) Molecular basis of joint development. *Jpn J Rheumatol* 9: 17–29

81   Erlacher L, Ng CK, Ullrich R, Krieger S, Luyten FP (1998) Presence of cartilage-derived morphogenetic proteins in articular cartilage and enhancement of matrix replacement *in vitro*. *Arthritis Rheum* 41: 263–273

82   Chubinskaya S, Merrihew C, Cs-Szabo G, Mollenhauer J, McCartney J, Rueger DC, Kuettner KE (2000) Human articular chondrocytes express osteogenic protein-1. *J Histochem Cytochem* 48: 239–250

83   Marinova-Mutafchieva L, Taylor P, Funa K, Maini RN, Zvaifler NJ (2000) Mesenchymal cells expressing bone morphogenetic protein receptors are present in the rheumatoid arthritis joint. *Arthritis Rheum* 43: 2046–2055

84   Yazaki Y, Matsunaga S, Onishi T, Nagamine T, Origuchi N, Yamamoto T, Ishidou Y, Imamura T, Sakou T (1998) Immunohistochemical localization of bone morphogenetic proteins and the receptors in epiphyseal growth plate. *Anticancer Res* 18: 2339–2344

85   Sakou T, Onishi T, Yamamoto T, Nagamine T, Sampath T, Ten Dijke P (1999) Localization of Smads, the TGF-beta family intracellular signaling components during endochondral ossification. *J Bone Miner Res* 14: 1145–1152

86   Serra R, Johnson M, Filvaroff EH, LaBorde J, Sheehan DM, Derynck R, Moses HL (1997) Expression of a truncated, kinase-defective TGF-beta type II receptor in mouse skeletal tissue promotes terminal chondrocyte differentiation and osteoarthritis. *J Cell Biol* 139: 451–452

87   Benya PD, Shaffer JD (1982) Dedifferentiated chondrocytes reexpress the differentiated collagen phenotype when cultured in agarose gels. *Cell* 30: 215–224

88   Vukicevic S, Luyten FP, Reddi AH (1989) Stimulation of the expression of osteogenic and chondrogenic phenotypes *in vitro* by osteogenin. *Proc Natl Acad Sci USA* 86: 8793–8797

89   Carrington JL, Chen P, Yanagishita M, Reddi AH (1991) Osteogenin (bone morphogenetic protein-3) stimulates cartilage formation by chick limb bud cells *in vitro*. *Dev Biol* 146: 406–415

90   Chen P, Carrington JL, Hammonds RG, Reddi AH (1991) Stimulation of chondrogenesis limb bud mesoderm cells by recombinant human bone morphogenetic protein 2B (BMP-2B) and modulation by transforming growth factor beta 1 and beta 2. *Exp Cell Res* 195: 509–515

91   Flechtenmacher J, Huch K, Thonar EJ, Mollenhauer JA, Davies SR, Schmid TM, Puhl W, Sampath TK, Aydelotte MB, Kuettner KE (1996) Recombinant human osteogenic protein 1 is a potent stimulator of the synthesis of cartilage proteoglycans and collagens by human articular chondrocytes. *Arthritis Rheum* 39: 1896–1904

92   Luyten FP, Yu YM, Yanagishita M, Vukicevic S, Hammonds RG, Reddi AH (1992) Nat-

ural bovine osteogenin and recombinant human bone morphogenetic protein-2B are equipotent in the maintenance of proteoglycans in bovine articular cartilage explant cultures. *J Biol Chem* 267: 3691–3695

93 Luyten FP, Chen P, Paralkar V, Reddi AH (1994) Recombinant bone morphogenetic protein-4, transforming growth factor-beta 1, and activin A enhance the cartilage phenotype of articular chondrocytes *in vitro*. *Exp Cell Res* 210: 224–229

94 Harrison ET Jr, Luyten FP, Reddi AH (1991) Osteogenin promotes reexpression of cartilage phenotype by dedifferentiated articular chondrocytes in serum-free medium. *Exp Cell Res* 192: 340–345

95 Harrison ET Jr, Luyten FP, Reddi AH (1992) Transforming growth factor-beta: its effect on phenotype reexpression by dedifferentiated chondrocytes in the presence and absence of osteogenin. *Cell Dev Biol* 28A: 445–448

96 Pittenger MF, Mackay AM, Beck SC, Jaiswal RK, Douglas R, Mosca JD, Moorman MA, Simonetti DW, Craig S, Marshak DR (1999) Multilineage potential of adult human mesenchymal stem cells. *Science* 284: 143–147

97 Nakahara H, Goldberg VM, Caplan AI (1991) Culture-expanded human periosteal-derived cells exhibit osteochondral potential *in vivo*. *J Orthop Res* 9: 465–476

98 De Bari C, Dell'Accio F, Luyten FP (2000) Human periosteum-derived cells maintain phenotypic stability and chondrogenic potential throughout expansion regardless of donor age. *Arthritis Rheum* 44: 85–95

99 De Bari C, Dell'Accio F, Tylzanowski P, Luyten FP (2001) Multipotent mesenchymal stem cells from adult human synovial membrane. *Arthritis Rheum* 44: 1928–1942

100 De Bari C, Dell'Accio F, Vandenabeele F, Vermeesch JR, Raymackers JM, Luyten FP (2003) Skeletal muscle repair by adult human mesenchymal stem cells from synovial membrane. *J Cell Biol* 160: 909–918

101 De Bari C, Dell'Accio F, Luyten FP (2004) Failure of *in vitro*-differentiated mesenchymal stem cells from the synovial membrane to form ectopic stable cartilage *in vivo*. *Arthritis Rheum* 50: 142–150

102 Wolfman NM, Hattersley G, Cox K, Celeste AJ, Nelson R, Yamaji N, Dube JL, DiBlasio-Smith E, Nove J, Song JJ et al (1997) Ectopic induction of tendon and ligament in rats by growth and differentiation factors 5, 6, and 7, members of the TGF-beta gene family. *J Clin Invest* 100: 321–330

103 Aspenberg P, Forslund C (1999) Enhanced tendon healing with GDF 5 and 6. *Acta Orthop Scand* 70: 51–54

104 Lories RJ, Derese I, Ceuppens JL, Luyten FP (2003) Bone morphogenetic proteins 2 and 6, expressed in arthritic synovium, are regulated by proinflammatory cytokines and differentially modulate fibroblast-like synoviocyte apoptosis. *Arthritis Rheum* 48: 2807–2818

105 Adams ME, Brandt KD (1991) Hypertrophic repair of canine articular cartilage in osteoarthritis after anterior cruciate ligament transection. *J Rheumatol* 18: 428–435

106 van der Kraan PM, Vitters EL, van Beuningen HM, van den Berg WB (1992) Proteoglycan synthesis and osteophyte formation in "metabolically" and "mechanically"

induced murine degenerative joint disease: an *in-vivo* autoradiographic study. *Int J Exp Pathol* 73: 335–350

107 Serra R, Johnson M, Filvaroff EH, LaBorde J, Sheehan DM, Derynck R, Moses HL (1997) Expression of a truncated, kinase-defective TGF-beta type II receptor in mouse skeletal tissue promotes terminal chondrocyte differentiation and osteoarthritis. *J Cell Biol* 139: 541–552

108 von der Mark K, Kirsch T, Nerlich A, Kuss A, Weseloh G, Gluckert K, Stoss H (1992) Type X collagen synthesis in human osteoarthritic cartilage. Indication of chondrocyte hypertrophy. *Arthritis Rheum* 35: 806–811

109 Weinstein M, Yang X, Deng C (2000) Functions of mammalian Smad genes revealed by targeted gene disruption in mice. *Cytokine Growth Factor Rev* 11: 49–58

110 van Beuningen HM, Glansbeek HL, van der Kraan PM, van den Berg WB (1998) Differential effects of local application of BMP-2 or TGF-beta 1 on both articular cartilage composition and osteophyte formation. *Osteoarthritis Cartilage* 6: 306–317

111 Glansbeek HL, van Beuningen HM, Vitters EL, Morris EA, van der Kraan PM, van den Berg WB (1997) Bone morphogenetic protein 2 stimulates articular cartilage proteoglycan synthesis *in vivo* but does not counteract interleukin-1alpha effects on proteoglycan synthesis and content. *Arthritis Rheum* 40: 1020–1102

112 van Beuningen HM, van der Kraan PM, Arntz OJ, van den Berg WB (1994) *in vivo* protection against interleukin-1-induced articular cartilage damage by transforming growth factor-beta 1: age-related differences. *Ann Rheum Dis* 53: 593–600

113 Tomita T, Nakase T, Kaneko M, Tsuboi H, Takahi K, Hashimoto J, Takano H, Myoui A, Shi K, Yoshikawa H, Ochi T (2000) Distributions of BMP-2 and BMP receptors in the osteophyte of patients with osteoarthritis. *Arthritis Rheum* 43: S350

114 Zoricic S, Maric I, Bobinac D, Vukicevic S (2003) Expression of bone morphogenetic proteins and cartilage-derived morphogenetic proteins during osteophyte formation in humans. *J Anat* 202: 269–277

115 Erlacher L, Ng CK, Ullrich R, Krieger S, Luyten FP (1998) Presence of cartilage-derived morphogenetic proteins in articular cartilage and enhancement of matrix replacement *in vitro*. *Arthritis Rheum* 41: 263–273

116 Chubinskaya S, Merrihew C, Cs-Szabo G, Mollenhauer J, McCartney J, Rueger DC, Kuettner KE (2000) Human articular chondrocytes express osteogenic protein-1. *J Histochem Cytochem* 48: 239–250

117 Scharstuhl A, Vitters EL, van der Kraan MP, van den Berg WB (2003) Reduction of osteophyte formation and synovial thickening by adenoviral overexpression of transforming growth factor beta/bone morphogenetic protein inhibitors during experimental osteoarthritis. *Arthritis Rheum* 48: 3442–3451

118 Braun J, Bollow M, Neure L, Seipelt E, Seyrekbasan F, Herbst H, Eggens U, Distler A, Sieper J (1995) Use of immunohistologic and *in situ* hybridization techniques in the examination of sacroiliac joint biopsy specimens from patients with ankylosing spondylitis. *Arthritis Rheum* 38: 499–505

119 Fukui N, Zhu Y, Maloney WJ, Clohisy J, Sandell LJ (2003) Stimulation of BMP-2

expression by pro-inflammatory cytokines IL-1 and TNF-alpha in normal and osteo-arthritic chondrocytes. *J Bone Joint Surg Am* 85-A (Suppl 3): 59–66

120 Bhatia M, Bonnet D, Wu D, Murdoch B, Wrana J, Gallacher L, Dick JE (1999) Bone morphogenetic proteins regulate the developmental program of human hematopoietic stem cells. *J Exp Med* 189: 1139–1148

121 Yoshikawa H, Takaoka K, Hamada H, Ono K (1985) Clinical significance of bone mor-phogenetic activity in osteosarcoma. A study of 20 cases. *Cancer* 56: 1682–1687

122 Yoshikawa H, Rettig WJ, Takaoka K, Alderman E, Rup B, Rosen V, Wozney JM, Lane JM, Huvos AG, Garin-Chesa P (1994) Expression of bone morphogenetic proteins in human osteosarcoma. Immunohistochemical detection with monoclonal antibody. *Cancer* 73: 85–91

123 Yoshikawa H, Takaoka K, Masuhara K, Ono K, Sakamoto Y (1988) Prognostic signif-icance of bone morphogenetic activity in osteosarcoma tissue. *Cancer* 61: 569–573

124 Guo W, Gorlick R, Ladanyi M, Meyers PA, Huvos AG, Bertino JR, Healey JH (1999) Expression of bone morphogenetic proteins and receptors in sarcomas. *Clin Orthop* 175–183

125 Mehdi R, Shimizu T, Yoshimura Y, Gomyo H, Takaoka K (2000) Expression of bone morphogenetic protein and its receptors in osteosarcoma and malignant fibrous histio-cytoma. *Jpn J Clin Oncol* 30: 272–275

126 Sulzbacher J, Birner P, Trieb K, Pichlbauer E, Lang S (2002) The expression of bone morphogenetic proteins in osteosarcoma and its relevance as a prognostic parameter. *J Clin Pathol* 55: 381–385

The role of bone morphogenetic proteins in developing and adult kidney

Fran Borovecki, Petra Simic, Lovorka Grgurevic and Slobodan Vukicevic

Department of Anatomy, School of Medicine, University of Zagreb, Salata 11, Zagreb 10000, Croatia

Introduction

The kidney has been identified as a major site of BMP-7 (BMP-7) synthesis during embryonal and postnatal development [1–4]. Gene knockout [5, 6] and *in vitro* experiments [4, 7] demonstrated the importance of BMP-7 in kidney development. Many developmental features are recapitulated during renal injury, and BMPs may be important in both preservation of function and resistance to injury [8, 9]. BMP-7 has a cytoprotective and anti-inflammatory effect in both acute and chronic renal failure [8, 9].

Kidney development

The kidney development commences with lateral outgrowth of the epithelial bud from the Wolfian duct. This event occurs at the 5th week of fetal gestation in humans or day 11 of gestation in embryonic mice (E11) [10]. The outgrowth of the ureteric bud is induced by interaction with the surrounding metanephric mesenchyme. Signals secreted by the metanephric mesenchyme propagate elongation of the ureteric bud and invasion of the metanephric mesenchyme, which in turn induces the mesenchyme to establish two cell fates – the stromal progenitor cells and the nephrogenic mesenchyme. Interaction with the ureteric bud promotes survival of the nephrogenic mesenchyme and induces the mesenchymal-epithelial transition which occurs in these cells. Subsequent formation of comma- and S-shaped bodies around the tip of the growing and branching ureteric bud leads to the formation of the glomerulus, proximal tubule, loop of Henle and distal tubule (Fig. 1). Finally, the fusion of the distal developing metanephric tubule with a ureteric bud branch completes formation of the nephron. The nephrogenic mesenchyme, on the other hand, during its transformation induces elongation of the ureteric bud resulting in the development of the renal collecting system.

Bone Morphogenetic Proteins: Regeneration of Bone and Beyond, edited by Slobodan Vukicevic
and Kuber T. Sampath
© 2004 Birkhäuser Verlag Basel/Switzerland

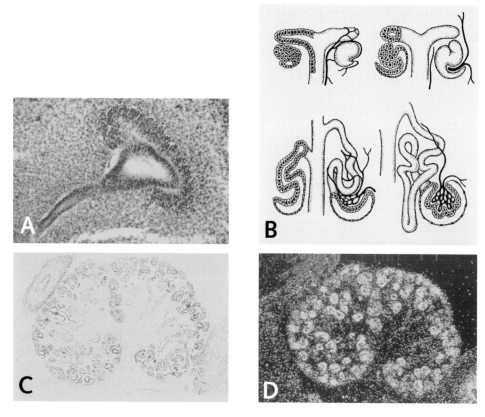

Figure 1
Schematic description of various stages in the development of the nephron
As a result of inductive interaction between the ureteric bud and the metanephric mesenchyme (magnification × 100) (A), a condensate is formed (human embryo five weeks of
gestation). It goes through the comma-shape (B) and S-shape body stages. This is followed
by tubule elongation and its connection with the nephric duct. When the blood vessels
invade the distal curve of the S-shaped body, the future mature glomeruli begin to form. (C)
Toluidin blue-stained bright field image of a section through the developing human (6 w)
kidney. (D) Dark field image of the same section with an antisense mRNA probe indicate
synthesis in kidney glomeruli, magnification × 12.5.

The reciprocal induction was documented by *in vitro* experiments when the
ureteric bud and the metanephric mesenchyme were cultured separately [11, 12].
The ureteric bud does not branch in the absence of the mesenchyme, and the mesenchyme dies without the ureteric bud. Although certain tissues (such as neural

214

tube, spinal cord and salivary glands) enable the metanephric mesenchyme to form kidney tubules, the ureteric bud branches only under instructions from the metanephric mesenchyme [10]. However, the extrinsic influences, namely growth factors and proto-oncogenes, control the proliferation and differentiation of the metanephric cells. Recent experiments have shown that soluble factors in the correct combination are capable of inducing growth and branching of the isolated ureteric bud.

Extensive studies of mammalian kidney development have revealed that several transcription factors play a pivotal role in the initial outgrowth of the ureteric bud from the Wolfian duct. These include Wilms tumor suppressor gene (WT-1) [13, 14], Pax2 [15, 16], Lim1 [17], Sall1 [18] and Eya1 [19]. The nuclear protein Formin [20] and the peptide growth factor Gdnf [21–23], as well as its receptor complex consisting of the receptor tyrosin kinase Ret [24, 25] and its co-receptor Gfrα1 [26], are also important in the initiation of nephrogenesis. Prior to the onset of ureteric bud outgrowth, Pax2, Lim1, Formin and Ret are expressed in the murine Wolfian duct, while Pax2, Sall1 and Gdnf are detected in the metanephric mesenchyme surrounding the Wolfian duct. At the initial stages of ureteric bud outgrowth, Wt1 is expressed in the metanephric mesenchyme [27]. Following the induction of the ureteric bud outgrowth and throughout nephrogenesis, Eya1, Lim1 and Formin are additionally expressed in induced mesenchyme and its epithelial derivatives, whereas Wt1 and Pax2 expression progressively diminishes as the nephrogenesis continues. Pax2 and Lim1 expression is maintained in ureteric bud/collecting duct branches continuously during renal development, whereas Ret expression becomes spatially restricted to the branch tips. Pax-2 is necessary for the mesenchymal aggregation and mesenchyme-to-epithelial transition during nephrogenesis, and it disappears after terminal differentiation of nephrons [15, 28]. WT-1 is a negative regulator of Pax-2 during kidney development [29]. Its expression is elevated in a variety of renal tumors [30]. Homozygous deletion in any of the aforementioned genes in mice causes failure of ureteric bud outgrowth and results in bilateral renal agenesis or severe renal dysgenesis with variable penetrance depending on the gene involved. Several secreted peptide growth factors have been shown to exert effects on collecting duct branching after the initial stages of ureteric bud morphogenesis. Members of the fibroblast growth factor (FGF) family are involved in controlling collecting duct morphogenesis in mice. One member of the family, FGF7, is expressed in stromal cells surrounding the ureteric bud [31] and treatment of whole rat embryonic kidney explants *ex vivo* with recombinant FGF7 induced expansion of the ureteric bud cell population in these tissues [32], indicating that FGF7 could play a role in the regulation of collecting duct cell proliferation. Homozygous null mutant mice exhibit a reduced number of ureteric bud branches and nephrogenic tubules at early stages of renal development, while at later stages demonstrate insufficiently developed renal papillae. Another member of the FGF family, FGF10, is known to be essential for lung branching morphogenesis in mice [33], but FGF10 knockout mice also possess smaller kidneys than wild-type animals with decrease in number of medullary collecting ducts, medullary dysplasia and dilatation of the renal pelvis. FGF2 induces full epithelial differentiation of isolated rat metanephric mesenchyme, if it is combined with a conditioned medium from ureteric bud cell cultures, or if FGF-2 pulse is followed by leukemia inhibitory factor (LIF) treatment [34]. FGFs and BMPs act synergistically during nephrogenesis to maintain competence of the metanephric mesenchyme. It has been shown that LIF and members of the IL-6 family, including cardiotrophin, oncostatin and ciliary neurotrophic factor (CNTF) are expressed in the ureter and can induce nephrogenesis in culture. This possibly explains why the LIF knockout has no obvious kidney phenotype [35]. Hepatocyte growth factor (HGF) and epidermal growth factor (EGF) stimulate branched tubule formation [36, 37], while activin A and TGF-β inhibit branched tubule formation in three-dimensional cultures [38–40]. Several growth factors, including GDNF, FGF7 and TGF-β have been shown to promote the expression of tissue inhibitors of metalloproteinases (TIMPs) [41]. TIMPs regulate the local proteolytic activity of secreted, zinc-requiring, extracellular matrix (ECM) proteinases called matrix metalloproteinases (MMPs), which alter the composition of ECM proteins, and thus modulate interactions between the ECM, cells and signaling molecules. Addition of exogenous TIMPs inhibits ureteric bud branching in cultured kidney explants [42], which indicates that growth factors might play a role in regulating the local activity of matrix-degrading proteases by controlling TIMP expression.

Among the growth factors whose importance has been demonstrated in kidney development, BMPs have been shown to play a significant role in embryonic nephrogenesis [43, 44], especially in renal branching morphogenesis and induction of metanephric mesenchyme. It has been shown that BMP-4 and BMP-5 provide essential roles in collecting duct system development *in vivo* [45, 46]. *In vivo* and *in vitro* studies have also indicated that BMP-2 and BMP-7 act as regulators of renal branching morphogenesis. In a manner similar to BMP-4 and BMP-5, BMP-2 and BMP-7 mRNA is expressed in the developing kidney (Fig. 2) in a temporal and spatial pattern that is consistent with roles in inductive tissue interactions during nephrogenesis [47].

BMP knock-out mice and kidney development

Mice lacking the BMP-7 gene died of uremia within 24 hrs following birth. One group reported the absence of tubules and immature glomeruli apparatus (S- and comma-shaped bodies) following the ingrowth of the ureteric bud into the metanephric mesenchyme in E-11 mice, suggesting that BMP-7 is necessary for the induction of the E-11 mesenchyme [6]. Another BMP-7 knockout phenotype suggested that unaltered kidney development progressed up to E-14 in BMP-7 null mice, which was, however, followed by a rapid disappearance of the metanephric

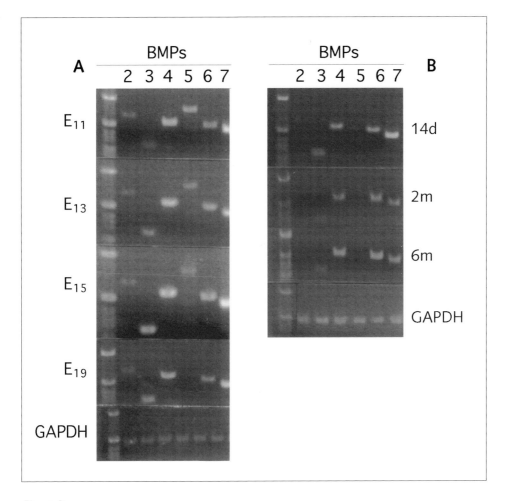

Figure 2

Expression of bone morphogenetic proteins in developing and adult mouse kidneys

(A) Whole kidney RNA was isolated, cDNA was synthesized and analyzed by RT-PCR. Reactions without cDNA were used as a negative control. GAPDH was used to normalize the reaction. At E11 of mouse development, BMPs 2-7 are expressed, with BMP-4 and BMP-7 being most abundant. BMP-3 and BMP-6 are gradually upregulated, while BMP-5 expression declines from E11 towards E19. (B) After two weeks, 2 and 6 months following delivery, BMP-7 is strongly expressed, while BMP-2 and BMP-5 appear low.

mesenchyme resulting in loss of kidney mass upon birth [5]. While this apparent discrepancy can be attributed to the variance observed in mouse genetics, the precise role of BMP-7 in metanephric differentiation remains unknown.

After the initial induction, BMP-7 and Wnt-4 are required for subsequent mesenchymal differentiation by maintaining the inductive response. Wnt-4 is a cysteine-rich signaling molecule expressed in pretubular cells of the metanephric mesenchyme at the base of the ureteric bud. Its expression is absolutely necessary for kidney development and is lost upon fusion of nephron with the collecting duct [48]. As cell proliferation and differentiation proceed, more and more molecules are involved in the regulation. BF-2 is the "winged helix" transcription factor expressed in stromal cells. It is necessary for regulation of the nephrogenesis in the induced cell population that is destined to make epithelium [49]. In mice lacking PDGF B or its receptor PDGFRβ, mesangial cells are absent thus disabling formation of the glomeruli [50].

BMP-7 is expressed in several tissues associated with inductive interactions and is required for proper nephrogenesis using gene targeting in mice [5, 6]. BMP-7 mRNA expression is the highest on day 13 of kidney development, which corresponds with its proposed role in nephrogenesis. In the normal kidney, the highest expression of BMP-7 mRNA could be seen in tubules of the outer medulla, in cells at the periphery of the glomerular tuft, adventitia of renal arteries and epithelial cells of the renal pelvis and the ureter [7]. During development, BMP-7 transcripts are most abundantly present, first, in the epithelium of the branching ureteric buds, and later in the glomeruli [1]. Most of the homozygous animals die the first postnatal day from acute renal failure. Their kidneys failed to develop normally, and they also have microopthalmia and various degrees of skeletal deformities. The kidney starts to develop, reciprocal interactions occur, but further development ceases by approximately 14 days *post coitum* accompanied by extensive apoptosis. Glomeruli and proximal convoluted tubules are well developed, so it seems that BMP-7 is absolutely necessary for development of distal convoluted tubules and maintenance of the kidney structure. Multiple cysts are observed in the kidneys of animals that survived for a few days [47]. In the central nervous system (CNS) and heart of the mutant animals, expression domains of the BMP family members completely overlap with that of BMP-7. It seems that at such places other BMP family members can substitute for BMP-7 [51].

It has been demonstrated that during kidney development, high doses of BMP-7 inhibit branching morphogenesis, whereas low doses are stimulatory [40]. Another study [52] showed that BMP-7 suppresses tubulogenesis and, in synergy with FGF–2, increases the cell population of stromal precursor cells in the developing kidney. These results indicate an important function for BMP-7 in the maintenance of blastemal tissue and hence the continuous growth of the kidney during development. In cultured embryonic kidneys, BMP-7 mRNA expression was demonstrated in several glomerular cell types, such as mesangial, epithelial and endothelial cells. Distal tubule MDCK cells also expressed BMP-7 mRNA but human proximal tubule HK-2 cells did not. Treatment with BMP-7 increased cellular proliferation of HK-2 cells, but not of the mesangial cells. These results suggest that BMP-7 is produced in the renal glomerulus and then travels to the proximal tubule to regulate the proliferation of the cells in this region of the nephron [52]. BMP-7 expression in the epithelial components of the kidney is not dependent on cell–cell or cell-BMP-7 interactions with the metanephric mesenchyme. Disruption of proteoglycan synthesis results in the loss of BMP-7 in the mesenchyme. It seems that BMP-7 expression in the metanephric mesenchyme is dependent on proteoglycans and proper protein glycosylation [53]. The current data support a model in which signaling from the ureter induces metanephric expression of Pax2 and WT-1. They subsequently activate the signaling molecules BMP-7 and Wnt-4, which promote tubulogenesis and expression of stromal precursor cells. Several other BMPs are expressed during kidney development and in the postnatal life.

BMP-4 is expressed in mesenchymal cells surrounding the Wolffian duct and the ureter stalk. It is important in the early morphogenesis of the kidney and urinary tract. It inhibits ectopic budding from the Wolffian duct or the ureter stalk by antagonizing inductive signals from the metanephric mesenchyme to the illegitimate sites on the Wolffian duct. Another function is to promote the elongation of the branching ureter within the metanephros. BMP-4 signaling can substitute for the surface ectoderm in supporting nephric duct morphogenesis [54]. BMP-4 heterozygous null-mutant mice display abnormalities of the genitourinary tract including hypoplasic kidneys, hydroureter, ectopic ureterovesical junction and double collecting system ([46, 55]; see the chapter by Martinovic et al.). In the organ culture of developing kidney, human recombinant BMP-4 diminishes the number of ureteric branches and changes the branching pattern *via* interfering with the differentiation of the metanephric mesenchyme [56]. In BMP-7 null-mutant mice, BMP-4 is expressed in the mesenchyme surrounding the ureteric bud in the early stages of development, then in the area of nephron development, and finally, its expression is limited to the Bowmann capsule [47]. Its expression reaches maximal value from day 15 to 17 of embryonal development suggesting its role in tubulogenesis.

Although BMP-2 and BMP-4 share 92% homology in amino acid sequence, mRNA transcripts for BMP-2 and BMP-4 are expressed in distinct domains within developing renal tissues at several stages of nephrogenesis [47]. Early embryonic lethality in BMP-2$^{-/-}$ mice and the normal development of the kidneys in BMP-2$^{+/-}$ mice represent a challenge in the quest to elucidate the function of BMP-2 in kidney development *in vivo* [57]. BMP-2 has been shown to inhibit renal branching morphogenesis in cultured embryonic kidney explants and in the mIMCD-3 cell culture model of collecting duct morphogenesis [40]. In the aforementioned models, BMP-2 inhibits cell proliferation and stimulates apoptosis in a Smad-dependent manner [58]. These findings suggest that a BMP-dependent pathway may act *in vivo* to inhibit branching morphogenesis, thereby providing spatial regulation of this process. BMP-2 and HGF function to control parallel pathways downstream of their respective cell surface receptors regulating the collecting duct morphogenesis [59]. In mesangial cells, BMP-2 inhibits PDGF-induced DNA synthesis and c-fos gene transcription [60]. BMP-2 expression is persistent during intrauterine and post-natal kidney development, while its expression is downregulated in adult kidneys.

Osteogenin (BMP-3) is mainly synthesized in developing lung and kidney [61]. In normal rat kidneys, BMP-3 mRNA expression is limited to areas of tubule development, and is not found in the glomeruli [62]. On the contrary, Dudley and Robertson have found BMP-3 mRNA in the glomerular area of the future nephron in BMP-7 null-mutant mice [47]. Gradually, BMP-3 mRNA expression is upregulated from day 13 to 17 of embryonal development, and then decreases. BMP-3 knockout mice do not have kidney abnormalities (see the chapter by Martinovic et al.).

BMP-5 expression is demonstrated in the cell layer adjacent to epithelial cells of the ureteric bud and in renal calices of the more mature kidneys in BMP-7 null-mutant mice [47]. In normal mouse embryos, BMP-5 expression is found in mesenchymal cells surrounding the ureter, but also in the renal calices at later stages of development. BMP-5 mRNA is expressed in mice embryonal kidneys from day 12 to day 17 kidney during the postnatal life. From the beginning of kidney development BMP-6 expression is upregulated, and the highest level is found in mature, adult kidneys.

BMP receptors in kidney development

In situ hybridization histochemistry experiments were used to determine localization of mRNAs for BMP type-I receptors (BMPR-IA and BMPR-IB) and the BMP type-II receptor (BMPR-II) in developing mouse metanephroi [63]. These experiments have shown that at E12.5 and E14.5 BMPR-IA and BMPR-IB are expressed in the tips and body of the branching ureter as well as mesenchymal condensates, developing vesicles and comma-shaped bodies. Localization of BMPR-II transcripts was similar, although expression of this receptor was not observed in the body of the ureter. At E17.5, transcripts for all of these BMP receptors were localized in the nephrogenic zone, including ureteric tips, vesicles, comma- and S-shaped bodies as well the body of the ureter and in tubules. The observed patterns of expression suggest that BMPs are important regulators of epithelial–mesenchymal interactions, nephron development and ureteric branching morphogenesis.

The role of Smads, the transcription factors that act as downstream signaling molecules which translate TGF-β signals into gene expression has not been studied extensively. RT-PCR data and *in situ* hybridization analysis showed that the receptor-regulated (R) Smads (Smad1, 2, 3, 5, and 8), the common partner Smad (Smad4), and the inhibitory (I) Smads (Smad6 and 7) were all expressed during mouse kidney development from E12 until the end of nephrogenesis at postnatal day 15 [64]. Although each Smad had a distinct spatial distribution during development, all were expressed by mesenchymal cells in the nephrogenic zone and were downregulated once these cells began to epithelialize. Smad4 was detected in uninduced mesenchymal cells and at ureteric bud tips. The bone morphogenetic-responsive R-Smads, Smad1, 5, and 8, were mainly expressed in the nephrogenic zone, whereas the TGF-β-responsive R-Smads were predominantly noted in the medullary interstitium. Expression of the I-Smad Smad7 was also seen in mesenchymal cells in the interstitium. The observed patterns of expression suggest that individual Smad or combinations of Smads play specific roles in cell-fate determination during kidney development.

BMP antagonists in kidney development

The role of specific BMP antagonists in kidney development has yielded limited advancements in understanding their possible role in nephrogenesis. Gremlin belongs to the family of BMP antagonists that includes the head-inducing factor Cerebrus and the tumor suppressor DAN. Gremlin is a secreted protein that blocks BMP-2, BMP-4 and BMP-7 signaling by binding BMPs and preventing them from interacting with their receptors [65]. Experiments using *Xenopus laevis* have shown the expression of gremlin can be detected in neural crest cell lineages and in the developing pronephros, suggesting a role in nephrogenesis. Furthermore, mice with targeted disruption of gremlin exhibit no apparent malformations during kidney development [66]. However, recent studies have shown that gremlin plays an important role in pathogenesis of diabetic nephropathy [67–69]. Recently, uterine sensitization-associated gene-1 (USAG-1), previously reported as gene of unknown function predominately expressed in sensitized endometrium of the rat uterus, has been shown to be abundantly expressed in the kidney and to function as a BMP antagonist [70]. Recombinant USAG-1 binds directly to BMPs and inhibits their actions. It is expressed in the developing metanephros at early stages of embryonal development. At later stages of embryogenesis, its expression is confined to renal tubules, while in adult kidneys its expression is further limited to distal tubules of the kidney, in a pattern similar to the localization of BMP-7.

Although BMPs have principally been shown to promote nephrogenesis, increasing evidence is emerging that a BMP-dependent inhibitory pathway is also active during kidney development. Critical importance of BMP-dependent inhibitory pathway during renal branching morphogenesis is becoming more and more evident through studies of the BMP type I receptor, ALK3 and Smad4. Alk3 mRNA is expressed in the metanephric mesenchyme and the ureteric bud [46, 71]. Animals with a mutational deficiency of ALK3 gene exhibit embryonic lethality and thus reveal little insight into the function of Alk3 during kidney development [72]. However, overexpression of a constitutively active ALK3 allele in collecting duct cells strongly inhibits tubule formation *in vitro*, which suggests that ALK3 acts in the BMP-dependent inhibitory pathway [73]. These findings complement the *in vivo* observations that overexpression of the constitutively active ALK3 allele in transgenic mice causes a variable phenotype including reduced ureteric bud branching and cystic malformation of the kidney. Smad4, which is a common mediator of Smad-dependent signaling also exerts an inhibitory role in kidney development [74]. In Smad4[+/-] mice, ureteric bud branching is increased and the inhibitory effect of recombinant BMP-2 is reversed *in vitro*. This suggests a role for Smad4 in the BMP-dependent inhibitory pathway.

The role of BMPs in the repair of adult kidney

Acute kidney failure

The finding that BMP-7 expression remains high both in the fetal and in postnatal life, and is available in the circulation suggests that BMP-7 may have a systemic function and a role in the repair and regeneration of the adult kidney [3, 8].

Acute renal failure represents a clinical condition with persistently high mortality (40–80%), despite technical advances in both critical care medicine and dialysis. The successful treatment of patients with acute renal failure who require dialysis remains one of the greatest challenges facing nephrology today [75]. This condition can be fully understood, and optimal treatment measures defined, only with knowledge of the underlying molecular and structural changes and events.

The damaged kidney is capable of complete repair and regeneration after acute injury and the process recapitulates features that occur during the development. It is assumed that regenerating cells take a step back, towards an earlier ontogenic stage, which makes the cells sensitive to embryonic stimuli [76, 77]. BMP-7 may be important in both preservation of function and resistance to injury [8].

The mechanisms controlling the cascade of cellular migration, growth and proliferation following acute renal failure undoubtedly comprise a number of autocrine and paracrine growth factors [78, 79], such as insulin-like growth factors (IGFs), epidermal growth factor (EGF), fibroblasts growth factor (FGF), transforming growth factors (TGF-α, TGF-β), and hepatocytes growth factor (HGF) [80–83]. Animal studies dealing with acute renal failure due to ischemic-reperfusion insult have indeed proven that administration of hr BMP-7 has a beneficial effect on the extent of injury and the regeneration of kidney function (Fig. 3) [8]. Bioavailability studies have shown that human BMP-7 has a serum half-life of about 30 min, and that significant amounts of ^{125}I-BMP-7 can be found in both kidney cortex and medulla shortly after iv administration [8].

Apart from being protective in ischemic acute renal failure, BMP-7 also influences the course of toxic kidney injury *in vitro*, as well as in acute nephrotoxic animal models utilizing administration of mercuric chloride and cisplatinum [84]. Both prophylactic and therapeutic systemic administration of BMP-7 to rats given mer-

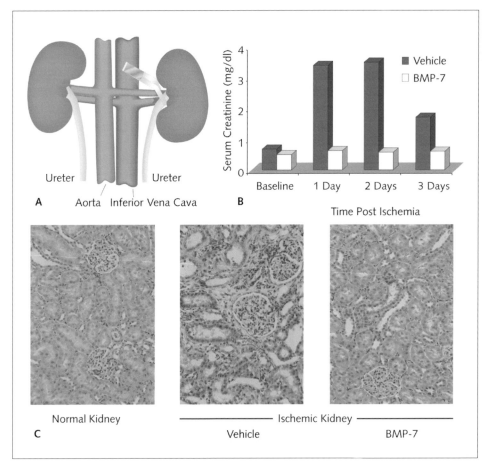

Figure 3
Therapeutic effect of BMP-7 on serum creatinine values and kidney regeneration in a rat model of ischemic acute renal failure
(A) Schematic depiction of the ischemic acute renal failure. (B) Serum creatinine levels in rats treated with vehicle and BMP-7. Vehicle (acetate buffer, pH 4.5) and BMP-7 were administered daily at 24-hour intervals beginning eight hours following 40 min ischemic injury. (data shown as mean ± SEM; P < 0.01, Student's t-test). (C) Histological images of a normal rat kidney, ischemic kidney and an ischemic kidney of rats treated with BMP-7 (magnification × 200).

curic chloride protected the kidney function and significantly extended the survival rate (Fig. 4). Similarly, BMP-7 protected the kidney function in rats treated with a high dose of cisplatinum.

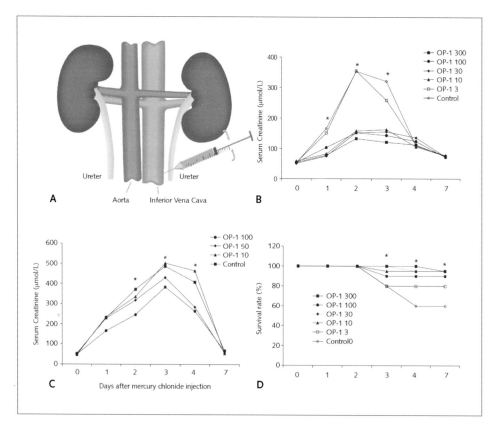

Figure 4
Effect of BMP-7 on the survival rate and serum creatinine values in rats subjected to acute toxic renal failure
*Animals were given mercuric chloride (4 mg/kg) in a bolus at the beginning of the experiment. Vehicle (acetate buffer, pH 4.5) and BMP-7 were administered daily at 24-hour intervals beginning on day 0, 10 minutes before the injection (data shown as mean ± SEM; P < 0.01, Student's t-test), or beginning 8 hours following the injection. (data shown as mean ± SEM; * P < 0.01, Student's t-test)*

Mercuric chloride exerts its toxic effects on kidney cells through a variety of mechanisms, the principle target being S3 segment of proximal tubules. Intracellular pathways contributing to cell damage by mercury are primarily the consequence of its high affinity for sulfhydril groups. Those protein groups are of utmost importance for cell function, since they are both located within active centers of various vital enzymes and they represent one of the main defense mechanisms against oxidative damage [85, 86]. Indeed, increased H_2O_2 production in mitochondria and heme

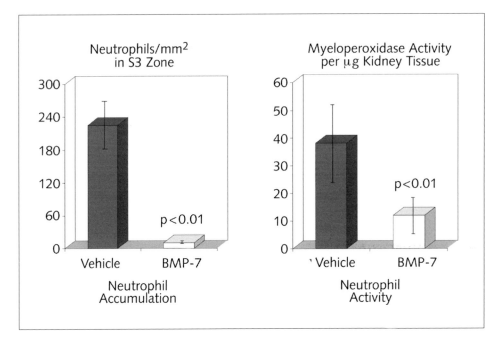

Figure 5
BMP-7 decreases inflammation in the rat model of acute ischemic renal failure Neutrophil accumulation and neutrophil activity, as measured by myeloperoxidase activity per µg of kidney tissue, are significantly decreased in animals treated with BMP-7, when compared to untreated controls with acute ischemic renal failure.

oxygenase induction have been demonstrated both *in vitro* and in tubular cells isolated from rats treated with $HgCl_2$ [87]. Apart from interfering with respiratory chain and oxidative phosphorylation enzymes, mercury was shown in numerous studies to cause oxidative injury with subsequent lipid peroxidation, DNA damage and protein oxidation [88]. Thus, in terms of cytoprotection, multiple pathways may need to be blocked as well, since this toxicant may activate multiple pathways. *In vitro* studies show that BMP-7 significantly promotes cell survival and proliferation in human primary proximal tubule cells treated with mercury chloride, while ineffective in intact cells. In rats with an ischemic-reperfusion kidney damage [8], BMP-7 was shown to ameliorate the course of injury through a variety of mechanisms, including inhibition of apoptosis, minimizing of infarction and cell necrosis and preventing intracellular adhesion molecule (ICAM) expression, thus suppressing the inflammatory response (Fig. 5) [8]. Whether the same mechanisms are responsible for its beneficial effects observed in nephrotoxic studies remains to be elucidated. However, the oxidative damage is a principal cause of cell injury and death in both mercury-induced and ischemic-reperfusion insult to the kidney. Considering the fact that BMP-7 has a characteristic cysteine-rich region in the carboxyterminal part of the polypeptide chain, it is conceivable that it might function as both mercury and/or free radical scavenger. On the other hand, the finding that BMP-7 is effective in promoting the proliferation and viability of renal tubular cells previously injured by mercury *in vitro*, while ineffective in intact cells, points to a difference in sensitivity to external stimuli between regenerating and intact cells. Indeed, the experiments dealing with liver regeneration [89] have shown that hepatocytes first need to be "primed" with either cytokines or reactive oxygen species in order to become fully competent to respond to growth factor stimuli. It is well established that kidney cells have a capacity for repair and function recovery after injury by recapitulating the molecular and cellular events that take place during nephrogenesis [81, 82] very similar to regenerating fractured bone [90, 91]. Since BMP-7 is a morphogenic protein involved in nephrogenesis during the embryogenesis, it may be postulated that injured cells exhibit *de novo* sensitivity to BMP-7 stimulation *in vitro*. During prenatal development of the mouse kidney, BMP-7 mRNA expression is most abundant on day 12, with a slow decline after day 15. It seems that there is a timeframe during nephrogenesis in which the presence of BMP-7 is required for normal kidney development. In nephrogenic mesenchyme tissue explant cultures, BMP-7 was shown to prevent apoptosis [92] and the same effect was observed *in vivo* in ischemic-reperfusion injury [8]. Kidney BMP-7 mRNA and protein are selectively downregulated in the medulla after acute ischemic renal injury (Fig. 6) [8], thus BMP-7 modulation may be a key element for kidney repair [93]. Whether BMP-7 has a direct growth-promoting function either through early genes activation or apoptosis inhibition in damaged tubular cells, or it simply serves as a functional free radical scavenger, remains to be determined. Collectively, these data suggest that BMP-7 reduces the severity of renal damage associated with ischemia/reperfusion and nephrotoxic agents, and, as such, may provide a basis for the treatment of acute renal failure.

Chronic renal failure

Progressive and permanent reduction in the glomerular filtration rate (GFR), which is associated with the loss of functional nephron units, leads to chronic renal failure (CRF).

The subject progresses to end-stage renal disease when the GFR continues to decline to less than 10% of normal values (5–10 ml/min). At this point, renal failure will rapidly progress to cause death unless the subject receives renal replacement therapy, i.e., chronic haemodialysis, continuous peritoneal dialysis or kidney transplantation, or therapy that delays the progression of chronic renal disease.

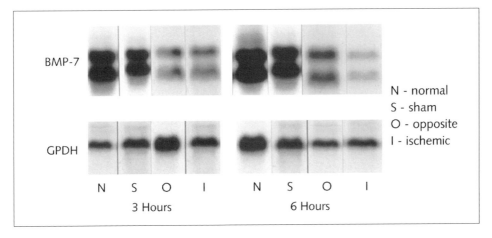

Figure 6
BMP-7 mRNA levels in the kidney in a model of ischemic injury
Northern blot analysis of BMP-7 mRNA expression of normal (N), sham operated (S), oppo-
site (O) and ischemic (I) kidneys in rat model of acute ischemic renal failure. The samples
were collected three and six hours following the surgery. GAPDH was used to normalize the
reaction. BMP-7 mRNA expression is clearly decreased in animals with acute ischemic renal
failure, when compared to normal and sham operated animals at three hours following oper-
ation, both in the ischemic and the opposite kidney. The expression is further decreased at
six hours following surgery in the ischemic kidney.

Remnant kidney model

The effect of systemically administered BMP-7 to delay or halt progression of end stage renal failure in a remnant kidney (5/6 nephrectomy) rat model was investigated (Fig. 7). Recombinant human BMP-7 at doses of 10 μg/kg was administered intravenously three times per week beginning two days following surgery and continued for 11 weeks. The effect of BMP-7 was monitored by serum creatinine values (Cr), glomerular filtration rate (GFR), and the survival rate. The results indicate that two weeks after the beginning of treatment, BMP-7 considerably decreased serum Cr values as compared to control animals. Rats treated with BMP-7 had better GFR and prolonged survival rate.

The higher GFR observed in BMP-7-treated rats and the hystomorphometric analysis (Figs 8 and 9) suggest that BMP-7 is capable of preventing rapid deterioration of the glomerular function in this model. In 18 weeks following nephrectomy, the survival rate was 88% in BMP-7-treated rats as compared to 32% in controls. The experiment was terminated 30 weeks following nephrectomy with 60% survivors in BMP-7-treated and 15% survivors in control rats, respectively. This result

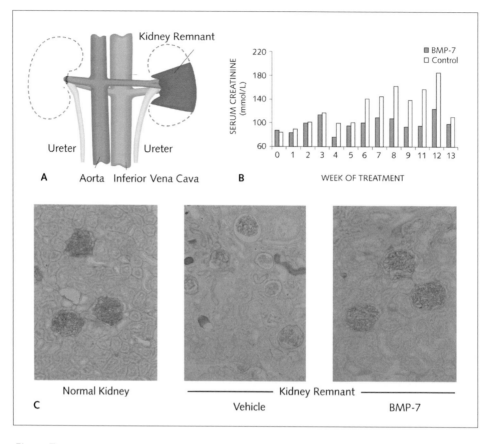

Figure 7

Therapeutic effect of BMP-7 on serum creatinine values and kidney regeneration in rats following 5/6 nephrectomy

(A) Schematic depiction of the 5/6 nephrectomy. One week following removal of 5/6 kidney mass, rats were subjected to i.v. application of BMP-7 (10 µg/kg) or a vehicle acetate buffer three times a week and serum creatinine values were measured throughout 13 weeks. Data are shown as average ± SEM; $p \leq 0.01$ for BMP-7 versus vehicle treated rats, Student's t-test in (B). Histological analysis of normal rat kidney, rat kidney following 5/6 nephrectomy and kidney following 5/6 nephrectomy treated with BMP-7 (C), magnification × 200.

suggests that BMP-7 can delay the progression of the terminal phase of chronic renal failure. Since the process of the chronic kidney failure in humans lasts over years, delaying the progression is critical for the treatment of chronic kidney diseases. BMP-7 might provide a potential therapeutic basis for the treatment of end-stage renal failure.

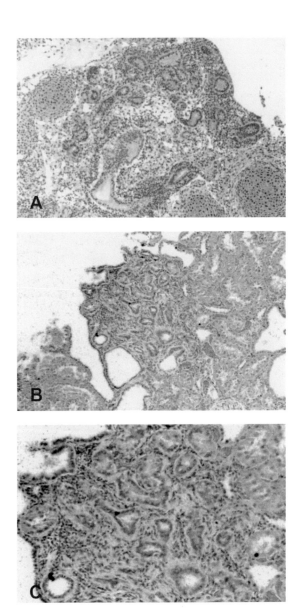

Figure 8
BMP-7 treatment in 5/6 kidney remnant model induces formation of embryonic nephron-like structures
(A) Histological image of rat kidney mesenchyme at E12.5. (B) Histological analysis of rat kidney following 5/6 nephrectomy with clearly visible newly induced embryonic-like structures, which comprise 3–4% percent of the entire kidney mass, magnification × 10. (C) Higher magnification (× 200) of the same segment of the kidney section.

%	Normal Histology	Mesangial Thickening	Glomerular Sclerosis & Loop Collapse	Scattered Sclerosis & Microaneurysms	Absence of Viable Glomeruli
Control (N=15)	2.58±0.22	27.3±2.4	26.5±3.5	34.7±4.2	8.9±0.7
BMP-7 (N=20)	11.4±1.1	58.6±3.2	14.7±1.3	11.8±4.1.1	2.5±0.2
P	<.01	<.01	<.02	<.01	<.01

Figure 9
Pathological changes in glomeruli of rat kidney following 5/6 nephrectomy
Histomorphometric analysis of kidneys of rats following 5/6 neprectmy reveals decreased glomarular sclerosis, microaneurysms and increased viability of the glomeruli in rats treated with BMP-7.

Unilateral ureteral obstruction

In another model mimicking chronic renal injury, human recombinant BMP-7 was systemically administered to rats with unilateral ureteral obstruction (UUO) and produced nearly complete protection for five days against tubulointerstitial fibrosis (Fig. 10) [9]. Tubulointerstitial fibrosis is a com-mon final pathway contributing to progression of many chronic kidney diseases [94–96]. UUO activates a cascade of events that produce tubulointerstitial fibrosis [97–99]. An early event in the damage cascade is angiotensin II upregulation, which stimulates tumor necrosis factor-α (TNF-α) production and TGF-β expression [100–104]. These cytokines activate nuclear factor κB (NF-κB), a crucial transcription factor in fibroblasts, macrophages and epithelial cells, involved in renal cellular transformation and apoptosis as well as interstitial inflammation and subsequent fibrosis. The damage cascade stimulated by UUO closely resembles that produced by several forms of renal injury [100, 105–107]. Suppression of this damage cascade might prevent fibrogenesis and preserve renal function. BMP-7 suppressed UUO-stimulated loss of the tubular epithelium due to apoptosis and prevented the transformation of renal cells into interstitial myofibroblasts [9]. This suggests that, whereas BMP-7 prevented tubular cell apoptosis as previously reported [8], it further appears to have maintained the phenotype of tubular cells and the interstitial fibroblasts. Both tubular cells and interstitial fibroblasts are subjected to phenotypic alterations as a result of UUO [96, 97, 108–110]. The preponderance of evidence is that phenotypic alteration of epithelial and fibroblastic cells to myofibroblasts is detrimental and leads to a progressive loss of renal function [97, 109–112]. BMP-7 administration was similar to but greater

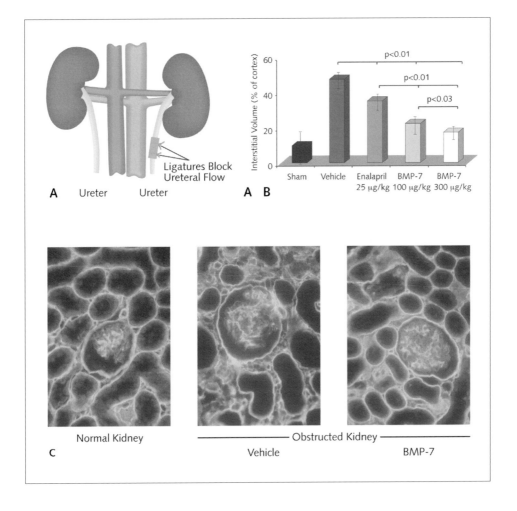

Figure 10
Therapeutic effect of BMP-7 on interstitial volume and interstitial fibrosis in rats unilateral ureteric obstruction
(A) Schematic depiction of the unilateral ureteric obstruction. Rats were subjected to unilateral ureteric obstruction, which was removed after five days. Rats were subjected to i.v. application of enalapril (25 mg/kg), BMP-7 (100 or 300 μg/kg) or a vehicle acetate buffer three times a week and interstitial volume was measured at the end of the experiment. Data are shown as average ± SEM; P ≤ 0.03 for BMP-7 versus vehicle treated rats, Student's t-test in (B). Histological analysis of type IV collagen in normal rat kidney, rat kidney following unilateral ureteric obstruction and kidney following unilateral ureteric obstruction treated with BMP-7 (C), magnification × 400.

than enalapril in its protective action against tubulointerstitial fibrosis [9]. In addition, BMP-7 preserved the tubular epithelial structure and prevented tubular atrophy. In comparison, ACE inhibition decreases the activity of the damage cascade by suppressing UUO stimulation of TGF-β, TNF-α, and NF-κB, which are mediated by angiotensin II [98, 99]. Approximately 50% of the stimulation of this damage cascade, after UUO, is due to angiotensin II [97], and data suggest that more than 50% is suppressed by BMP-7. Thus, BMP-7 may function as a renal homeostasis signal by providing a survival signal to epithelial cells, protecting the tubular epithelial cell phenotype, and suppressing gene activation associated with injury.

Diabetic nephropathy

Diabetic nephropathy is the leading cause of end-stage renal failure in Western society, accounting for approximately 40% of new patients starting renal replacement therapy. A majority of patients suffering from diabetic nephropathy have type II diabetes mellitus. Morphological traits of diabetic nephropathy include progressive thickening of the glomerular basement membrane and expansion of the mesangial compartment due to accumulation of extracellular matrix. Transdifferentiation of epithelial cells to fibroblastic phenotype has been implicated as the major pathological event which promotes tubulointerstitial fibrosis, a hallmark of progressive renal failure [69]. Hyperglycemia and increased glomerular pressure, main pathophysiological findings in the kidney of diabetic patients, are major factors which set the stage for ensuing degradation of the renal tissues and consequent fibrosis. Both factors induce release of cytokines and growth factors by a variety of renal cells, predominantly mesangial cells. These growth factors in turn give rise to renal hypertrophy. Glomerular growth and hypertrophy of the tubulointerstitium precede glomerulosclerosis and tubulointerstitial hypertrophy, which are observed at later stages of the disease. In the model of STZ-induced diabetes in rats, BMP-7 therapy partially reversed renal hypertrophy, restored GFR to normal and decreased proteinuria 32 weeks following a single dose of diabetes-inducing agent [113]. These data correlate with the finding that BMP-7 protein and mRNA are significantly decreased by 15 weeks after STZ treatment in rat and declined further by 30 weeks. The levels of BMP-7 expression were less than 10% of controls at week 30 [114]. Anti-inflammatory properties have been demonstrated in the model of ischemic acute renal failure. It has also been shown that BMP-7 reduces basal and TNF-α-stimulated expression of interleukin-6, interleukin-1β, monocyte chemotactic protein 1 (MCP-1) and interleukin-8 [115]. High levels of MCP-1 have been found in diseased renal tissue supporting the role of this chemokine in renal inflammation and interstitial fibrosis.

A specific BMP antagonist, gremlin, has also been implicated in initiation and progression of diabetic nephropathy. Although gremlin is expressed during kidney development, it is not present in normal adult kidney [69]. However, it is re-expressed in *in vitro* and *in vivo* models of diabetic nephropathy [68]. Mesangial cells cultured in conditions of high glucose concentration express gremlin, and addition of TGF-β to serum-restricted mesangial cells increases gremlin mRNA levels [116]. Addition of TGF-β neutralizing antibodies causes a reduction in hyperglycemia-induced gremlin expression.

Renal osteodystrophy

Chronic renal disease influences the skeletal system through extensive range of skeletal abnormalities which are the result of impaired kidney function and secondary hyperparathyroidism. These skeletal deviations include both states of high bone turnover and states of low bone turnover. High-turnover renal osteodystrophy, also known as osteitis fibrosa, is characterized by increasing numbers of osteoblasts as well as increased bone formation rate and osteoid accumulation. At the same time, increased number of osteoclasts and increased bone resorption can also be observed [117]. These findings are accompanied by fibrosis in the peritrabecular bone marrow space of unknown origin [118]. Accumulation of fibrous tissue in the bone marrow leads in these patients to anemia and decreased response to erythropoietin [119, 120].

Maladapted response of secondary increase of parathyroid hormone (PTH) activity in response to initial decrease in skeletal remodeling seems to be the underlying mechanism by which the renal osteodystrophy occurs [121]. This explanation of the pathophysiology is strongly supported by the fact that patients with chronic renal disease, in which hyperthyroidism is prevented, develop adynamic bone disorder [122]. BMP-7 expression in adult rat kidney is decreased in conditions such as acute renal ischemia and diabetic nephropathy [7–9, 123]. This hints to the possibility that initial decrease in bone turnover could in part be caused by BMP deficiency and decreased osteoblast differentiation. This in turn leads to hyperparathyroidism, a misguided attempt on the part of the organism to counteract the low remodeling rates in the face of CDK. Because PTH is not an osteoblast differentiation factor [124, 125], increased levels of PTH stimulate an abnormal phenotype of osteoblastic cells with fibroblast-like properties that accumulate in the peritrabecular space and produce marrow fibrosis. Exogenously administered BMP-7 resulted in elimination of these cells, most likely by inducing terminal differentiation into the osteoblastic phenotype in CKD mice model of secondary hyperparathyroidism [126]. Even in the presence of high levels of PTH, BMP-7 was able to promote remodeling, increased bone formation and to decrease bone resorption.

Adynamic bone disorder, a low-turnover variant of the renal osteodystrophy, is an important complication of the chronic renal disease. In the animal model of chronic renal disease accompanied by adynamic bone disorder which is characterized by absence of secondary hyperparathyroidism, treatment with BMP-7 reversed the abnormal bone histomorphometry to normal values [127]. Untreated animals developed hyperphosphatemia, secondary hyperparathyroidism, and a mild osteodystrophy, as expected. With restriction of dietary phosphorus and administration of calcitriol calcium, phosphorus and parathyroid hormone levels were maintained within the normal range. Significant adynamic bone disorder developed in the untreated group characterized by significant depression in osteoblast number, perimeter, bone formation rate, and mineral apposition rate when compared with the control group. The animals treated with BMP-7 exhibited a significant decrease in the phosphorus plasma levels when compared to untreated animals [127].

Renal fibrosis

Advanced chronic renal failure is characterized by progressive renal fibrosis which ultimately leads to complete collapse of the renal function [128]. Scarring of the kidney due to progressive fibrosis occurs during a variety of diseases, such as hypertension, primary glomerulopathies, autoimmune disease, diabetes mellitus, toxic injury or congenital malformations. Renal fibrosis is distinguished by the triad of glomerulosclerosis, interstitial fibrosis and tubular atrophy, implicating common mechanisms which are independent of the underlying primary disease [129, 130]. Although it has been assumed that renal fibroblasts play the major role in the occurrence of fibrosis, recent studies have highlighted importance of tubular epithelial cells in progression of chronic renal disease. Tubular epithelial cells are capable of conversion between the epithelial and mesenchymal phenotypes, due to their unique plasticity [131, 132]. The process of epithelial-to-mesenchymal transition (EMT) is present both in physiological conditions during embryonic development [133] and in pathophysiological setting, for instance in tumor progression and organ fibrosis [134]. During this transition cells lose epithelial characteristics and differentiate to acquire hallmarks of a mesenchymal phenotype. Transforming growth factor β1 (TGF-β1) has been identified as the main inducer of EMT in the kidney and other organ systems. TGF-β1 contributes to progressive glomerular and interstitial fibrosis by increasing gene expression of several proteins of the extracellular matrix (ECM) and acts to reduce degradation of these proteins [135, 136]. *In vitro* studies on cultured mesangial cells have shown that TGF-β1 increased cell-associated collagen type IV and fibronectin, soluble collagen type IV, thrombospondin and connective tissue growth factor (CTGF) [137]. Treatment with recombinant human BMP-7 impaired the increase in the aforementioned ECM proteins and CTFG. It has been shown that BMP-7 reverses the TGF-β1-induced ETM by reinduction of E-cadherin, an important cell adhesion molecule. Due to the fact that TGF-β1 and BMP-7 mediate their effect through the Smad signaling pathway, interactions on the level of these second messenger molecules have also been implicated in the reversal of TGF-β1-induced EMT. While TGF-β1 directly inhibits E-cadherin expression and induces EMT in a Smad3-dependent manner, BMP-7 enhances E-cadherin expression *via* Smad5 and restores the epithelial phenotype [131]. *In vivo* studies on the TGF-β1-induced *de novo* EMT model have shown that systemic administration of human recombinant BMP-7 results in repair of severely damaged renal tubular epithelial cells and ultimately leads to reversal of chronic renal injury [131].

Conclusion

BMPs may have important functions in kidney development and renal diseases. BMP-7 regulates kidney mesenchyme differentiation and preserves renal function by preventing inflammation and fibrosis following ischemia, nephrectomy, ureteral obstruction and diabetes.

References

1   Helder MN, Ozkaynak E, Sampath TK, Luyten FP, Latin V, Oppermann H, Vukicevic S (1995) Expression pattern of osteogenic protein-1 (bone morphogenetic protein-7) in human and mouse development. *J Histochem Cytochem* 43: 1035–1044

2   Vukicevic S, Stavljenic A, Pecina M (1995) Discovery and clinical applications of bone morphogenetic proteins. *Eur J Clin Chem Clin Biochem* 33: 661–671

3   Ozkaynak E, Schnegelsberg PN, Opperman H (1991) Murine osteogenic protein -1 (OP-1): high levels of mRNA in kidney. *Biochem Biophys Res Commun* 179: 116–123

4   Vukicevic S, Kopp JB, Luyten FB, Sampath TK (1996) Induction of nephrogenic mesenchyme by osteogenic protein-1 (bone morphogenetic protein 7). *Proc Natl Acad Sci USA* 93: 9021–9026

5   Dudley AT, Lyons KM, Robertson EJ (1995) A requirement for bone morphogenetic protein-7 during development of the mammalian kidney and eye. *Genes Dev* 9: 2795–2807

6   Luo O, Hofmann A, Bronckers JJ, Sohocki M, Bradley A, Karsenty G (1995) BMP-7 is an inducer of nephrogenesis and is also required for eye development and skeletal patterning. *Genes Dev* 9: 2808–2820

7   Simon M, Maresh JG, Harris SE, Hernandez JD, Arar M, Olson MS, Abboud HE (1999) Expression of bone morphogenetic protein-7 mRNA in normal and ischemic adult rat kidney. *Am J Physiol* 276: 382–389

8   Vukicevic S, Basic V, Rogic D, Basic N, Shih M, Shepard A, Jin D, Dattatreyamurty B, Jones W, Dorai H et al (1998) Osteogenic protein-1 (bone morphogenetic protein-7) reduces severity of injury after ischemic acute renal failure in rat. *J Clin Invest* 102: 202–214

9   Hruska KA, Guo G, Wozniak M, Martin D, Miller S, Liapis H, Loveday K, Klahr S,

Sampath TK, Morrissey J (2000) Osteogenic protein-1 prevents renal fibrogenesis associated with ureteral obstruction. *Am J Physiol Renal Physiol* 279: F130–F143

10 Saxen L (1987) *Organogenesis of the kidney.* Cambridge Univ. Press, Cambridge

11 Grobstein C (1953) Inductive epithelio-mesenchymal interactions in cultured organ rudiments of the mouse. *Science* 118: 52–55

12 Grobstein C (1956) Trans-filter induction of tubules in mouse metanephrogenic mesenchyme. *Exp Cell Res* 10: 434–440

13 Kreidberg JA, Sariola H, Loring JM, Maeda M, Pelletier J, Housman D, Jaenisch R (1993) WT-1 is required for early kidney development. *Cell* 74: 679–691

14 Lee SB, Huang K, Palmer R, Truong VB, Herzlinger D, Kolquist KA, Wong J, Paulding C, Yoon SK, Gerald W et al (1999) The Wilms tumor suppressor WT1 encodes a transcriptional activator of amphiregulin. *Cell* 98: 663–673

15 Rothenpieler UW, Dressler GR (1993) Pax-2 is required for mesenchyme-to-epithelium conversion during kidney development. *Development* 119: 711–720

16 Dressler GR, Deutsch U, Chowdhury K, Nornes HO, Gruss P (1990) Pax-2, a new murine paired-box-containing gene and its expression in the developing excretory system. *Development* 109: 787–795

17 Karavanov AA, Karavanova I, Perantoni A, Dawid IB (1998) Expression pattern of the rat Lim-1 homeobox gene suggests a dual role during kidney development. *Int J Dev Biol* 42: 61–66

18 Brophy PD, Ostrom L, Lang KM, Dressler GR (2001) Regulation of ureteric bud outgrowth by Pax2-dependent activation of the glial derived neurotrophic factor gene. *Development* 128: 4747–4756

19 Xu PX, Adams J, Peters H, Brown MC, Heaney S, Maas R (1999) Eya1-deficient mice lack ears and kidneys and show abnormal apoptosis of organ primordia. *Nat Genet* 23: 113–117

20 Chan DC, Wynshaw-Boris A, Leder P (1995) Formin isoforms are differentially expressed in the mouse embryo and are required for normal expression of Fgf-4 and Shh in the limb bud. *Development* 121: 3151–3162

21 Moore MW, Klein RD, Farinas I, Sauer H, Armanini M, Philips H, Reichardt LF, Ryan AM, Carver-Moore K, Rosenthal A (1996) Renal and neuronal abnormalities in mice lacking GDNF. *Nature* 382: 76–79

22 Jing S, Wen D, Yu Y, Holst PL, Luo Y, Fang M, Tamir R, Antonio L, Hu Z, Cupples R et al (1996) GDNF-induced activation of the Ret protein tyrosin kinase is mediated by GDNFR- alpha, a novel receptor for GDNF. *Cell* 85: 1113–1124

23 Pichel JG, Shen L, Sheng HZ, Granholm AC, Drago J, Grinberg A, Lee EJ, Huang SP, Saarma M, Hoffer BJ et al (1996) Defects in enteric innervation and kidney development in mice lacking GDNF. *Nature* 382: 73–76

24 Suchardt A, D'Agati V, Larsson-Blomberg L, Constantini F, Pachinis V (1994) Defects in kidney and enteric nervous system of mice lacking the tyrosin kinase receptor Ret. *Nature* 367: 380–383

25 Vega QC, Worby CA, Lechner MS, Dixon JE, Dressler GR (1996) Glial cell line-derived

neurotrophic factor activates RET and promotes kidney morphogenesis. *Proc Natl Acad Sci USA* 93: 10657–10661

26   Cacalano G, Farinas I, Wang LC, Hagler K, Forgie A, Moore M, Armanini M, Phillips H, Ryan AM, Reichardt LF et al (1998) GFRalpha1 is an essential receptor component for GDNF in the developing nervous system and kidney. *Neuron* 21: 53–62

27   Pelletier J, Schalling M, Buckler AJ, Rogers A, Haber DA, Housman D (1991) Expression of the Wilms' tumor gene WT-1 in the murine urogenital system. *Genes Dev* 5: 1345–1356

28   Torres M, Gomez Pardo E, Dressler GR, Gruss P (1995) Pax-2 controls multiple steps of urogenital development. *Development* 121: 4057–4065

29   Ryan G, Steele-Perkins V, Morris J, Rauscher FJ, Dressler GR (1995) Repression of Pax-2 by WT-1 during normal kidney development. *Development* 121: 867–875

30   Dressler GR, Wilkinson JE, Rothenpieler UW, patterson LT, Williams-Simons L, Westphal H (1993) Deregulation of Pax-2 expression in transgenic mice generates severe kidney abnormalities. *Nature* 362: 65–67

31   Finch PW, Cunha GR, Rubin JS, Wong J, Ron D (1995) Pattern of keratinocyte growth factor and keratinocyte growth factor receptor expression during mouse fetal development suggests a role in mediating morphogenetic mesenchymal-epithelial interactions. *Dev Dyn* 203: 223–240

32   Qiao J, Uzzo R, Obara-Ishihara T, Degenstein L, Fuchs E, Herzlinger D (1999) FGF-7 modulates ureteric bud growth and nephron number in the developing kidney. *Development* 126: 547–554

33   Sekine K, Ohuchi H, Fujiwara M, Yamasaki M, Yoshizawa T, Sato T, Yagishita N, Matsui D, Koga Y, Itoh N et al (1999) Fgf10 is essential for limb and lung formation. *Nat Genet* 21: 138–141

34   Plisov SY, Yoshino K, Dove LF, Higinbotham KG, Rubin JS, Perantoni AO (2001) TGF beta 2, LIF and FGF2 cooperate to induce nephrogenesis. *Development* 128: 1045–1057

35   Barasch J, Yang J, Ware CB, Taga T, Yoshida K, Erdjument-Bromage H, Tempst P, Parravicini E, Malach S, Aranoff T, Oliver JA (1999) Mesenchymal to epithelial conversion in rat metanephros is induced by LIF. *Cell* 99: 377–386

36   Montesano R, Matsumoto K, Nakamura T, Orci L (1991) Identification of a fibroblast-derived epithelial morphogen as hepatocyte growth factor. *Cell* 67: 901–908

37   Barros EJG, Santos OFP, Matsumoto K, Nakamura T, Nigam SK (1995) Differential tubulogenic and branching morphogenetic activities of growth factors: implications for epithelial tissue development. *Proc Natl Acad Sci USA* 92: 4412–4416

38   Piscione TD, Yager TD, Gupta IR, Grinfeld B, Pei Y, Attisano L, Wrana J, Rosenblum ND (1997) BMP-2 and BMP-7 exert direct and opposite effects on renal branching morphogenesis. *Am J Physiol* 273: F961–F975

39   Santos OFP, Nigam SK (1993) HGF-induced tubulogenesis and branching of epithelial cells is modulated by extracellular matrix and TGF-β. *Dev Biol* 160: 293–302

40   Ritvos O, Tuuri T, Eramaa M, Sainio K, Hilden K, Saxen L, Gilbert SF (1995) Activin

disrupts epithelial branching morphogenesis in developing glandular organs of the mouse. *Mech Dev* 50: 229–245

41 Sakurai H, Barros EJ, Tsukamoto T, Barasch J, Nigam SK (1997) An *in vitro* tubulogenesis system using cell lines derived from the embryonic kidney shows dependence on multiple soluble growth factors. *Proc Natl Acad Sci USA* 94: 6279–6284

42 Barasch J, Yang J, Qiao JY, Tempst P, Erdjument-Bromage H, Leung W, Oliver JA (1999) Tissue inhibitor of metalloproteinase-2   stimulates mesenchymal growth and regulates epithelial branching during morphogenesis of the rat metanephros. *J Clin Invest* 103: 1299–1307

43 Hogan BLM (1996) Bone morphogenetic proteins: multifunctional regulators of vertebrate development. *Genes & Develop* 10: 1580–1594

44 Piscione TD, Rosenblum ND (2002) The molecular control of renal branching morphogenesis: current knowledge and emerging insights. *Differentiation* 70: 227–246

45 King JA, Marker PC, Seung KJ, Kingsley DM (1994) BMP5 and the molecular, skeletal, and soft-tissue alterations in short ear mice. *Dev Biol* 166: 112–122

46 Miyazaki Y, Oshima K, Fogo A, Hogan BL, Ichikawa I (2000) Bone morphogenetic protein 4 regulates the budding site and elongation of the mouse ureter. *J Clin Invest* 105: 863–873

47 Miyazaki Y, Oshima K, Fogo A, Ichikawa I (2003) Evidence that bone morphogenetic protein 4 has multiple biological functions during kidney and urinary tract development. *Kidney Int* 63: 835–844

48 Dudley AT, Robertson EJ (1997) Overlapping expression domains of bone morphogenetic protein family members potentially account for limited tissue defects in BMP-7 deficient embryos. *Dev Dyn* 208: 349–362

49 Stark K, Vainio S, Vassileva G, McMahon AP (1994) Epithelial transformation of metanephric mesenchyme in the developing kidney regulated by Wnt-4. *Nature* 372: 679–683

50 Hatini V, Huh SO, Hertzlinger D, Soares VC, Lai E (1996) Essential role of stromal mesenchyme in kidney morphogenesis revealed by targeting disruption of Winged Helix transcription factor BF-2. *Genes Dev* 10: 1467–1478

51 Leveen P, Pekny M, Gebre-Medhin S, Swolin B, Larsson E, Betzholtz C (1994) Mice deficient for PDGF B show renal, cardiovascular and haematological abnormalities. *Genes Dev* 8: 1875–1887

52 Karsenty G, Luo G, Hofmann C, Bradley A (1996) BMP-7 is required for nephrogenesis, eye development, and skeletal patterning. *Ann NY Acad Sci* 785: 98–107

53 Kitten AM, Kreisberg JI, Olson MS (1999) Expression of osteogenic protein-1 mRNA in cultured kidney. *J Cell Physiol* 181: 410–415

54 Godin RE, Takaesu NT, Robertson EJ, Dudley AT (1998) Regulation of BMP-7 expression during kidney development. *Development* 125: 3473–3482

55 Obara-Ishihara T, Kuhlman J, Niswander L, Herzlinger D (1999) The surface ectoderm is essential for nephric duct formation in intermediate mesoderm. *Development* 126: 1103–1108

56 Raatikainen-Ahokas A, Hytonen M, Tehnunen A, Sainio K, Sariola H (2000) BMP-4 affects the differentiation of metanephric mesenchyme and reveals an early anterior-posterior axis of the embryonic kidney. *Dev Dyn* 217: 146–158

57 Zhang H, Bradley A (1996) Mice deficient for BMP2 are nonviable and have defects in amnion/chorion and cardiac development. *Development* 122: 2977–2986

58 Piscione TD, Phan T, Rosenblum ND (2001) BMP7 controls collecting tubule cell proliferation and apoptosis *via* Smad1-dependent and -independent pathways. *Am J Physiol* 280: F19–F33

59 Gupta IR, Macias-Silva M, Kim S, Zhou X, Piscione TD, Whiteside C, Wrana JL, Rosenblum ND (2000) BMP-2/ALK3 and HGF signal in parallel to regulate renal collecting duct morphogenesis. *J Cell Sci* 113: 269–278

60 Ghosh Choundhury G, Kim YS, Simon M, Wozney J, Harris S, Ghosh Choundhury N, Abboud HE (1999) Bone morphogenetic protein 2 inhibits platelet-derived growth factor-induced c-fos gene transcription and DNA synthesis in mesangial cells. Involvement of mitogen-activated protein kinase. *J Biol Chem* 274: 10897–10902

61 Vukicevic S, Helder MN, Luyten FP (1994) Developing human lung and kidney are major sites for synthesis of bone morphogenetic protein-3 (osteogenin). *J Histochem Cytochem* 42: 869–875

62 Takahashi H, Ikeda T (1996) Transcripts for two members of the transforming growth factor-beta superfamily BMP-3 and BMP-7 are expressed in developing rat embryos. *Dev Dyn* 207: 439–449

63 Martinez G, Loveland KL, Clark AT, Dziadek M, Bertram JF (2001) Expression of bone morphogenetic protein receptors in the developing mouse metanephros. *Exp Nephrol* 9: 372–379

64 Vrljicak P, Myburgh D, Ryan AK, van Rooijen MA, Mummery CL, Gupta IR (2004) Smad expression during kidney development. *Am J Physiol Renal Physiol* 286: F625–633

65 Hsu DR, Economides AN, Wang X, Eimon PM, Harland RM (1998) The *Xenopus* dorsalizing factor Gremlin identifies a novel family of secreted proteins that antagonize BMP activities. *Mol Cell* 1: 673–683

66 Khokha MK, Hsu D, Brunet LJ, Dionne MS, Harland RM (2003) Gremlin is the BMP antagonist required for maintenance of Shh and Fgf signals during limb patterning. *Nat Genet* 34: 303–307

67 McMahon R, Murphy M, Clarkson M, Taal M, Mackenzie HS, Godson C, Martin F, Brady HR (2000) IHG-2, a mesangial cell gene induced by high glucose, is human gremlin. Regulation by extracellular glucose concentration, cyclic mechanical strain, and transforming growth factor-beta1. *J Biol Chem* 275: 9901–9904

68 Lappin DW, McMahon R, Murphy M, Brady HR (2002) Gremlin: an example of the re-emergence of developmental programmes in diabetic nephropathy. *Nephrol Dial Transplant* 9: 65–67

69 Dolan V, Hensey C, Brady HR (2003) Diabetic nephropathy: renal development gone awry? *Pediatr Nephrol* 18: 75–84

70 Yanagita M, Oka M, Watabe T, Iguchi H, Niida A, Takahashi S, Akiyama T, Miyazono K, Yanagisawa M, Sakurai T (2004) USAG-1: a bone morphogenetic protein antagonist abundantly expressed in the kidney. *Biochem Biophys Res Commun* 316: 490–500

71 Dewulf N, Verschueren K, Lonnoy O, Moren A, Grimsby S, Vande Spiegle K, Miyazono K, Huylebroeck D, Ten Dijke P (1995) Distinct spatial and temporal expression patterns of two type 1 receptors for bone morphogenetic proteins during mouse embryogenesis. *Endocrinology* 136: 2652–2663

72 Mishina YA, Susuki A, Ueno N, Behringer RR (1995) BMPr encodes a type I bone morphogenetic protein receptor that is essential for gastrulation during mouse embryogenesis. *Genes Dev* 9: 3027–3037

73 Gupta IR, Macias-Silva M, Kim S, Zhou X, Piscione TD, Whiteside C, Wrana JL, Rosenblum ND (2000) HGF Rescues BMP-2-mediated inhibition of renal collecting duct morphogenesis without interrupting Smad1 dependent signaling. *J Cell Sci* 113: 269–278

74 Piscione TD, Cella C, Rosenblum ND (2001) ALK3 and SMAD4 regulate collecting duct morphogenesis *in vivo* [abstract]. *J Am Soc Nephrol* 12: 525A

75 Thadhani R, Pascual M, Bonventre JV (1996) Acute renal failure. *N Engl J Med* 334: 1448–1460

76 Humes HD, MacKay SM, Funke AJ, Buffington DA (1997) Acute renal failure: growth factors, cell therapy and gene therapy. *Proc Assoc Am Physicians* 109: 547–557

77 Hirschberg R, Ding H (1998) Growth factors and acute renal failure. *Semin Nephrol* 18: 191–207

78 Humes DH, Liu S (1994) Cellular and molecular basis of renal repair in acute renal failure. *J Am Soc Nephrol* 5: 1–11

79 Witzgall R, Brown D, Schwarz C, Bonventre JV (1994) Localization of proliferating cell number antigen, vimentin, c-Fos and clusterin in the post-ischemic kidney: evidence for a heterogenous genetic response among nephron segments and a large pool of mitoticaly active and differentiated cells. *J Clin Invest* 93: 2175–2188

80 Hirschberg R, Kopple JD (1989) Evidence that insulin-like growth factor I increases renal plasma flow and glomerular filtration rate in fasted rats. *J Clin Invest* 83: 326–330

81 Andersson G, Jennische E (1988) IGF-I immunoreactivity is expressed by regenerating renal tubule cells after ischaemic injury in the rat. *Acta Physiol Scand* 132: 453–457

82 Sugimura K, Goto T, Kasai S, Tsuchida K, Takemoto Y, Yamagami S (1998) The activation of serum hepatocyte growth factor in acute renal failure. *Nephron* 76: 364–365

83 Coimbra TM, Cieslinski DA, Humes HD (1990) Epidermal growth factor accelerates renal repair in mercuric chloride nephrotoxicity. *Am J Physiol* 259: 438–443

84 Weinberg JM (1993) The cellular basis of nephrotoxicity. In: Schrier RW, Gottschalk CW (eds): *Diseases of the kidney*. Little, Brown and Company, Boston, 1031–1098

85 Guillermina G, Adriana TM, Monica EM (1989) The implications of renal glutathione level in mercuric chloride nephrotoxicity. *Toxicology* 58: 187–195

86 Houser MT, Milner LS, Kolbeck PC, Wei SH, Stohs SJ (1992) Glutathione monoethyl ester moderates mercuric chloride-induced acute renal failure. *Nephron* 61: 449–455

87 Nath KA, Croatt AJ, Likely S, Behrens TW, Warden D (1996) Renal oxidant injury and oxidant response induced by mercury. *Kidney Int* 50: 1032–1043

88 Southard J, Nitisewojo P, Green DE (1974) Mercurial toxicity and the perturbation of the mitochondrial control system. *Fed Proc* 33: 2147–2153

89 Fausto N (2000) Liver regeneration. *J Hepathol* 32: 19–31

90 Reddi AH, Huggins C (1972) Biochemical sequences in the transformation of normal fibroblasts in adolescent rats. *Proc Natl Acad Sci USA* 69: 1601–1605

91 Luyten FP, Cunningham NS, Vukicevic S, Paralkar V, Ripamonti U, Reddi AH (1992) Advances in osteogenin and related bone morphogenetic proteins in bone induction and repair. *Acta Orthop Belg* 58: 263–267

92 Dudley AT, Godin RE, Robertson EJ (1999) Interaction between FGF and BMP signaling pathways regulates development of metanephric mesenchyme. *Genes Dev* 15: 1601–1613

93 Almanzar MM, Kendall FS, Philip DH, Piqueras AI, Jones WK, Charette MF, Paredes AL (1998) Osteogenic protein-1 mRNA expression is selectively modulated after acute ischemic renal injury. *J Am Soc Nephrol* 9: 1456–1463

94 Couser WG (1993) Mediators of immune glomerular injury. *Clin Invest* 71: 8–11

95 Couser WG (1993) Pathogenesis of glomerulonephritis. *Kidney Int* 44: S519–S526

96 Johnson RJ, Hugo C, Hasley C, Pichler RH, Bassuk J, Thomas S, Suga S, Couser WG, Shankland SJ (1998) Mechanisms of progressive glomerulonecrosis and tubulointerstitial fibrosis. *Clin Exp Nephrol* 2: 307–312

97 Fern RJ, Yesko CM, Thornhill BA, Kim H-Y, Smithies O, Chevalier RL (1999) Reduced angiotensinogen expression attenuates renal interstitial fibrosis in obstructive nephropathy in mice. *J Clin Invest* 103: 39–46

98 Klahr SS (1998) Nephrology forum: obstructive nephropathy. *Kidney Int* 54: 286–300

99 Klahr S, Morrissey J (1998) Angiotensin II and gene expression in the kidney. *Am J Kidney Dis* 31: 171–176

100 Border WA, Noble NA (1998) Interactions of transforming growth factor-beta and angiotensin II in renal fibrosis. *Hypertension* 31: 181–188

101 Douglas JG, Romero M, Hopfer U (1990) Signaling mechanisms coupled to the angiotensin receptor of proximal tubular epithelium. *Kidney Int* 30: S43–S47

102 Kaneto H, Morrissey J, Klahr S (1993) Increased expression of TGF-β 1 mRNA in the obstructed kidney of rats with unilateral ureteral ligation. *Kidney Int* 44: 313–321

103 Kaneto H, Morrissey J, McCracken R, Ishidoya S, Reyes A, Klahr S (1996) The expression of mRNA for tumor necrosis factor increases in the obstructed kidney of rats soon after unilateral ureteral ligation. *Nephrology* 2: 161–166

104 Kalahr S, Ishidoya S, Morrissey J (1995) Role of angiotensin II in the tubulointerstitial fibrosis of obstructive nephropathy. *Am J Kidney Dis* 26: 141–146

105 Johnson RJ, Alpers CE, Yoshimura A, Lombardi D, Pritzl P, Floege J, Schwartz SM (1992) Renal injury from angiotensin II-mediated hypertension. *Hypertension* 19: 464–474

106 Kagami S, Border WA, Miller DE, Noble NA (1994) Angiotensin II stimulates extracel-

lular matrix protein synthesis through induction of transforming growth factor-beta expression in rat glomerular mesangial cells. *J Clin Invest* 93: 2431–2437

107 Yoo KH, Thornhill BA, Wolstenholme JT, Chevalier RL (1998) Tissue-specific regulation of growth factors and clusterin by angiotensin II. *Am J Hypertens* 11: 715–722

108 Chevalier RL, Kim A, Thornhill BA, Wolstenholme JT (1999) Recovery following relief of unilateral ureteral obstruction in the neonatal rat. *Kidney Int* 55: 793–807

109 Nagle RB, Johnson ME, Jervis HR (1976) Proliferation of renal interstitial cells following injury induced by ureteral obstruction. *Lab Invest* 35: 18–22

110 Ng YY, Huang TP, Yang WC, Chen ZP, Yang AH, Mu W, Nikolic-Paterson DJ, Atkins RC, Lan HY (1998) Tubular epithelial-myofibroblast transdifferentiation in progressive tubulointerstitial fibrosis in 5/6 nephrectomized rats. *Kidney Int* 54: 864–876

111 Ishidoya S, Morrissey J, McCracken R, Reyes A, Klahr S (1995) Angiotensin II receptor antagonist ameliorates renal tubulointerstitial fibrosis caused by unilateral ureteral obstruction. *Kidney Int* 47: 1285–1294

112 Ishidoya S, Morrissey J, McCracken R, Klahr S (1996) Delayed treatment with enalapril halts tubulointerstitial fibrosis in rats with obstructive nephropathy. *Kidney Int* 49: 1110–1119

113 Wang S, Chen Q, Simon TC, Strebeck F, Chaudhary L, Morrissey J, Liapis H, Klahr S, Hruska KA (2003) Bone morphogenic protein-7 (BMP-7), a novel therapy for diabetic nephropathy. *Kidney Int* 63: 2037–2049

114 Wang SN, Lapage J, Hirschberg R (2001) Loss of tubular bone morphogenetic protein-7 in diabetic nephropathy. *J Am Soc Nephrol* 12: 2392–2399

115 Gould SE, Day M, Jones SS, Dorai H (2002) BMP-7 regulates chemokine, cytokine, and hemodynamic gene expression in proximal tubule cells. *Kidney Int* 61: 51–60

116 Lappin DW, Hensey C, McMahon R, Godson C, Brady HR (2000) Gremlins, glomeruli and diabetic nephropathy. *Curr Opin Nephrol Hypertens* 9: 469–472

117 Malluche HH, Ritz E, Lange HP, Kutschera L, Hodgson M, Seiffert U, Schoeppe W (1976) Bone histology in incipient and advanced renal failure. *Kidney Int* 9: 355–362

118 Nomura S, Ogawa Y, Osawa G, Katagiri M, Harada T, Nagahana H (1996) Myelofibrosis secondary to renal osteodystrophy. *Nephron* 72: 683–687

119 Rao DS, Shih MS, Mohini R (1993) Effect of serum parathyroid hormone and bone marrow fibrosis on the response to erythropoietin in uremia. *N Engl J Med* 328: 171–175

120 Gallieni M, Corsi C, Brancaccio D (2000) Hyperparathyroidism and anemia in renal failure. *Am J Nephrol* 20: 89–96

121 Hruska KA, Saab G, Chaudhary LR, Quinn CO, Lund RJ, Surendran K (2004) Kidney-bone, bone-kidney, and cell-cell communications in renal osteodystrophy. *Semin Nephrol* 24: 25–38

122 Goodman WG, Ramirez JA, Belin T (1994) Development of adynamic bone in patients with secondary hyperparathyroidism after intermittent calcitriol therapy. *Kidney Int* 46: 1160–1166

123 Hruska KA (2000) Pathophysiology of renal osteodystrophy. *Pediatr Nephrol* 14: 636–640

124 Isogai Y, Akatsu T, Ishizuya T, Yamaguchi A, Hori M, Takahashi N, Suda T (1996) Parathyroid hormone regulates osteoblast differentiation positively or negatively depending on the differentiation stages. *J Bone Miner Res* 11: 1384–1393

125 Chaudhary LR, Avioli LV (1998) Identification and activation of mitogen-activated protein (MAP) kinase in normal human osteoblastic and bone marrow stromal cells: Attenuation of MAP kinase activation by cAMP, parathyroid hormone and forskolin. *Mol Cell Biochem* 178: 59–68

126 Gonzalez EA, Lund RJ, Martin KJ, McCartney JE, Tondravi MM, Sampath TK, Hruska KA (2002) Treatment of a murine model of high-turnover renal osteodystrophy by exogenous BMP-7. *Kidney Int* 61: 1322–1331

127 Lund RJ, Davies MR, Brown AJ, Hruska KA (2004) Successful treatment of an adynamic bone disorder with bone morphogenetic protein-7 in a renal ablation model. *J Am Soc Nephrol* 15: 359–369

128 Zeisberg M, Muller GA, Kalluri R (2004) Are there endogenous molecules that protect kidneys from injury? The case for bone morphogenic protein-7 (BMP-7). *Nephrol Dial Transplant* 19: 759–761

129 Remuzzi G, Bertani T (1998) Pathophysiology of progressive nephropathies. *N Engl J Med* 339: 1448–1456

130 Brenner BM (2002) Remission of renal disease: recounting the challenge, acquiring the goal. *J Clin Invest* 110: 1753–1758

131 Zeisberg M, Hanai J, Sugimoto H, Mammoto T, Charytan D, Strutz F, Kalluri R (2003) BMP-7 counteracts TGF-beta1-induced epithelial-to-mesenchymal transition and reverses chronic renal injury. *Nat Med* 9: 964–968

132 Zeisberg M, Kalluri R (2004) The role of epithelial-to-mesenchymal transition in renal fibrosis. *J Mol Med* 82: 175–181

133 Hay ED (1995) An overview of epithelio-mesenchymal transformation. *Acta Anat* 154: 8–20

134 Thiery JP (2002) Epithelial-mesenchymal transitions in tumour progression. *Nat Rev Cancer* 2: 442–454

135 Tomooka S, Border WA, Marshall BC, Noble NA (1992) Glomerular matrix accumulation is linked to inhibition of the plasmin protease system. *Kidney Int* 42: 1462–1469

136 Wilson HM, Reid FJ, Brown PA, Power DA, Haites NE, Booth NA (1993) Effect of transforming growth factor-beta 1 on plasminogen activators and plasminogen activator inhibitor-1 in renal glomerular cells. *Exp Nephrol* 1: 343–350

137 Wang S, Hirschberg R (2003) BMP7 antagonizes TGF-beta-dependent fibrogenesis in mesangial cells. *Am J Physiol Renal Physiol* 284: F1006–1013

Bone morphogenetic proteins in the nervous system

Pamela Lein[1] and Dennis Higgins[2]

[1]Center for Research on Occupational and Environmental Toxicology, Oregon Health & Sciences University, Portland, Oregon, USA; [2]Department of Pharmacology and Toxicology, State University of New York, Buffalo, 3455 Main Street, Buffalo, NY 14214, USA

Introduction

Bone morphogenetic proteins (BMPs) belong to the TGF-β superfamily of growth factors and morphogens. BMPs regulate neurulation, the sequence of morphogenetic events that specify neural tissue. Thereafter, BMPs and the closely related Growth/Differentiation Factors (GDFs) are prominently expressed in the central and peripheral nervous systems as well as in target tissues of sensory and motor neurons. The actions of BMPs and GDFs on neural tissue are both profound and diverse. They have been implicated in crucial developmental events such as: specification of neural and glial cell lineages, neural cell survival and proliferation, dorsalventral patterning, segmentation, axonal guidance, determination of neurotransmitter phenotype, regulation of dendritic growth and synapse formation. In addition, BMPs and GDFs are neuroprotective in mature animals in models of stroke and Parkinson's disease. In a previous article, we surveyed the effects of BMPs and GDFs on neural tissue [1]. In this review, we update summaries of some of the more rapidly advancing research areas. Other recent reviews have examined neural actions of TGF-β [2] and the roles of BMPs and their antagonists in neural induction [3, 4], neural crest development [5] and dorsal-ventral patterning [6, 7]. Therefore, these topics are not considered here.

Effects of BMPs and GDF on neural cell fate decisions

In the vertebrate nervous system, all neuronal and glial cells are derived from a common neural stem cell through a sequential process involving initial subdivision into regionally specific groups, followed by proliferation and then differentiation into distinct neuronal and glial cell types [8–10]. En route to becoming a differentiated postmitotic neural cell, a neural stem cell must make three fundamental decisions [11]. First, it determines its positional identity (dorsal–ventral, rostral–caudal and left–right). Second, it decides whether to proliferate (self-renew) or undergo mitotic arrest. Third, it assumes a particular cell fate based on inherited or external signals. The effect of external signals on cell fate determination varies according to developmental stage, resulting sequentially in apoptosis, neurogenesis and then gliogenesis. Experimental evidence indicates that BMP and GDF signaling influences all three of these decisions. Thus, BMPs induce dorsal precursor fates, including roof plate, choroid plexus epithelium and neural crest [12–15], and induce apoptosis or terminal differentiation in the developing neural tube and olfactory epithelium [14, 16–19]. BMP stimulation of CNS precursors *in vitro* leads to differentiation into neurons [20] and glia [21] and neural crest derivatives such as smooth muscle, peripheral neurons and Schwann cells [14, 22, 23]. The role of BMPs and GDFs in determining positional identity, has recently been reviewed [6, 7, 24], as has the literature documenting the influence of BMPs and GDFs on neuronal *versus* glial cell fate decisions in various regions of the central and peripheral nervous systems [1, 11, 25]. Thus, the remainder of this discussion will summarize the recent evidence that BMPs and GDFs coordinate the decision of precursors to either self-renew or arrest and the rapid advances in elucidating the molecular mechanisms underlying the diverse and seemingly paradoxical effects of BMP and GDF on cell fate decisions.

Neural stem cells are broadly defined as multipotent, self-renewing progenitor cells. A critical role for BMPs in maintaining stem cell self-renewal and multipotency is suggested by recent *in vitro* studies of embryonic stem (ES) cells. Early stem cell experiments demonstrated that production of mouse ES cell lines in culture depended on the addition of leukemia inhibitory factor (LIF) in addition to serum and/or growth on feeder cells. One of the necessary signaling pathways for self-renewal in ES cells is activation of the transcription factor STAT3 subsequent to binding of LIF and related cytokines to the gp130 receptor [26]. However, in the absence of serum, ES cells differentiate predominantly into neural phenotypes, suggesting that LIF alone is not sufficient to maintain ES cell self-renewal. The observation that neural phenotypes develop in the absence of serum suggested that factors known to inhibit neurogenesis might mediate the effects of serum. It is well-established that BMPs inhibit neural induction in the early embryo [1]. Moreover, BMPs have been shown to antagonize [27], and BMP antagonists to promote [28], neural differentiation of ES cells. However, exposure of ES cells to BMPs alone promotes differentiation into non-neural fates [29–31], suggesting that BMPs are unlikely candidate self-renewal factors. In contrast, Smith and colleagues showed in an elegant series of experiments that exposure of ES cells to BMPs 2 or 4 in combination with LIF is sufficient to sustain self-renewal and preserve multipotency, embryo colonization capacity and germline transmission properties [32]. GDF-6, but not TGF-β1 or activin, similarly supported ES cell self-renewal in the presence of LIF, suggesting that this activity is restricted to BMP receptor ligands.

BMPs may contribute to ES cell self-renewal *via* upregulation of *Id* (Inhibitor of differentiation) genes, which encode negative bHLH factors that have the HLH domain, but lack a basic DNA binding domain [33–35]. Id proteins inhibit the function of bHLH transcription factors, including the proneural bHLH transcription factors Mash-1, Neurogenin and NeuroD [36]. In ES cells, *Id* genes are transcriptionally activated *via* Smad activation [32, 37], and ES cells transfected with *Id1*, *Id2* and *Id3* maintain self-renewal in the presence of LIF alone [32]. Upon withdrawal of LIF, Id-expressing ES cells differentiate but do not give rise to neural lineages. Conversely, removal of the floxed Id construct resulted in neuronal differentiation upon removal of LIF. These findings, in conjunction with the observation that the neurogenic transcription factor *Mash-1* is expressed in undifferentiated ES cells [32], suggest that BMP-induced upregulation of Id expression blocks neuronal lineage-specific transcription factors, thereby enabling the self-renewal response to LIF signaling.

Similarly, BMPs have been shown to promote expansion of neural progenitor cells in the developing spinal cord by inhibiting neuronal differentiation *via* upregulation of Zic genes, which encode zinc finger proteins homologous to *Drosophila* Odd-paired [38]. In the developing chick spinal cord, *Zic1* is increasingly restricted to the dorsal axis of the neural tube. Misexpression of *Zic1* on the ventral spinal cord inhibits neuronal differentiation, whereas overexpression in the dorsal spinal cord causes significant increase in the number of dorsal progenitors. Inhibition of neuronal differentiation by *Zic1* appears to be mediated by upregulation of *Notch1*. Notch signaling is an evolutionarily conserved mechanism that controls differentiation and proliferation of many tissues [39, 40]. Notch signaling activates transcription of *Hes-1* (vertebrate homologs of *Drosophila* hairy and Enhancer of split) and *Hes-5*, both of which encode negative bHLH transcription factors that inhibit transcription activity of neurogenic bHLH proteins by competitive binding to their heterodimer partners, E12 and E47 [41–43]. Expression of *Zic1* enhances *Notch1* and *Hes-1* expression [38], suggesting a mechanism similar to that evoked to explain the effects of BMPs on ES cell self-renewal: BMPs promote expansion of neural progenitor cells by inhibiting expression of neurogenic genes.

Interestingly, BMP signaling has been shown to exert both stimulatory and inhibitory effects on self-renewal. Smith and colleagues demonstrated that BMP induces activation of p38 in ES cells, but using pharmacological inhibitors of p38, determined that this is not required for self-renewal and may even have a negative impact by promoting apoptosis [32]. Since p38 has been shown to antagonize BMP action by inhibitory phosphorylation of Smad1 [44], these findings suggest that self-renewal stimuli may simultaneously activate counterbalancing signals that modulate responsiveness to apoptosis- or differentiation-inducing stimuli and thus act to constrain unrestricted expansion of the stem cell pool. Recent reports by McKay and colleagues suggest an additional mechanism to explain the shift of BMP influence from proliferation to termination in the developing nervous system [11]. Their model, which they refer to as induction-termination, is derived from observations that *in vivo*, neural precursor cells always express *Bmpr-1a*, but *Bmpr-1b* is not expressed until embryonic day 9 and is restricted to the dorsal neural tube surrounding the source of BMP ligands [14]. Activation of BMPR-IA induces dorsal identity and proliferation in both transgenic mice and in cultured stem cells. BMPR-IA signaling simultaneously induces expression of a second BMP receptor, *Bmpr1b*. Activation of BMPR-IB drives mitotic arrest *via* the CDK inhibitor p21^{cip1} and results initially in apoptosis in early gestation embryos and then in neuronal differentiation in mid-gestation embryos [14]. Dominant-negative experiments demonstrated that BMPR-IB activation is both necessary and sufficient for terminal differentiation of neural stem cells that are dorsalized consequent to BMPR-IA activation [14]. Interestingly, two closely related receptors, Alk1 and Alk5, have similarly been shown to mediate opposing effects in response to TGF-β: Alk1 promotes endothelial cell proliferation whereas Alk5 promotes cell-cycle arrest [45]. Thus self-renewal *versus* mitotic arrest is a function of the changing stoichiometry between two receptors in response to a single ligand. Such a mechanism could explain how BMPs cause proliferation in some contexts and differentiation in others.

One of the more perplexing questions raised by observations of BMP effects on neural cell fate determination is why the response of neural progenitor cells to BMPs switches from neurogenesis during mid-gestation to gliogenesis during late gestation and after birth. Analysis of stem cell behavior *in vitro* strongly suggests that transitions in differentiation are due to the changing properties of stem cell populations over time rather than delayed differentiation of separate committed progenitor populations. First, clones of acutely dissociated E10 cortical stem cells give rise to clones containing mostly neurons, whereas clones of older stem cells remain multipotent, but give rise to higher proportions of glia. This progression is recapitulated in individual clones [46]. Second, precursors cultured from progressively older rat cortical tissue respond to BMPs with apoptosis (E13), neurogenesis plus gliogenesis (E16) and finally gliogenesis alone (perinatal) [21, 47, 48]. While committed neuronal or glial progenitors can be isolated and maintained in culture [49], committed oligodendrocyte progenitors can be re-instructed to multipotency using BMPs [50]. Thus, neuronal or glial commitment may remain subject to environmental cues that either promote or repress lineage commitment [50, 51].

These observations suggest that the diverse responses of stem cells to BMPs are due to other signals that modulate responses to BMPs [48, 52] and/or the activation of different BMP signal transduction pathways [53]. With respect to the latter, it has recently been reported that BMP-induced differentiation of smooth muscle cells from neural stem cells requires Smad activation, whereas glial differentiation in response to BMPs occurs *via* activation of a novel signaling pathway involving FRAP (FKPB12/rapamysin-associated protein) and STAT proteins [53]. With respect to the former possibility, the opposite effects of BMPs on neurogenesis at midgestation and late gestation have been attributed, at least in part, to expression and activity of the neurogenic bHLH protein neurogenin [52]. Neurogenin is abundantly expressed in neural precursor cells in midgestation, but is very low in late gestation when BMPs suppress neurogenesis and promote gliogenesis, most likely in cooperation with LIF cytokines [21, 54, 55]. Binding of BMP ligands to BMP receptors on neural precursor cells results in phosphorylation of Smads, which translocate to the nucleus and bind to p300 [54]. The resultant Smad/p300 complex has been shown to form a complex with STAT3 to effectively induce astrocyte differentiation [54]. However, the Smad/p300 complex has also been shown to associate with neurogenin [52], and this enhances the transcriptional activity and neurogenic potential of neurogenin, implying that the Smad/p300 complex acts as a coactivator for neurogenin. Moreover, when neurogenin expression is high, as in neural precursor cells at midgestation, neurogenin sequesters Smad/p300 and prevents its interaction with STAT3 [52]. In summary, these findings suggest that BMPs can differentially regulate cell fate decisions by activating distinct cytoplasmic signals, and that conserved paradigms of BMP signaling are repeated in various contexts throughout the development of the nervous system.

Regulation of neurotransmitter phenotype by BMPs

Some of the first studies of the effects of BMPs on neurons reported changes in expression of enzymes involved in norepinephrine synthesis [56–58] and in the abundance of mRNAs for neuropeptides in sympathetic ganglia [59]. Subsequent studies have borne out these observations and identified additional transmitter phenotypes that are affected by BMPs. In addition, the range of cells whose transmitter phenotype is known to be affected by BMPs has been greatly expanded. It now includes most types of neurons in the peripheral nervous system, including sympathetic [56–58], parasympathetic [60], and sensory neurons [61], as well as monoaminergic neurons in the vertebrate central nervous system [62] and peptidergic neurons in *Drosophila* [63, 64]. Thus, effects on neurotransmitter phenotypes represent one of the most profound and widespread actions of BMPs, GDFs and activins.

Sympathetic neurons are derived from the neural crest and their acquisition of the adrenergic phenotype is driven primarily by environmental interactions. Four lines of evidence indicate that BMPs induce the noradrenergic phenotype in sympathetic neurons. First, when migrating neural crest cells aggregate to form sympathetic ganglia, they are exposed to several BMPs (-2, -4, -7) that are synthesized in the dorsal aorta [12, 58]. Second, these same BMPs have been found to induce the expression of tyrosine hydroxylase (TH), the rate-limiting enzyme in norepinephrine synthesis. In mass avian cultures, this effect is observed under standard culture conditions [56–58], but it requires lowered oxygen pressure plus forskolin exposure in clonal cultures of mammalian neural crest cells [65]. The reason for these differences is unclear, but the oxygen concentration (~5%) that maximally facilitates BMP effects in mammalian neural crest is closer to physiological levels than are levels in standard tissue culture incubators. Third, in addition to inducing TH *in vitro*, it was also found that overexpression of BMP-4 in the vicinity of sympathetic ganglia *in ovo* caused ectopic cells to begin expressing TH [58, 66]. Finally, implantation of beads releasing noggin, a BMP inhibitor, prevented the development of TH-positive neurons in sympathetic ganglia [66].

How do BMPs induce the adrenergic phenotype? MASH1 is a transcription factor that is required for the development of autonomic neurons [67]. BMPs regulate the expression of MASH1 [67] and also the homeodomain transcription factor Phox 2b [68, 69], which is expressed in all noradrenergic neurons and is required for their development. Downstream from MASH1 and Phox 2b are other transcription factors, including Phox 2a, dHand and GATA3 [62, 70]. However, the pathway is not linear and downstream elements such as Phox 2a can influence upstream elements such as Phox 2b [62]. The actions of the Phox 2 genes on dopamine-beta hydroxylase are probably direct because they bind to regulatory sequences in its promoter [71]. Evidence for the regulation of the TH-promoter by Phox 2 genes is controversial [62].

MASH1 and the Phox 2 genes are also required for the development of dopaminergic and noradrenergic neurons in the central nervous system [68, 72, 73], and so it might be anticipated that BMPs would affect the differentiation of these neurons. Consistent with this possibility, it has been found that BMPs are required for the generation of noradrenergic neurons in zebra fish hindbrain [74] and the mammalian locus coeruleus [75] and for the differentiation of dopaminergic neurons in *C. elegans* [76]. In addition, BMP-2, -5, -6, -7 and activin stimulate the differentiation of dopaminergic neurons in cultures derived from ventral mesencephalon [77, 78] and basal forebrain ventricular zone [79], respectively, and BMP-2, -4 and -6 induce a dopaminergic phenotype in cultured striatal neurons [80].

BMPs also affect transmitter synthesis in non-catecholaminergic cells. BMP-2 promotes the survival and differential of GABAergic striatal neurons [81], and BMP-9 increases cholinergic function in embryonic forebrain cells, both *in vitro* and *in vivo* [82]. Surprisingly, BMPs also promote cholinergic differentiation in the parasympathetic ciliary ganglion and this effect is dependent on many of the same genes (MASH, Phox 2 a and b) that regulate TH expression in sympathetic neurons [60]. However, unlike sympathetic neurons, ciliary neurons do not express dHand and in the absence of this gene, cholinergic differentiation is favored over catecholaminergic development. Thus, in the peripheral nervous system, the expression of dHand qualitatively alters the response to BMPs.

The development of ciliary neurons is also regulated by target-derived activin and follistatin. Activin from the choroid induces somatostatin in ciliary neurons projecting to this target, whereas ciliary neurons projecting to the iris encounter both activin and its inhibitor follistatin, which prevents expression of somatostatin [83]. Activin and BMPs also alter peptide expression in dorsal root sensory neurons by increasing the synthesis of CGRP (calcitonin gene related product) [61]. One of the main targets of CGRP-containing neurons is skin, and activin and BMPs were detected in this tissue as well as other target tissues of CGRP containing axons [84]. Thus, target-derived members of the TGF-β family are likely candidates for regulators of sensory peptide transmitters and their activity may also be regulated by endogenous inhibitors.

Effects of BMPs on the growth of neural processes

One of the most prominent effects of BMPs in neural tissue is the stimulation of dendritic growth. This has been observed in a variety of systems including cultured sympathetic [85], hippocampal [86], cerebral cortical [87], and striatal neurons [88], as well as in intraocular transplants of spinal neurons [89]. In sympathetic neurons, the magnitude of the effect is exceptionally large [85]. BMP-7 induces a 40-fold increase in total dendritic length when compared to cultures grown with nerve growth factor alone, and the size of the dendritic arbor formed during a four-week exposure to BMP-7 *in vitro* is comparable to that which is formed during the first month of postnatal development *in vivo*. BMP-7 is thus a sufficient stimulus to allow generation of a normal dendritic arbor. In other cells, the effect is more modest, but still substantial. For example, the dendritic length of hippocampal neurons increases more than two-fold in the presence of BMP-7 [86]. The effect on dendritic length appears to represent a specific morphogenetic change because BMPs do not affect either cell survival or axonal length in these cultures [86, 88, 90, 91]. In this respect, BMPs differ from NGF, which acts non-selectively to increase neuronal survival and volume as well as axonal and dendritic length [85].

Dendrites are the primary site of synapse formation in the vertebrate nervous system and increased dendritic growth might be expected to lead to increases in the number of synaptic contacts. This has been observed in cultures of hippocampal and sympathetic neurons [85, 86], suggesting that BMPs have the potential to modify neural circuitry. Increased dendritic growth has been observed with several BMPs (-2, -4, -5, -6, -7) but not with CDMPs, TGF-β, or activin [92, 93] and it is accompanied with increases in MAP-2 [86, 92], a protein required for dendritic growth [94]. In sympathetic neurons, BMP-induced dendritic growth requires Smad1 [95] and the activation of an intracellular FGF receptor [96]. In addition, BMP-induced dendritic growth is inhibited by proteasome inhibitors and is associated with activation of the Tlx-2 [95]. Gruendler et al. [97] reported that Smad1 regulates the proteasomal degradation of SNIP1, which is a repressor of the master transcriptional coactivator CBP/p300 [98] and CPB/p300 has been implicated in Smad1 signaling. Thus, the inhibitory effects of proteasome inhibitors may be due to a blockage of SNIP1 degradation. In this respect, it is interesting to note the BMPs also stimulate the degradation of another transcriptional regulator, MASH1, in olfactory neurons [99], suggesting that regulated proteolysis of nuclear proteins is an important aspect of BMP-TGF-β signaling in neurons as well as epithelial cells [100].

BMP-induced dendritic growth has been studied primarily in tissue culture and so the cellular sources of BMPs *in vivo* are unknown. However, glial-derived BMPs are probably important during some stages of development. In the perinatal period, glial Schwann cells in rat sympathetic ganglia make large amounts of BMP-5, -6 and -7 *in vivo* [101]. Moreover, co-culture with Schwann cells promotes dendritic growth and the effects of these glial cells on dendritic growth are reduced by exposure to antibodies to BMP-6 and -7 or to endogenous BMP antagonists such as follistatin and noggin. It is therefore likely that ganglionic Schwann cells are a physiological source of dendrite-promoting BMPs in sympathetic ganglia during the first few postnatal months. However, after that time BMP levels fall in ganglia and it is likely that target-derived factors become more important [102]. In this respect, it is important to note many targets of sympathetic neurons secrete significant amounts of BMPs in adulthood and that BMPs can be retrogradely transported [102].

Dendritic arbors are complex and are subject to both positive and negative regulatory influences. In sympathetic ganglia, the dendrite-promoting activities of BMPs are antagonized by neuropeptide transmitters, such as PACAP and VIP, which are released from preganglionic neurons [103], as well as by retinoic acid, a morphogen that is synthesized in sympathetic ganglia [104]. In addition, agents that activate the ERK1/ERK2 pathway can interfere with BMP-induced dendritic growth [91]. A negative interaction between these pathways has been observed in several tissues and probably reflects phosphorylation of Smad1 at four MAPK sites in the linker region [105]. Nerve growth factor (NGF) is the most important survival factor for sympathetic neurons and one of the several signaling pathways that it activates is the ERK1/2 pathway [91]. Moreover, inhibition of this pathway augments BMP-induced dendritic growth, suggesting that one of the signaling cascades activated by NGF can act in an antagonistic manner in sympathetic neurons and reduce the dendritic growth induced by other NGF-sensitive pathways.

Interferon-γ [90] and members of the IL-6 cytokine family [106, 107] block the dendrite-promoting effects of BMPs and also cause retraction of existing dendrites. These properties are important because axotomy and inflammation induce the synthesis of these cytokines [108, 109] and axotomy causes dendritic retraction [110]. It is, therefore, likely that some of the regressive effects of axonal injury are mediated by cytokine-induced changes in the responsiveness to BMPs. The mechanism by which cytokines block BMP action in sympathetic ganglia is unclear. However, IL-6 related cytokines activate Stat 3 in sympathetic neurons [106] and Nakashima et al. [54] reported that Stat3 and Smad1 both bind to the p300 transcriptional activator and act synergistically in neuroepithelial cultures.

In addition to stimulating dendritic growth, BMPs have also been found to affect axons (reviewed in [1]). However, their most prominent effects appear to be on axonal guidance rather than axonal growth *per se*. Initially Colavita et al. [111] reported that mutation of the UNC-129 gene, which encodes a member of the TGF-β superfamily, interfered with pioneer axonal guidance in *C. elegans*. Subsequently it was found that BMPs mediate the repulsion of commissural axons by the roof plate [112] and that noggin-mutant mice show defects in the projections of spinal motor and dorsal root ganglion axons as well as in the guidance of cranial nerve VII [113]. In addition, Liu et al. [114] observed that in mice with targeted mutation of one of the BMP receptors (BMPR-IB), axons arising from ventrally located retinal ganglion cells failed to enter the head of the optic nerve and instead made abrupt turns away from it. In most cases it is not known whether the guidance errors reflect direct effects of BMPs on growth cones or indirect effects arising from BMP-induced changes in guidance molecules such as netrins or ephrins. However, with commissural neurons it was possible to demonstrate that BMPs cause growth cones on isolated neurons to undergo a rapid collapse [112]. Thus, under some circumstances, BMPs can act directly as repellents, just as they can act as chemoattractants [115]. Correct commissural guidance is dependent on expression of both BMP-7 and GDF-7 [116]. Moreover, BMP-7 and GDF-7 can form heterodimers *in vitro* and under these conditions, GDF-7 enhances the effects of BMP-7 on growth cones. It is, therefore, possible that it is the heterodimer that is involved in axonal guidance.

BMP regulation of synaptic growth

Precise regulation of synapse formation and subsequent synaptic expansion, retraction and remodeling throughout life is critical for normal neural function. However, the molecular machinery that controls synaptic growth is poorly understood. Recent evidence from forward [117] and reverse [63] genetic studies of synapse development in the *Drosophila* neuromuscular junction (NMJ) indicates a critical role for BMP signaling in synaptic growth. *Drosophila* NMJ synapses are initially established during late embryonic stages and then grow significantly during larval development as evidenced by increases in branching, bouton number, and active zones per bouton [118–120]. This synaptic growth is thought to help maintain constant synaptic efficacy as the muscle volume increases up to 100-fold during larval growth. Mutations in the *wishful thinking* (*wit*) gene, the *Drosophila* homolog of the vertebrate BMP type II receptor, greatly reduce the size of NMJ synapses relative to the size of the muscles [117]. This mutation appears to selectively target synaptic growth in the NMJ since other aspects of neural development, including general organization of neural centers, motor neuron specification, axonal outgrowth and target development, appear normal in *wit* mutants. Interestingly, synapse size is close to normal in *wit* mutants at the end of embryogenesis, but becomes increasingly smaller relative to wildtype NMJ synapses as larvae grow [117], suggesting that *wit* is required not for initial formation of synapses but rather for synaptic growth. Consistent with defective synaptic growth, mutant larvae also exhibit severe defects in evoked junctional potentials and a lower frequency of spontaneous vesicle release, which most likely result from ultrastructural defects in presynaptic active zones [117]. Two lines of evidence suggest that Wit functions as a presynaptic receptor that regulates synaptic size in the *Drosophila* NMJ [117]. First, *in situ* hybridization reveals *wit* expression in motor neurons, but not in muscle. Second, the NMJ phenotype is specifically rescued by transgenic expression of *wit* in motor neurons, but not in muscle.

These observations suggest a mechanism for coordinating synapse growth with muscle growth: muscle-derived BMP ligands act as a retrograde signal that binds to Wit receptors on motor neurons activating BMP signaling pathways resulting in transcriptional regulation of genes that modulate synapse assembly and activity. Consistent with this hypothesis are observations that mutations in *mad*, the *Drosophila* homolog of Smad1, or in *saxophone* (*sax*) and *thickveins* (*tkv*), the *Drosophila* genes that encode type I BMP receptors, phenocopy the *wit* mutation [117, 121], and that transgenic expression of constitutively active Sax and Tkv can rescue the NMJ phenotype in *wit* mutants [117]. Furthermore, *wit*, *sax* and *tkv* mutant embryos show a specific loss of phosphorylated Mad, within motor neurons, but not in other tissues [63, 121]. Another critical piece of evidence in support of the hypothesis that retrograde BMP signaling modulates NMJ synaptic growth is the finding that muscle cells express the *Drosophila* gene *glass bottom boat* (*gbb*), which encodes a BMP ligand, and that mutations in *gbb* also reduce NMJ synapse size, decrease neurotransmitter release and result in aberrant presynaptic ultrastructure [122]. Transgenic expression of *gbb* in muscle rescues key aspects of the gbb mutant phenotype, but notably defects in neurotransmission can be rescued by expression of Gbb either pre- or postsynaptically. *In vitro* experiments with S2 cells indicate that Gbb binds Wit receptors resulting in increased Mad phosphorylation [122]. That this interaction is important physiologically is suggested by demonstrations that inhibition of retrograde axonal transport in Drosophila larvae by overexpression of dominant negative homologs of dynein, effectively eliminates accumulation of phosphorylated Mad in motor neurons and results in small NMJ synapses [122].

In summary, these recent findings suggest a critical role for BMP signaling in regulating *Drosophila* synapse assembly and/or maintenance, and raise the possibility that like TGF-β [123, 124], BMP signaling regulates additional aspects of synaptic plasticity, such as long-term facilitation and memory. These studies also raise numerous questions, the most significant of which focus on the regulation of Gbb delivery to the presynaptic Wit receptor, the nature of the retrograde signal, the gene targets downstream of the Wit receptor, and whether BMP signaling similarly regulates synaptic growth in vertebrates.

BMPs in adult brain and potential therapeutic uses

BMPs are prominently expressed in many areas of the developing nervous system [1]. Overall expression is lower in the adult nervous system. However, BMPs can still be detected in many brain structures. For example, the adult hippocampus expresses significant levels of BMP-5, BMP-6, GDF-1 and GDF-10; the neocortex, BMP-5, BMP-6 and GDF-1; the cerebellum, BMP-5 and BMP-6; the striatum, BMP-5 and BMP-7; the spinal cord, BMP-7; and the brainstem, BMP-5 and BMP-6 [125–129]. BMP-7 has also been detected in the plexiform and ganglion layers of adult human retina [130].

BMP-7 and GDF-15 are highly expressed in the choroid plexus [131–133] and there is evidence that both are secreted into the cerebrospinal fluid (CSF) [134, 135]. In fact, BMP-7 levels are surprisingly high (~ 30 ng/ml) and this concentration is sufficient to elicit near maximal responses in bioassays [134]. In this respect BMPs appear to be unique: most other trophic factors are present at such low levels in CSF that their biological activity is undetectable. The role of BMP-7 and GDF-15 in CSF is unknown. There is evidence that GDF-15 may penetrate from the ventricles into the parenchyma during development, but this appears less prominent in adults [131, 135] and there are no data on BMP-7 penetration. However, it is important to note that Lim et al. [136] found that adult subventricular zone (SVZ) cells express BMPs and BMPRs and that adjacent ependymal cells produce noggin. In SVZ cells cultured from adult brains, exogenous BMPs or overexpression of constitutively active type I BMPRs inhibits neurogenesis and promotes glial differentiation, in agreement with observations reported by others [137]. In contrast, exogenous noggin promotes neurogenesis and inhibits glial differentiation. *In vivo*, overexpression of BMP-7 in the ependyma inhibits neurogenesis while stimulating generation of glial cell types and ectopic expression of noggin in the striatum promotes neuronal differentiation of SVZ cells grafted to the striatum. These data suggest that the production of noggin by the ependyma creates a neurogenic environment in the adjacent SVZ by blocking endogenous BMP signaling or possibly CSF mediated signaling. Similarly, it has been recently reported that a novel BMP antagonist, neurogenesin-1, promotes neuronal differentiation in neural stem cells of the adult hippocampus [138].

Lewen et al. [139] reported that mild cerebral contusion induced increased expression of mRNAs for BMPR-II and ActR-1 and Charytoniuk et al. [140] observed that ischemia also increased expression of BMPR-II transcripts. In addition, injury and ischemia have been found to increase the expression of BMP ligands, including: activin in the hippocampus [141, 142]; BMP-7 in the spinal cord [143]; and GDF-15 in the cerebral cortex [131]. These observations have led to suggestions that BMPs might be involved in neural regeneration; they have also prompted exploration of potential therapeutic uses of exogenous BMPs. Initial studies focused on the effects of BMP-7 (also known as osteogenic protein-1 or OP-1) because it had previously been found to be protective in a rat model of myocardial ischemia and reperfusion [144]. Perides et al. [145] found that BMP-7 reduced cerebral infarct area and mortality in 12 to 14-day old rats subjected to bilateral carotid artery ligation followed by hypoxia. Similarly, Lin et al. [146] and Wang et al. [147] found that prior administration of BMP-7 or BMP-6 reduced motor impairment caused by middle cerebral artery ligation. These observations indicate that BMPs can be neuroprotective, but their relevance to stroke was unclear because the BMPs were administered prior to the event.

Kawamata et al. [148] administered BMP-7 intracisternally one and four days after middle cerebral artery occlusion and observed a sustained increase in behavior recovery and a subsequent study revealed that the window for initial intervention could be as long as three days [149]. More recently, Chang et al. [141] examined the effects of intravenous administration of BMP-7 24 h after middle cerebral artery occulsion. BMP-7 crossed the blood brain barrier and entered the ischemic area. Behaviorally, BMP-7 caused a decrease in body asymmetry from days 7–14 and an increase in locomotor activity at later times. The finding that a single intravenous dose of BMP-7 can produce long-lasting behavioral improvement suggests that its potential for treating stroke should be further explored, as should its mechanism of action. In studies where BMPs were given before the ischemic event, a decreased infarct size was observed [145]. This action may be related to the ability of BMPs to regulate current flow through ionotropic glutamate receptors [130]. However, in studies where administration was delayed for 24 h, infarct size was not reduced [148], presumably because most of the cell death occurred within the first day. Under these conditions, the ability of BMPs to protect cerebral cortical dendrites from glutamate-induced damage [150] and to stimulate dendritic growth [85, 86] are likely to contribute to the enhanced recovery.

The protective action of BMPs also extends to other parts of the nervous system. GDF-5 protects dopaminergic neurons exposed to MPP$^+$ [151] or 6-hydroxydopamine [152] and GDF-15 reduces cell death in nigrostriatal neurons treated with 6-hydroxydopamine [135]. The survival of dopaminergic neurons that have been grafted into lesioned striatum [153, 154] is also increased by either BMP-2 or GDF-5. Thus, BMPs have neuroprotective effects in animal models of Parkinson's disease. In addition, striatal neurons were protected by activin A in a quinolinic lesion model of Huntington's disease [155] and this ligand also reduced excitotoxin-induced cell death in the hippocampus [156].

Acknowledgements
This work was supported by grants from the National Science Foundation (DH, IBN0121210) and NINDS (PL, R01 NS46649).

References

1   Lein P, Drahushak KM, Higgins D (2002) Bone morphogenetic proteins in the nervous system. In: S Vukicevic, KT Sampath (eds): *Bone morphogenetic proteins: From laboratory to clinical practice*. Birkhäuser Verlag, Basel, Switzerland, 289–319

2   Krieglstein K, Strelau J, Schober A, Sullivan A, Unsicker K (2002) TGF-beta and the regulation of neuron survival and death. *J Physiol Paris* 96: 25–30

3   Munoz-Sanjuan I, Brivanlou AH (2002) Neural induction, the default model and embryonic stem cells. *Nat Rev Neurosci* 3: 271–280

4   Schier AF (2003) Nodal signaling in vertebrate development. *Annu Rev Cell Dev Biol* 19: 589–621

5   Knecht AK, Bronner-Fraser M (2002) Induction of the neural crest: a multigene process. *Nat Rev Genet* 3: 453–461

6   Helms AW, Johnson JE (2003) Specification of dorsal spinal cord interneurons. *Curr Opin Neurobiol* 13: 42–49

7   Jessell TM (2000) Neuronal specification in the spinal cord: inductive signals and transcriptional codes. *Nat Rev Genet* 1: 20–29

8   Weiss S, Reynolds BA, Vescovi AL, Morshead C, Craig CG, van der Kooy D (1996) Is there a neural stem cell in the mammalian forebrain? *Trends Neurosci* 19: 387–393

9   McKay R (1997) Stem cells in the central nervous system. *Science* 276: 66–71

10   Edlund T, Jessell TM (1999) Progression from extrinsic to intrinsic signaling in cell fate specification: a view from the nervous system. *Cell* 96: 211–224

11   Panchision DM, McKay RD (2002) The control of neural stem cells by morphogenic signals. *Curr Opin Genet Dev* 12: 478–487

12   Shah NM, Groves AK, Anderson DJ (1996) Alternative neural crest cell fates are instructively promoted by TGFbeta superfamily members. *Cell* 85: 331–343

13   Sela-Donenfeld D, Kalcheim C (1999) Regulation of the onset of neural crest migration by coordinated activity of BMP4 and Noggin in the dorsal neural tube. *Development* 126: 4749–4762

14   Panchision DM, Pickel JM, Studer L, Lee SH, Turner PA, Hazel TG, McKay RD (2001) Sequential actions of BMP receptors control neural precursor cell production and fate. *Genes Dev* 15: 2094–2110

15   Liem KF, Tremml G, Jessell TM (1997) A role for the roof plate and its resident TGF-beta-related proteins in neuronal patterning in the dorsal spinal cord. *Cell* 91: 127–138

16   Graham A, Francis-West P, Brickell P, Lumsden A (1994) The signalling molecule BMP4 mediates apoptosis in the rhombencephalic neural crest. *Nature* 372: 684–686

17   Furuta Y, Piston DW, Hogan BL (1997) Bone morphogenetic proteins (BMPs) as regulators of dorsal forebrain development. *Development* 124: 2203–2212

18   Li W, Cogswell CA, LoTurco JJ (1998) Neuronal differentiation of precursors in the neocortical ventricular zone is triggered by BMP. *J Neurosci* 18: 8853–8862

19   Wu HH, Ivkovic S, Murray RC, Jaramillo S, Lyons KM, Johnson JE, Calof AL (2003) Autoregulation of neurogenesis by GDF11. *Neuron* 37: 197–207

20   Mabie PC, Mehler MF, Kessler JA (1999) Multiple roles of bone morphogenetic protein signaling in the regulation of cortical cell number and phenotype. *J Neurosci* 19: 7077–7088

21   Gross RE, Mehler MF, Mabie PC, Zang Z, Santschi L, Kessler JA (1996) Bone morphogenetic proteins promote astroglial lineage commitment by mammalian subventricular zone progenitor cells. *Neuron* 17: 595–606

22 Molne M, Studer L, Tabar V, Ting YT, Eiden MV, McKay RD (2000) Early cortical precursors do not undergo LIF-mediated astrocytic differentiation. *J Neurosci* Res 59: 301–311

23 Mujtaba T, Mayer-Proschel M, Rao MS (1998) A common neural progenitor for the CNS and PNS. *Dev Biol* 200: 1–15

24 Altmann CR, Brivanlou AH (2001) Neural patterning in the vertebrate embryo. *Int Rev Cytol* 203: 447–482

25 Anderson DJ (2001) Stem cells and pattern formation in the nervous system: the possible *versus* the actual. *Neuron* 30: 19–35

26 Smith AG (2001) Embryo-derived stem cells: of mice and men. *Annu Rev Cell Dev Biol* 17: 435–462

27 Tropepe V, Hitoshi S, Sirard C, Mak TW, Rossant J, van der Kooy D (2001) Direct neural fate specification from embryonic stem cells: a primitive mammalian neural stem cell stage acquired through a default mechanism. *Neuron* 30: 65–78

28 Gratsch TE, O'Shea KS (2002) Noggin and chordin have distinct activities in promoting lineage commitment of mouse embryonic stem (ES) cells. *Dev Biol* 245: 83–94

29 Ying QL, Stavridis M, Griffiths D, Li M, Smith A (2003) Conversion of embryonic stem cells to neuroectodermal precursors in adherent monoculture. *Nat Biotechnol* 21: 183–186

30 Wiles MV, Johansson BM (1999) Embryonic stem cell development in a chemically defined medium. *Exp Cell Res* 247: 241–248

31 Johansson BM, Wiles MV (1995) Evidence for involvement of activin A and bone morphogenetic protein 4 in mammalian mesoderm and hematopoietic development. *Mol Cell Biol* 15: 141–151

32 Ying QL, Nichols J, Chambers I, Smith A (2003) BMP induction of Id proteins suppresses differentiation and sustains embryonic stem cell self-renewal in collaboration with STAT3. *Cell* 115: 281–292

33 Norton JD, Deed RW, Craggs G, Sablitzky F (1998) Id helix-loop-helix proteins in cell growth and differentiation. *Trends Cell Biol* 8: 58–65

34 Christy BA, Sanders LK, Lau LF, Copeland NG, Jenkins NA, Nathans D (1991) An Id-related helix-loop-helix protein encoded by a growth factor-inducible gene. *Proc Natl Acad Sci USA* 88: 1815–1819

35 Benezra R, Davis RL, Lockshon D, Turner DL, Weintraub H (1990) The protein Id: a negative regulator of helix-loop-helix DNA binding proteins. *Cell* 61: 49–59

36 Ross SE, Greenberg ME, Stiles CD (2003) Basic helix-loop-helix factors in cortical development. *Neuron* 39: 13–25

37 Hollnagel A, Oehlmann V, Heymer J, Ruther U, Nordheim A (1999) Id genes are direct targets of bone morphogenetic protein induction in embryonic stem cells. *J Biol Chem* 274: 19838–19845

38 Aruga J, Tohmonda T, Homma S, Mikoshiba K (2002) Zic1 promotes the expansion of dorsal neural progenitors in spinal cord by inhibiting neuronal differentiation. *Dev Biol* 244: 329–341

39 Weinmaster G (1997) The ins and outs of notch signaling. *Mol Cell Neurosci* 9: 91–102

40 Artavanis-Tsakonas S, Rand MD, Lake RJ (1999) Notch signaling: cell fate control and signal integration in development. *Science* 284: 770–776

41 Sasai Y, Kageyama R, Tagawa Y, Shigemoto R, Nakanishi S (1992) Two mammalian helix-loop-helix factors structurally related to Drosophila hairy and Enhancer of split. *Genes Dev* 6: 2620–2634

42 Ohtsuka T, Ishibashi M, Gradwohl G, Nakanishi S, Guillemot F, Kageyama R (1999) Hes1 and Hes5 as notch effectors in mammalian neuronal differentiation. *EMBO J* 18: 2196–2207

43 Akazawa C, Sasai Y, Nakanishi S, Kageyama R (1992) Molecular characterization of a rat negative regulator with a basic helix-loop-helix structure predominantly expressed in the developing nervous system. *J Biol Chem* 267: 21879–21885

44 Kretzschmar M, Doody J, Massague J (1997) Opposing BMP and EGF signalling pathways converge on the TGF-beta family mediator Smad1. *Nature* 389: 618–622

45 Goumans MJ, Valdimarsdottir G, Itoh S, Rosendahl A, Sideras P, ten Dijke P (2002) Balancing the activation state of the endothelium *via* two distinct TGF-beta type I receptors. *EMBO J* 21: 1743–1753

46 Qian X, Shen Q, Goderie SK, He W, Capela A, Davis AA, Temple S (2000) Timing of CNS cell generation: a programmed sequence of neuron and glial cell production from isolated murine cortical stem cells. *Neuron* 28: 69–80

47 Mehler MF, Mabie PC, Zhu G, Gokhan S, Kessler JA (2000) Developmental changes in progenitor cell responsiveness to bone morphogenetic proteins differentially modulate progressive CNS lineage fate. *Dev Neurosci* 22: 74–85

48 White PM, Morrison SJ, Orimoto K, Kubu CJ, Verdi JM, Anderson DJ (2001) Neural crest stem cells undergo cell-intrinsic developmental changes in sensitivity to instructive differentiation signals. *Neuron* 29: 57–71

49 Mujtaba T, Piper DR, Kalyani A, Groves AK, Lucero MT, Rao MS (1999) Lineage-restricted neural precursors can be isolated from both the mouse neural tube and cultured ES cells. *Dev Biol* 214: 113–127

50 Kondo T, Raff M (2000) Oligodendrocyte precursor cells reprogrammed to become multipotential CNS stem cells. *Science* 289: 1754–1757

51 Morrow T, Song MR, Ghosh A (2001) Sequential specification of neurons and glia by developmentally regulated extracellular factors. *Development* 128: 3585–3594

52 Sun Y, Nadal-Vicens M, Misono S, Lin MZ, Zubiaga A, Hua X, Fan G, Greenberg ME (2001) Neurogenin promotes neurogenesis and inhibits glial differentiation by independent mechanisms. *Cell* 104: 365–376

53 Rajan P, Panchision DM, Newell LF, McKay RD (2003) BMPs signal alternately through a SMAD or FRAP-STAT pathway to regulate fate choice in CNS stem cells. *J Cell Biol* 161: 911–921

54 Nakashima K, Yanagisawa M, Arakawa H, Kimura N, Hisatsune T, Kawabata M, Miyazono K, Taga T (1999) Synergistic signaling in fetal brain by STAT3-Smad1 complex bridged by p300 [see comments]. *Science* 284: 479–482

55 Nakashima K, Yanagisawa M, Arakawa H, Taga T (1999) Astrocyte differentiation mediated by LIF in cooperation with BMP2. *FEBS Lett* 457: 43–46

56 Varley JE, Wehby RG, Rueger DC, Maxwell GD (1995) Number of adrenergic and islet-1 immunoreactive cells is increased in avian trunk neural crest cultures in the presence of human recombinant osteogenic protein-1. *Dev Dyn* 203: 434–447

57 Varley JE, Maxwell GD (1996) BMP-2 and BMP-4, but not BMP-6, increase the number of adrenergic cells which develop in quail trunk neural crest cultures. *Exp Neurol* 140: 84–94

58 Reissmann E, Ernsberger U, Francis-West PH, Rueger D, Brickell PM, Rohrer H (1996) Involvement of bone morphogenetic protein-4 and bone morphogenetic protein-7 in the differentiation of the adrenergic phenotype in developing sympathetic neurons. *Development* 122: 2079–2088

59 Fann MJ, Patterson PH (1994) Depolarization differentially regulates the effects of bone morphogenetic protein (BMP)-2, BMP-6, and activin A on sympathetic neuronal phenotype. *J Neurochem* 63: 2074–2079

60 Muller F, Rohrer H (2002) Molecular control of ciliary neuron development: BMPs and downstream transcriptional control in the parasympathetic lineage. *Development* 129: 5707–5717

61 Ai X, Cappuzzello J, Hall AK (1999) Activin and bone morphogenetic proteins induce calcitonin gene-related peptide in embryonic sensory neurons *in vitro*. *Mol Cell Neurosci* 14: 506–518

62 Goridis C, Rohrer H (2002) Specification of catecholaminergic and serotonergic neurons. *Nat Rev Neurosci* 3: 531–541

63 Marques G, Bao H, Haerry TE, Shimell MJ, Duchek P, Zhang B, O'Connor MB (2002) The Drosophila BMP type II receptor Wishful Thinking regulates neuromuscular synapse morphology and function. *Neuron* 33: 529–543

64 Allan DW, St Pierre SE, Miguel-Aliaga I, Thor S (2003) Specification of neuropeptide cell identity by the integration of retrograde BMP signaling and a combinatorial transcription factor code. *Cell* 113: 73–86

65 Morrison SJ, Csete M, Groves AK, Melega W, Wold B, Anderson DJ (2000) Culture in reduced levels of oxygen promotes clonogenic sympathoadrenal differentiation by isolated neural crest stem cells. *J Neurosci* 20: 7370–7376

66 Schneider C, Wicht H, Enderich J, Wegner M, Rohrer H (1999) Bone morphogenetic proteins are required *in vivo* for the generation of sympathetic neurons. *Neuron* 24: 861–870

67 Lo L, Tiveron MC, Anderson DJ (1998) MASH1 activates expression of the paired homeodomain transcription factor Phox2a, and couples pan-neuronal and subtype-specific components of autonomic neuronal identity. *Development* 125: 609–620

68 Pattyn A, Morin X, Cremer H, Goridis C, Brunet JF (1999) The homeobox gene Phox2b is essential for the development of autonomic neural crest derivatives. Nature 399: 366–370

69 Stanke M, Junghans D, Geissen M, Goridis C, Ernsberger U, Rohrer H (1999) The

Phox2 homeodomain proteins are sufficient to promote the development of sympathetic neurons. *Development* 126: 4087–4094

70 Howard MJ, Stanke M, Schneider C, Wu X, Rohrer H (2000) The transcription factor dHAND is a downstream effector of BMPs in sympathetic neuron specification [In Process Citation]. *Development* 127: 4073–4081

71 Kim HS, Seo H, Yang C, Brunet JF, Kim KS (1998) Noradrenergic-specific transcription of the dopamine beta-hydroxylase gene requires synergy of multiple cis-acting elements including at least two Phox2a-binding sites. *J Neurosci* 18: 8247–8260

72 Hirsch MR, Tiveron MC, Guillemot F, Brunet JF, Goridis C (1998) Control of noradrenergic differentiation and Phox2a expression by MASH1 in the central and peripheral nervous system. *Development* 125: 599–608

73 Morin X, Cremer H, Hirsch MR, Kapur RP, Goridis C, Brunet JF (1997) Defects in sensory and autonomic ganglia and absence of locus coeruleus in mice deficient for the homeobox gene Phox2a. *Neuron* 18: 411–423

74 Guo S, Brush J, Teraoka H, Goddard A, Wilson SW, Mullins MC, Rosenthal A (1999) Development of noradrenergic neurons in the zebrafish hindbrain requires BMP, FGF8, and the homeodomain protein soulless/Phox2a. *Neuron* 24: 555–566

75 Vogel-Hopker A, Rohrer H (2002) The specification of noradrenergic locus coeruleus (LC) neurones depends on bone morphogenetic proteins (BMPs). *Development* 129: 983–991

76 Lints R, Emmons SW (1999) Patterning of dopaminergic neurotransmitter identity among *Caenorhabditis elegans* ray sensory neurons by a TGFbeta family signaling pathway and a Hox gene. *Development* 126: 5819–5831

77 Reiriz J, Espejo M, Ventura F, Ambrosio S, Alberch J (1999) Bone morphogenetic protein-2 promotes dissociated effects on the number and differentiation of cultured ventral mesencephalic dopaminergic neurons. *J Neurobiol* 38: 161–170

78 Brederlau A, Faigle R, Kaplan P, Odin P, Funa K (2002) Bone morphogenetic proteins but not growth differentiation factors induce dopaminergic differentiation in mesencephalic precursors. *Mol Cell Neurosci* 21: 367–378

79 Daadi M, Arcellana-Panlilio MY, Weiss S (1998) Activin co-operates with fibroblast growth factor 2 to regulate tyrosine hydroxylase expression in the basal forebrain ventricular zone progenitors. *Neuroscience* 86: 867–880

80 Stull ND, Jung JW, Iacovitti L (2001) Induction of a dopaminergic phenotype in cultured striatal neurons by bone morphogenetic proteins. *Brain Res Dev Brain Res* 130: 91–98

81 Hattori A, Katayama M, Iwasaki S, Ishii K, Tsujimoto M, Kohno M (1999) Bone morphogenetic protein-2 promotes survival and differentiation of striatal GABAergic neurons in the absence of glial cell proliferation. *J Neurochem* 72: 2264–2271

82 Lopez-Coviella I, Berse B, Krauss R, Thies RS, Blusztajn JK (2000) Induction and maintenance of the neuronal cholinergic phenotype in the central nervous system by BMP-9. *Science* 289: 313–316

83   Darland DC, Nishi R (1998) Activin A and follistatin influence expression of somato-statin in the ciliary ganglion *in vivo*. *Dev Biol* 202: 293–303

84   Hall AK, Burke RM, Anand M, Dinsio KJ (2002) Activin and bone morphogenetic proteins are present in perinatal sensory neuron target tissues that induce neuropeptides. *J Neurobiol* 52: 52–60

85   Lein P, Johnson M, Guo X, Rueger D, Higgins D (1995) Osteogenic protein-1 induces dendritic growth in rat sympathetic neurons. *Neuron* 15: 597–605

86   Withers GS, Higgins D, Charette M, Banker G (2000) Bone morphogenetic protein-7 enhances dendritic growth and receptivity to innervation in cultured hippocampal neurons. *Eur J Neurosci* 12: 106–116

87   Le Roux P, Behar S, Higgins D, Charette M (1999) OP-1 enhances dendritic growth from cerebral cortical neurons *in vitro*. *Exp Neurol* 160: 151–163

88   Gratacos E, Checa N, Alberch J (2001) Bone morphogenetic protein-2, but not bone morphogenetic protein-7, promotes dendritic growth and calbindin phenotype in cultured rat striatal neurons. *Neuroscience* 104: 783–790

89   Granholm AC, Sanders LA, Ickes B, Albeck D, Hoffer BJ, Young DA, Kaplan PL (1999) Effects of osteogenic protein-1 (OP-1) treatment on fetal spinal cord transplants to the anterior chamber of the eye. *Cell Transplant* 8: 75–85

90   Kim IJ, Beck HN, Lein PJ, Higgins D (2002) Interferon gamma induces retrograde dendritic retraction and inhibits synapse formation. *J Neurosci* 22: 4530–4539

91   Kim IJ, Drahushuk K, Kim W, Lein PJ, Andres DA, Higgins D (2004) Extracellular signal-regulated kinases regulate dendritic growth in rat sympathetic neurons. *J Neurosci* 24: in press

92   Guo X, Rueger D, Higgins D (1998) Osteogenic protein-1 and related bone morphogenetic proteins regulate dendritic growth and the expression of microtubule-associated protein-2 in rat sympathetic neurons. *Neurosci Lett* 245: 131–134

93   Beck HN, Drahushuk K, Jacoby DB, Higgins D, Lein PJ (2001) Bone morphogenetic protein-5 (BMP-5) promotes dendritic growth in cultured sympathetic neurons. *BMC Neurosci* 2: 12

94   Caceres A, Mautino J, Kosik KS (1992) Suppression of MAP2 in cultured cerebellar macroneurons inhibits minor neurite formation. *Neuron* 9: 607–618

95   Guo X, Lin Y, Horbinski C, Drahushuk KM, Kim IJ, Kaplan PL, Lein P, Wang T, Higgins D (2001) Dendritic growth induced by BMP-7 requires Smad1 and proteasome activity. *J Neurobiol* 48: 120–130

96   Horbinski C, Stachowiak EK, Chandrasekaran V, Miuzukoshi E, Higgins D, Stachowiak MK (2002) Bone morphogenetic protein-7 stimulates initial dendritic growth in sympathetic neurons through an intracellular fibroblast growth factor signaling pathway. *J Neurochem* 80: 54–63

97   Gruendler C, Lin Y, Farley J, Wang T (2001) Proteasomal degradation of Smad1 induced by bone morphogenetic proteins. *J Biol Chem* 276: 46533–46543

98   Kim RH, Wang D, Tsang M, Martin J, Huff C, de Caestecker MP, Parks WT, Meng X, Lechleider RJ, Wang T, et al (2000) A novel smad nuclear interacting protein, SNIP1, suppresses p300-dependent TGF-beta signal transduction. *Genes Dev* 14: 1605–1616

99   Shou J, Rim PC, Calof AL (1999) BMPs inhibit neurogenesis by a mechanism involving degradation of a transcription factor [see comments]. *Nat Neurosci* 2: 339–345

100 Sun Y, Liu X, Ng-Eaton E, Lodish HF, Weinberg RA (1999) SnoN and Ski protooncoproteins are rapidly degraded in response to transforming growth factor beta signaling. *Proc Natl Acad Sci USA* 96: 12442–12447

101 Lein PJ, Beck HN, Chandrasekaran V, Gallagher PJ, Chen HL, Lin Y, Guo X, Kaplan PL, Tiedge H, Higgins D (2002) Glia induce dendritic growth in cultured sympathetic neurons by modulating the balance between bone morphogenetic proteins (BMPs) and BMP antagonists. *J Neurosci* 22: 10377–10387

102 Lein PJ, Chen HL, Beck HN, Dorsaneo D, Hedges AM, Gonsiorek E, Yost B, Higgins D, Morales M, Hoffer BJ (2003) Target-derived BMPs regulate dendritic growth in sympathetic neurons. *Soc Neurosci Abstr* 24: 127

103 Drahushuk K, Connell TD, Higgins D (2002) Pituitary adenylate cyclase-activating polypeptide and vasoactive intestinal peptide inhibit dendritic growth in cultured sympathetic neurons. *J Neurosci* 22: 6560–6569

104 Chandrasekaran V, Zhai Y, Wagner M, Kaplan PL, Napoli JL, Higgins D (2000) Retinoic acid regulates the morphological development of sympathetic neurons. *J Neurobiol* 42: 383–393

105 Massague J (2003) Integration of Smad and MAPK pathways: a link and a linker revisited. *Genes Dev* 17: 2993–2997

106 Guo X, Metzler-Northrup J, Lein P, Rueger D, Higgins D (1997) Leukemia inhibitory factor and ciliary neurotrophic factor regulate dendritic growth in cultures of rat sympathetic neurons. *Brain Res Dev Brain Res* 104: 101–110

107 Guo X, Chandrasekaran V, Lein P, Kaplan PL, Higgins D (1999) Leukemia inhibitory factor and ciliary neurotrophic factor cause dendritic retraction in cultured rat sympathetic neurons. *J Neurosci* 19: 2113–2121

108 Banner LR, Patterson PH (1994) Major changes in the expression of the mRNAs for cholinergic differentiation factor/leukemia inhibitory factor and its receptor after injury to adult peripheral nerves and ganglia. *Proc Natl Acad Sci USA* 91: 7109–7113

109 Sun Y, Landis SC, Zigmond RE (1996) Signals triggering the induction of leukemia inhibitory factor in sympathetic superior cervical ganglia and their nerve trunks after axonal injury. *Mol Cell Neurosci* 7: 152–163

110 Purves D, Snider WD, Voyvodic JT (1988) Trophic regulation of nerve cell morphology and innervation in the autonomic nervous system. *Nature* 336: 123–128

111 Colavita A, Krishna S, Zheng H, Padgett RW, Culotti JG (1998) Pioneer axon guidance by UNC-129, a *C. elegans* TGF-beta. *Science* 281: 706–709

112 Augsburger A, Schuchardt A, Hoskins S, Dodd J, Butler S (1999) BMPs as mediators of roof plate repulsion of commissural neurons. *Neuron* 24: 127–141

113 Dionne MS, Brunet LJ, Eimon PM, Harland RM (2002) Noggin is required for correct guidance of dorsal root ganglion axons. *Dev Biol* 251: 283–293

114 Liu J, Wilson S, Reh T (2003) BMP receptor 1b is required for axon guidance and cell survival in the developing retina. *Dev Biol* 256: 34–48

115 Postlethwaite AE, Raghow R, Stricklin G, Ballou L, Sampath TK (1994) Osteogenic protein-1, a bone morphogenic protein member of the TGF-beta superfamily, shares chemotactic but not fibrogenic properties with TGF-beta. *J Cell Physiol* 161: 562–570

116 Butler SJ , Dodd J (2003) A role for BMP heterodimers in roof plate-mediated repulsion of commissural axons. *Neuron* 38: 389–401

117 Aberle H, Haghighi AP, Fetter RD, McCabe BD, Magalhaes TR, Goodman CS (2002) wishful thinking encodes a BMP type II receptor that regulates synaptic growth in *Drosophila*. *Neuron* 33: 545–558

118 Zito K, Parnas D, Fetter RD, Isacoff EY, Goodman CS (1999) Watching a synapse grow: noninvasive confocal imaging of synaptic growth in *Drosophila*. *Neuron* 22: 719–729

119 Schuster CM, Davis GW, Fetter RD, Goodman CS (1996) Genetic dissection of structural and functional components of synaptic plasticity. I. Fasciclin II controls synaptic stabilization and growth. *Neuron* 17: 641–654

120 Schuster CM, Davis GW, Fetter RD, Goodman CS (1996) Genetic dissection of structural and functional components of synaptic plasticity. II. Fasciclin II controls presynaptic structural plasticity. *Neuron* 17: 655–667

121 Rawson JM, Lee M, Kennedy EL, Selleck SB (2003) Drosophila neuromuscular synapse assembly and function require the TGF-beta type I receptor saxophone and the transcription factor Mad. *J Neurobiol* 55: 134–150

122 McCabe BD, Marques G, Haghighi AP, Fetter RD, Crotty ML, Haerry TE, Goodman CS, O'Connor MB (2003) The BMP homolog Gbb provides a retrograde signal that regulates synaptic growth at the *Drosophila* neuromuscular junction. *Neuron* 39: 241–254

123 Zhang F, Endo S, Cleary LJ, Eskin A, Byrne JH (1997) Role of transforming growth factor-beta in long-term synaptic facilitation in Aplysia. *Science* 275: 1318–1320

124 Chin J, Angers A, Cleary LJ, Eskin A, Byrne JH (1999) TGF-beta1 in Aplysia: role in long-term changes in the excitability of sensory neurons and distribution of TbetaR-II-like immunoreactivity. *Learn Mem* 6: 317–330

125 Tomizawa K, Matsui H, Kondo E, Miyamoto K, Tokuda M, Itano T, Nagahata S, Akagi T, Hatase O (1995) Developmental alteration and neuron-specific expression of bone morphogenetic protein-6 (BMP-6) mRNA in rodent brain. *Brain Res Mol Brain Res* 28: 122–128

126 Mehler MF, Mabie PC, Zhang D, Kessler JA (1997) Bone morphogenetic proteins in the nervous system. *Trends Neurosci* 20: 309–317

127 Soderstrom S, Ebendal T (1999) Localized expression of BMP and GDF mRNA in the rodent brain. *J Neurosci Res* 56: 482–492

128 Martinez G, Carnazza ML, Di Giacomo C, Sorrenti V, Vanella A (2001) Expression of bone morphogenetic protein-6 and transforming growth factor-beta1 in the rat brain after a mild and reversible ischemic damage. *Brain Res* 894: 1–11

129 Angley C, Kumar M, Dinsio KJ, Hall AK, Siegel RE (2003) Signaling by bone morphogenetic proteins and Smad1 modulates the postnatal differentiation of cerebellar cells. *J Neurosci* 23: 260–268

130 Shen W, Finnegan S, Lein PJ, Sullivan S, Slaughter M, Higgins D (2004) Bone morphogenetic proteins regulate ionotropic glutamate receptors. *Submitted*

131 Schober A, Bottner M, Strelau J, Kinscherf R, Bonaterra GA, Barth M, Schilling L, Fairlie WD, Breit SN, Unsicker K (2001) Expression of growth differentiation factor-15/macrophage inhibitory cytokine-1 (GDF-15/MIC-1) in the perinatal, adult, and injured rat brain. *J Comp Neurol* 439: 32–45

132 Johanson CE, Palm DE, Primiano MJ, McMillan PN, Chan P, Knuckey NW, Stopa EG (2000) Choroid plexus recovery after transient forebrain ischemia: role of growth factors and other repair mechanisms. *Cell Mol Neurobiol* 20: 197–216

133 Helder MN, Ozkaynak E, Sampath KT, Luyten FP, Latin V, Oppermann H, Vukicevic S (1995) Expression pattern of osteogenic protein-1 (bone morphogenetic protein- 7) in human and mouse development. *J Histochem Cytochem* 43: 1035–1044

134 Dattatreyamurty B, Roux E, Horbinski C, Kaplan PL, Robak LA, Beck HN, Lein P, Higgins D, Chandrasekaran V (2001) Cerebrospinal fluid contains biologically active bone morphogenetic protein-7. *Exp Neurol* 172: 273–281

135 Strelau J, Sullivan A, Bottner M, Lingor P, Falkenstein E, Suter-Crazzolara C, Galter D, Jaszai J, Krieglstein K, Unsicker K (2000) Growth/differentiation factor-15/macrophage inhibitory cytokine-1 is a novel trophic factor for midbrain dopaminergic neurons *in vivo. J Neurosci* 20: 8597–8603

136 Lim DA, Tramontin AD, Trevejo JM, Herrera DG, Garcia-Verdugo JM, Alvarez-Buylla A (2000) Noggin antagonizes BMP signaling to create a niche for adult neurogenesis. *Neuron* 28: 713–726

137 Zhu G, Mehler MF, Mabie PC, Kessler JA (1999) Developmental changes in progenitor cell responsiveness to cytokines. *J Neurosci Res* 56: 131–145

138 Ueki T, Tanaka M, Yamashita K, Mikawa S, Qiu Z, Maragakis NJ, Hevner RF, Miura N, Sugimura H, Sato K (2003) A novel secretory factor, Neurogenesin-1, provides neurogenic environmental cues for neural stem cells in the adult hippocampus. *J Neurosci* 23: 11732–11740

139 Lewen A, Soderstrom S, Hillered L, Ebendal T (1997) Expression of serine/threonine kinase receptors in traumatic brain injury. *Neuroreport* 8: 475–479.

140 Charytoniuk DA, Traiffort E, Pinard E, Issertial O, Seylaz J, Ruat M (2000) Distribution of bone morphogenetic protein and bone morphogenetic protein receptor transcripts in the rodent nervous system and up-regulation of bone morphogenetic protein receptor type II in hippocampal dentate gyrus in a rat model of global cerebral ischemia [In Process Citation]. *Neuroscience* 100: 33–43

141 Chang CF, Lin SZ, Chiang YH, Morales M, Chou J, Lein P, Chen HL, Hoffer BJ, Wang Y (2003) Intravenous administration of bone morphogenetic protein-7 after ischemia improves motor function in stroke rats. *Stroke* 34: 558–564

142 Lai M, Gluckman P, Dragunow M, Hughes PE (1997) Focal brain injury increases activin betaA mRNA expression in hippocampal neurons. *Neuroreport* 8: 2691–2694

143 Setoguchi T, Yone K, Matsuoka E, Takenouchi H, Nakashima K, Sakou T, Komiya S,

Izumo S (2001) Traumatic injury-induced BMP7 expression in the adult rat spinal cord. *Brain Res* 921: 219–225

144 Lefer AM, Tsao PS, Ma XL, Sampath TK (1992) Anti-ischaemic and endothelial protective actions of recombinant human osteogenic protein (hOP-1). *J Mol Cell Cardiol* 24: 585–593

145 Perides G, Jensen FE, Edgecomb P, Rueger DC, Charness ME (1995) Neuroprotective effect of human osteogenic protein-1 in a rat model of cerebral hypoxia/ischemia. *Neurosci Lett* 187: 21–24

146 Lin SZ, Hoffer BJ, Kaplan P, Wang Y (1999) Osteogenic protein-1 protects against cerebral infarction induced by MCA ligation in adult rats. *Stroke* 30: 126–133

147 Wang Y, Chang CF, Morales M, Chou J, Chen HL, Chiang YH, Lin SZ, Cadet JL, Deng X, Wang JY et al (2001) Bone morphogenetic protein-6 reduces ischemia-induced brain damage in rats. *Stroke* 32: 2170–2178

148 Kawamata T, Ren J, Chan TC, Charette M, Finklestein SP (1998) Intracisternal osteogenic protein-1 enhances functional recovery following focal stroke. *Neuroreport* 9: 1441–1445

149 Ren J, Kaplan PL, Charette MF, Speller H, Finklestein SP (2000) Time window of intracisternal osteogenic protein-1 in enhancing functional recovery after stroke. *Neuropharmacology* 39: 860–865

150 Esquenazi S, Monnerie H, Kaplan P, Le Roux P (2002) BMP-7 and excess glutamate: opposing effects on dendrite growth from cerebral cortical neurons *in vitro*. *Exp Neurol* 176: 41–54

151 Krieglstein K, Suter-Crazzolara C, Hotten G, Pohl J, Unsicker K (1995) Trophic and protective effects of growth/differentiation factor 5, a member of the transforming growth factor-beta superfamily, on midbrain dopaminergic neurons. *J Neurosci Res* 42: 724–732

152 Sullivan AM, Opacka-Juffry J, Hotten G, Pohl J, Blunt SB (1997) Growth/differentiation factor 5 protects nigrostriatal dopaminergic neurones in a rat model of Parkinson's disease. *Neurosci Lett* 233: 73–76

153 Sullivan AM, Pohl J, Blunt SB (1998) Growth/differentiation factor 5 and glial cell line-derived neurotrophic factor enhance survival and function of dopaminergic grafts in a rat model of Parkinson's disease. *Eur J Neurosci* 10: 3681–3688

154 Espejo M, Cutillas B, Ventura F, Ambrosio S (1999) Exposure of foetal mesencephalic cells to bone morphogenetic protein-2 enhances the survival of dopaminergic neurones in rat striatal grafts. *Neurosci Lett* 275: 13–16

155 Hughes PE, Alexi T, Williams CE, Clark RG, Gluckman PD (1999) Administration of recombinant human Activin-A has powerful neurotrophic effects on select striatal phenotypes in the quinolinic acid lesion model of Huntington's disease. *Neuroscience* 92: 197–209

156 Tretter YP, Hertel M, Munz B, ten Bruggencate G, Werner S, Alzheimer C (2000) Induction of activin A is essential for the neuroprotective action of basic fibroblast growth factor *in vivo*. *Nat Med* 6: 812–815

Bone morphogenetic proteins and cancer

Joachim H. Clement[1] and Stefan Wölfl[2]

[1]Department of Internal Medicine II (Oncology, Hematology, Gastroenterology, Hepatology, Infectiology), Friedrich Schiller University Jena, Erlanger Allee 101, D-07740 Jena; [2]Department of Pharmacy and Molecular Biotechnology, University of Heidelberg, Im Neuenheimer Feld 364, D-69120 Heidelberg, Germany

Introduction

During early embryonic development several bone morphogenetic proteins (BMPs) are involved in processes such as determination, morphogenesis, proliferation, differentiation and apoptosis [1]. The interplay between BMPs, especially BMP-2 and BMP-4 with its antagonists Noggin and Chordin, is fundamental for mesoderm induction and further formation and thus overcoming the default state of neuroectoderm formation [2, 3]. Constantly high levels of BMPs in early *Xenopus* embryos maintain the ventral phenotype and keep the embryos alive for a long time. This is one hint that BMPs are necessary and sufficient for cell survival. The regular interplay between the processes of proliferation, apoptosis and differentiation is no longer correctly regulated during tumor initiation and progression (Fig. 1). Biology of tumor cells changes and they become more and more dedifferentiated and/or transdifferentiated. Finally, the tumor cells gain new functions and lose tissue-specific behavior and common restrictions, like proliferation control. It is generally accepted that the initial events of tumor formation occur at the genome level. Alterations in the DNA sequence of genes involved in the control of proliferation, cell-cycle regulation or apoptosis lead to malfunctions of the gene products and the beginning of a multi-step process ending in the formation of a tumor with the ability to form metastases (Fig. 2).

The focus of this review is tumor-related alterations of the BMP signalling network, although it is difficult to dissect BMPs clearly from transforming growth factor β (TGF-β) and activins. In the first part we present observed alterations on the genome and the expression level and their relevance for tumor diagnosis and prognosis. Furthermore, tumor-specific alterations of the biological role of various members of the BMP signalling network and their effect on the proliferation of tumor cells and the surrounding tissue will be described. Based on the present knowledge, we will try to elucidate the importance of BMPs and their activities in cancer research and therapy.

Bone Morphogenetic Proteins: Regeneration of Bone and Beyond, edited by Slobodan Vukicevic and Kuber T. Sampath
© 2004 Birkhäuser Verlag Basel/Switzerland

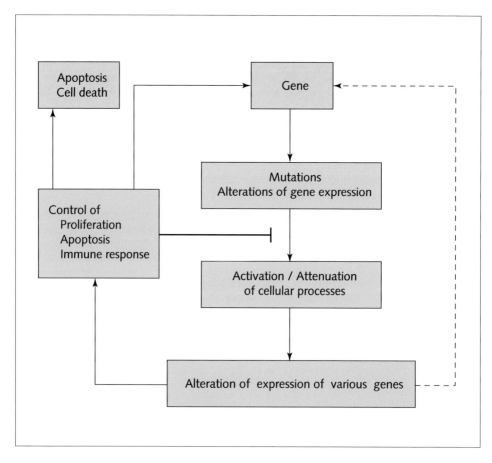

Figure 1
Alteration of gene expression interferes with control of cellular metabolism.
Genes are targets for mutations as well as altered regulatory mechanisms, e.g. enhanced transcriptional activity resulting in irregular levels of the gene products. In consequence, cellular processes are activated or attenuated which causes further variations in gene expression. Usually, this disturbance of normal cell physiology is corrected by the control mechanisms that guide proliferation, apoptosis and other processes. During tumor initiation and progression key regulators of the cellular metabolism, e.g. p53, are inactivated.

Genetic alterations of BMPs, their receptors and signalling molecules

In contrast to pro-proliferative pathways, e.g. the epidermal growth factor (EGF) receptor pathway or the platelet-derived growth factor (PDGF) receptor pathway, the BMP signalling pathway is rarely affected by genetic alterations like mutations

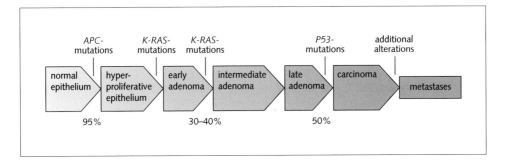

Figure 2
Multi-step carcinogenesis in colorectal cancer.
During development of colorectal cancer distinct genetic alterations affect tumor-suppressor
genes (APC, adenomatous polyposis coli protein; p53) and oncogenes (K-Ras) and acceler-
ate tumor progression. Modified from [134].

or deletions/insertions with regard to tumorigenesis (Tab. 1). The most prominent molecule of the BMP signalling network found to be altered in many tumors is Smad4 (Smad proteins are vertebrate homologues of Sma and Mad). It is deleted in more than 50% of pancreatic carcinoma [7], in about 20% of colorectal carcinoma [8] and in some cases of other tumor entities like seminoma germ cell tumors [9].

The receptor-linked Smad1 and Smad5, which are predominantly responsible for BMP-derived signals, as well as Smad3 or the inhibitory Smad6 seem to be unchanged in solid tumors, e.g. tumors of the breast, colon, lung or pancreas [6]. A screening of seminomas elucidated no mutations for Smad1, Smad2, Smad3, Smad5, Smad6 or Smad7 [9]. Neither homozygous deletions nor point mutations in the Smad5 gene were found in 40 primary gastric tumors and 51 cell lines derived from diverse types of human cancer, including 20 cell lines resistant to the growth-inhibitory effects of TGF-β1 [10]. Smad2 on the other hand is a rare target for tumor-associated mutations [5, 6]. This has severe consequences, because cells lacking Smad2 may escape from growth inhibition and the non-functional Smad2 might promote cancer cell progression [11]. In the juvenile polyposis syndrome, which has a high incidence for gastrointestinal tumors, no germline mutations could be detected in Smad1, Smad2, Smad3 or Smad5 [12]; only in the gene of the BMP receptor IA (BMPR-IA) were alterations detected [4]. Karoui et al. [13] reported that the BMPR-IA gene was affected by mutations in a significant proportion of juvenile polyposis cases, but not in a subset of colorectal tumors [13]. An analysis of pancreatic tumors and their cell lines, as well as breast cancer cell lines, also does not indicate a high incidence of BMPR-IA (ALK3) and BMPR-IB (ALK6) mutations [14].

Table 1 - Members of the BMP signalling pathway affected by mutations in tumors

Gene	Organ	Genetic alteration	Refs.
BMPR-IA	Colorectal	Juvenile polyposis coli	[4]
Smad2	Colon, breast, lung	Mutations (rare)	[5, 6]
Smad4	Pancreas	Deleted in more than 50%	[7]
	Colorectal	Deleted in 20%	[8]
	Seminoma	C-terminal mutations	[9]

The expression of BMPs and their receptors is frequently altered in tumors

The expression of several BMPs, their receptors and intracellular signalling molecules is described as altered in solid tumors in comparison to unaffected tissue. An overview is presented in Table 2.

Alterations of the expression patterns of the BMPs are phenomena which can be observed in tumors of epithelial (lung, breast) and mesodermal (kidney, bone) origin, which points to an important role in both cell types. An upregulation of BMP production is also seen in adenomas as well as in carcinomas. The first reports on changed mRNA and/or protein content of BMPs in malignant tumors in comparison to normal tissue or benign lesions came from groups working on the salivary gland and prostate. In the meantime, tumors from various tissues were tested for BMP expression showing that the members of the BMP family are upregulated in almost all cases. But there are also conflicting results. This might be due to the methods applied for expression analysis on the mRNA and protein level. There are several reports using conventional reverse transcriptase (RT)-PCR, Northern blotting, Western blotting or immunohistochemistry demonstrating an elevated presence of BMP-2 mRNA and protein in breast tumor tissue. Using quantitative gene expression analysis Reinholz et al. [29] presented data that BMP-2 is downregulated in non-invasive, invasive and liver metastases compared to normal breast tissue. Immunohistochemical studies showed that BMP-2 is predominantly detectable in the cytoplasm of the tumor cells, which might be due to the fact that the amounts of BMPs in the extracellular matrix are too low to be detected by this method. The majority of the cytoplasmic fraction should represent the various forms of precursor monomeric and dimeric BMPs. An important point that must be considered looking at BMP-2 in immunohistochemical studies is that some of the antibodies used also recognized BMP-4. Thus, the results obtained may be in part BMP-4 signals. Approaches focusing on BMP-4 showed an increase in all tissues studied until now. The expression analyses of BMP-5, BMP-6 and BMP-7, a structurally related group of BMPs, showed increased levels of these proteins in tumor tissues – with the exception of BMP-7 – in nephroblastomas and controversial results in prostate cancer. In selected subgroups of tumors like EGF receptor-negative invasive ductal breast carcinomas, there might be divergent results as shown for BMP-6, a gene regulated *via* the EGF receptor signalling pathway [43].

BMP receptors BMPR-IA, BMPR-IB and BMPR-II had a tendency towards upregulation in tumors, although some data showed that the expression of these proteins remains unchanged or reduced during tumor development, indicating that subgroups of tumors should be taken into consideration.

The expression patterns of the signal transduction molecules of the Smad-family is confusing. Receptor Smads, Smad2, Smad3 and Smad 5, seem to be downregulated during chemically induced skin carcinogenesis [52] and, in contrast, upregulated in colorectal cancer [54]. Smad1 was described as downregulated and the inhibitory Smad, Smad7, as upregulated. The expression of the BMP antagonist Gremlin was reduced in brain tumors [57]. A chordin-like protein, which was claimed a putative BMP antagonist, was downregulated in follicular carcinomas of the thyroid [46].

Altered expression affects biological function of tumor cells

First reports published 15 years ago show that BMPs are detectable in osteosarcomas and chondrosarcomas [61]. Because of the elevated expression of BMPs in bone-related tumors these tissues were a useful source from which to isolate and clone members of the BMP family [62–65]. The originally described important functions of BMPs in bone formation and remodelling, and the elevated expression of these proteins in bone-derived or -related tumor tissues, raised the question of whether BMPs might be more generally involved in tumor formation. In addition, the observation that tumors from other organs tend to calcify or synthesize chondroid or osseous structures also might be connected to BMPs. Finally, several tumors metastasize in bone. Increased BMP expression might be a prerequisite for tumor cells to settle within the bone. Elevated BMP levels in the target tissues might favor these areas for preferential metastasis formation. Therefore, the detection of BMPs in primary tumors as well as in metastases was believed to be a good hint for an involvement of these proteins in the process of tumorigenesis.

BMPs – indicators for metastasis formation?

The putative connection between the altered expression of BMPs and metastasis formation was studied intensively. Analysis of prostate tumor tissue in comparison with normal tissue as well as of prostate tumor cell lines revealed that the expression of BMPs is upregulated and might be important in the pathogenesis of

Table 2 - Members of the BMP signalling pathway showed altered expression in tumor tissues

Gene	Mode	Organ	Subtype	Refs.
BMP-2	Increased expression	Salivary gland	Pleomorphic adenocarcinoma	[15, 16]
		Stomach	Poorly differentiated	[17]
		Pancreas		[18]
		Oral epithelium	Squamous	[19]
		Lung	NSCLC	[20]
		Breast		[21, 22]
		Prostate		[23, 24]
		Lung, pleura	Mesothelioma	[25]
		Colorectal	Adenocarcinoma	[26]
		Skin	Pilomatricoma	[27]
		Rectum	Adenocarcinoma	[28]
	Reduced expression	Breast		[29]
		Colon	Microadenomas of familial adenomatous polyposis	[30]
BMP-2/4	Increased expression		Osteosarcoma	[31]
			Malignant histiocytoma	
			Liposarcoma	
			Leiomyosarcoma	
			Schwannoma	
BMP-3	Increased expression	Prostate	Adenocarcinoma	[24]
BMP-4	Increased expression	Salivary gland	Pleomorphic adenoma	[16]
		Stomach	Poorly differentiated	[17]
		Prostate		[24]
		Oral epithelium	Squamous	[19]
		Colon		[32]
		Ovary		[33]
		Pituitary	Prolactinoma	[34]
BMP-5	Increased expression	Oral epithelium	Squamous	[19]
		Colon	Adenocarcinoma	[35]
BMP-6	Increased expression	Prostate		[23, 24, 36–38]
		Breast cancer		[39]
		Colon	Adenocarcinoma	[35]
		Salivary gland	Pleomorphic adenoma	[16]
		Salivary gland	Acinic cell carcinoma	[40]
		Esophagus		[41]
		Pituitary	Non-functioning adenoma	[42]
	Reduced expression	Breast	EGF receptor-reduced	[43]
BMP-7	Increased expression	Salivary gland	Pleomorphic adenoma	[16]
		Pituitary	Non-functioning adenoma	[42]
		Prostate	Adenocarcinoma	[23]
		Breast	Invasive-ductal	[114]

Table 2 (continued)

Gene	Mode	Organ	Subtype	Refs.
BMP-7	Reduced expression	Kidney	Nephroblastoma	[44]
		Prostate		[42, 45]
CDMP-1	Increased expression	Salivary gland	Pleomorphic adenoma	[16]
GDF-10	Reduced expression	Thyroid	Follicular carcinoma	[46]
GDF15	Increased expression	Prostate	Adenocarcinoma	[135–137]
PLAB	Reduced expression	Prostate		[47]
MIC-1			Benign hyperplasia	[48]
PDF				
BMPR-IA	Increased expression	Pancreas		[18]
		Oral epithelium	Squamos	[19]
	Reduced expression	Prostate	Poorly differentiated	[49]
BMPR-IB	Increased expression	Brain	Glioma	[50]
	Reduced expression	Prostate	Poorly differentiated	[49, 51]
BMPR-II	Increased expression	Pancreas		[18]
	Reduced expression	Prostate	Poorly differentiated	[49]
Smad1	Reduced expression	Skin		[52]
		Cervix		[53]
Smad2	Increased expression	Colorectal		[54]
	Reduced expression	Skin		[52]
Smad3	Increased expression	Colorectal		[54]
	Reduced expression	Skin		[52]
Smad4	Reduced expression	Skin		[52]
Smad5	Increased expression	Colorectal		[54, 55]
	Reduced expression	Skin		[52]
Smad7	Increased expression	Thyroid	Anaplastic carcinoma	[56]
		Skin		[52]
Gremlin	Reduced expression	Brain		[57]
Chordin-like	Reduced expression	Thyroid	Follicular carcinoma	[46]
BAMBI	Increased expression	Colorectal, liver		[58]
SnoN	Reduced expression	Breast	Estrogen-receptor positive	[59]
Ski	Increased expression	Melanoma		[60]

BAMBI, BMP and activin membrane-bound inhibitor; BMP. bone morphogenetic protein; BMPR, bonoe morphogenetic protein receptor; CDMP, cartilage-derived morphogenetic protein; GDF, growth and differentiation factor; MIC-1, macrophage inflammatory peptide 1; PDF, prostate derived factor; PLAB, placental bone morphogenetic protein; Smad, vertebrate homologues of Sma and Mad; SnoN, Ski-related novel protein N osteoblastic metastases [23, 24]. Focusing on BMP-6, they could show that the BMP-6 mRNA was highly expressed at the primary sites of prostate adenocarcinomas. The overexpression of BMP-6 in prostate cancer was significantly correlated with the appearance of metastases [37, 38]. At the metastasis site, BMP-6 expression levels were similar in bone metastasis tissue and normal bone tissue [66]. On the other hand, looking at BMP-7, overexpression of this BMP in bone metastasis tissue indicated that these cells originated from the prostate, because similar high levels of BMP-7 mRNA were also detected at the primary site [66]. So far, the connection between BMP expression and formation of metastases is not yet fully understood. Lee et al. [67] demonstrated that the expression of BMP-2, BMP-4 and BMP-6 contributed to the formation of osteoblastic metastases induced by the prostate cancer cell line LAPC-9, but not to osteolytic lesions by PC-3 in a xenograft model indicating the involvement of other tumor cell and/or tissue-derived factors. An interesting and challenging question is whether BMP levels are continuously high all the way through, from the primary site to the metastatic site. Jin et al. [19] could show that BMP-2/-4 as well as BMP-5 protein levels are elevated in tumor cells detected in lymph nodes, an early step in metastasis formation. A biphasic expression was observed for growth and differentiation factor GDF-15, which is down-regulated during the progression of primary prostate cancer, but re-expressed in osseous metastases [33].

BMPs are involved in calcification and ossification of tumor areas

Several groups could show with immunohistochemistry that BMP proteins are present in the cytoplasm of tumor cells, which might be due to the fact that the antibodies used detect the precursor as well as the mature protein. According to Israel et al. [68], most of the BMP protein is located in the cytoplasm and only 10% in the extracellular space. First evidence that BMPs might be involved in calcification came from studies on pleomorphic adenomas of the salivary gland [16, 69]. The BMPs secreted from the tumor cells increase the local level of BMPs and influence the tumor cells and the surrounding tissue. Xenograft studies revealed that BMP-over-expressing cells from various origin were able to survive and even proliferate at the site of application [70–74]. Gitelman et al. [71] reported that tumors derived from vgr-1 (BMP-6)-overexpressing cells were dense, firm and fibrotic, whereas control cells produce tumors which were hemorrhagic, necrotic and friable. In addition, the higher levels of BMPs, secreted by the tumor cells, recruited cells from the surrounding tissue. These cells might be the source for the deposition of chondrogenic and osseous material, not the tumor cells. In this respect, Hirata et al. [74] described a massive ossification and replacement of the injected cells after several weeks when analyzing a xenograft model with BMP-2-transfected skin fibroblasts. Tokunaga et al. [75] showed that BMPs might be involved in the development of bone structures in an osteosarcoma xenograft model. As mentioned above, they found the host cells, not the human tumor cells responsible for that observation. During mixed tumor formation, BMP-2 is overexpressed in proliferating epithelial cells resulting in the development of ectopic bone [73]. Finally, Hatano et al. [72] showed that osteosarcomas transplanted into nude mice gave rise to two different types of bone, by tumor cells and an admixture of tumor and non-tumor cells. Studies in tumor tissues from patient samples revealed that the bone-inducing activity of tumors might be due to an altered protein synthesis. Kiyozuka et al. [25] reported that normal mesothelioma and tumor cells expressed BMP-2 mRNA, but BMP-2 protein expression, was restricted to the bone-forming area in malignant mesothelioma. Ossification of tumor areas as a result of BMP action was also described in primary colon adenocarcinomas [35] and in primary rectal carcinomas [28], and also in metastases of colon adenocarcinomas in the lung [26]. In some cases, mixed tumors formed by tumor cells and mesenchymal cells were the origin of ossification. In one case, tumors adjacent to bone form new bony structures [76]. Osteogenic or chondrogenic activity is not restricted to BMP-overexpressing malignant tumors, but is also observed with benign tumors. Tateyama et al. [39] described high BMP-6 expression in the vicinity of chondroid matrix in complex adenomas and benign mixed tumors in the mammary glands of dogs, which is localized to myoepithelial cells. But BMP expression is not always essential for bone formation in tumor cells [77]. The observations described above were made with xenograft models or already established primary tumors. Blessing et al. [78] could show that chemical skin carcinogenesis could be prevented by induced expression of BMP-4. The same observation was made with BMP-6. Overexpression of BMP-6 inhibited induction of tumor formation in the skin, which might be based on an augmented apoptosis and a downregulation of the transcription of AP-1 family members [79]. This is supported by the observation that the development of microadenomas of familial adenomatous polyposis patients is accompanied by a loss of BMP-2 expression and thus marked BMP-2 as a tumor suppressor, which promotes apoptosis in mature colonic epithelial cells [30].

Upregulation of BMPs is induced by upstream mediators

The altered expression of BMPs in cancer progression is dependent on regulatory factors. On the genetic level, the 5'-flanking region of the BMP-6 gene was found to be demethylated, especially around the Sp-1-binding site, which might be an epigenetic phenomenon contributing to activation of this gene [80]. Expression of BMPs is regulated by factors such as Sonic hedgehog, nuclear factor κB (NF-κB) and Wnt signalling. Sonic hedgehog induces BMP-4 expression in basal cell carcinoma [81]. Inhibition of NF-κB signalling activity in the osteosarcoma cell line SaOS-2 results in the induction of BMP-4 and BMP-7 with a marked decrease in cell proliferation [82]. The Wnt and the BMP signalling pathways interact with each other. Oncogenic β-catenin is absolutely required for BMP-4 expression and secretion by colon cancer cells. Furthermore, BMP-4 is overexpressed in the presence of mutant APC gene [32]. A membrane-bound antagonist of BMP and activin signalling, BAMBI (BMP and activin membrane-bound inhibitor), was upregulated by β-catenin and blocked growth arrest in colorectal and hepatocellular carcinoma cells [58]. Finally, Glypican-3 modulated cell proliferation in hepatocellular carcinoma by inhibiting BMP-7 [83].

Overexpressed BMPs affect the tumor cells in an auto- and paracrine fashion

What happened with the tumor cells themselves? Application of a single dose of BMP-2 led to a time-dependent alteration of gene expression [22, 84]. Among the affected genes, some genes like the ID genes are directly involved in cell-cycle control and thus in the control of proliferation and differentiation. The BMP-dependent suppression of proliferation is controversial. As expected, BMPs as members of the TGF-β superfamily act as antiproliferative agents in some systems. In this regard, recombinant BMP-2 did not stimulate the proliferation of tumor cells from various origins in a tumor-colony-forming assay [85] and BMP-2 reduced proliferation of breast cancer cell lines, depending on the specific nature of the cell line [21]. The growth-inhibitory effect of BMP-2 was dose-dependent on gastric cancer cells [86] and seemed to mediate cell-cycle arrest in the G_1 phase and upregulation of p21. Pouliot et al. [87] demonstrated that disruption of BMP receptor II-dependent signalling led to reduced proliferation of breast cancer cells. In contrast, inactivation of TGF-β receptor II attenuated BMP-2 mediated anti-proliferative effects in the breast cancer cell line MDA-MB-231 [88]. They speculated that BMP-2 may be an autocrine negative regulator of breast cancer proliferation. The biological effects of the BMPs might be dependent on tissue as well as on the composition of various BMPs and their antagonists or the receptor repertoire of the cells. Consequently, BMP-2, BMP-4 and BMP-7 led to the upregulation of the expression of p21 and p27. Furthermore, these factors induced association of p21 with cyclin D1 and cyclin E [89], downregulation of cyclin-dependent kinases (Cdk2, Cdk6) and decrease of Skp2 expression, required for ubiquitin-dependent p27 degradation [90]. Finally, BMP-2, BMP-4 and BMP-7 induced hypomethylation of the retinoblastoma protein [89, 91, 92], which resulted in an increase of the G_1 fraction of cells. The activation of the p21 promoter required both Smad1 and Smad4 and was induced by either BMP-2 or a constitutively active BMP receptor type I [93]. p27 contributed to the BMP-2-induced growth arrest and neuronal differentiation of neuroblastoma [90]. Surprisingly, overexpression of the BMP antagonist Gremlin increased the level of p21, too, but unlike BMP-2, reduced the level of phosphorylated p42/44 mitogen-activated protein (MAP) kinase [94]. The inhibition of p38 MAP kinase did not abrogate BMP-4 signalling [95]. BMP-2 treatment decreased the association of PTEN (phosphatase and tensin homologue deleted on chromosome 10) with two proteins in its degradation pathway and thus caused increased PTEN levels [96]. The important role of the BMP/Smad signalling pathway in tumorigenesis is further strengthened by the fact that dominant-negative mutants of BMP receptor II, TGF-β receptor II and Smad4 reduced tumorigenicity whereas overexpression of Smad7 supports tumor progression [97]. Kleeff et al. [18] showed that mutated Smad4, in contrast to wild-type Smad4, allowed BMP-2 to act as a mitogen in human pancreatic cancer. This finding points towards the importance of other pathways which are activated by BMPR-II and Smad4, but not by Smad7. Elevated levels of the BMP antagonist Gremlin reduced the MAP kinase phosphorylation and, in consequence, inhibited tumorigenesis.

BMPs interfere with steroid hormone activity

The expression of BMPs as well as their biological action in carcinogenesis is coupled with the activity of other factors, especially steroid hormones. In prostate cancer cells, BMP-2 increased the proliferation in the absence of androgen [51]. Estrogen-induced growth of the breast cancer cell line MCF-7 was inhibited by BMP-2. The same holds true for the basal proliferation rates of MCF-7 and MDA-MB-231 [89, 98]. Estrogen suppresses BMP-2-induced activation of the Smad pathway and BMP-mediated gene expression [99]. This effect is dependent on the direct physical interaction of Smad4 with the estrogen receptors ERα/ERβ [34, 97]. The expression of the BMP receptors is differentially regulated by androgen. BMPR-IB is reduced markedly in the presence of androgen, whereas BMPR-IA is not affected [51]. Besides BMP-2, other members of the BMP family are affected by steroid hormones. The expression of BMP-7 is regulated by androgen in prostate adenocarcinoma [100], whereas Ong et al. [101] could show upregulation of BMP-6 by estrogen *via* ERα in tumor cell lines, but not in osteoblast-like cells. Retinoic acid was shown to induce apoptosis in concert with BMP signalling. Cooperation of retinoic acid and BMP-4 led to an induction of p27 and a subsequent activation of retinoblastoma protein and ultimately apoptosis [102]. Retinoic acid induced BMP-2 expression, which is necessary and sufficient for apoptosis of retinoid-sensitive medulloblastoma cells. Expression of BMP-2 by retinoid-sensitive cells was sufficient to induce apoptosis in surrounding retinoid-resistant cells [103]. The synergistical action of ATRA (α-*trans* retinoic acid) and BMP-6 differentiated human neuroblastoma cells [104].

Tumor progression and cell migration

The downregulation of cadherin-4 might be a prerequisite for an increased mobility of the tumor cells which could be important for the beginning of invasion and subsequent metastasis formation at distant sites [105]. SIP1, a Smad-interacting protein, participates in the Smad-dependent regulation of gene expression. It was described as highly expressed in several E-cadherin-negative human carcinoma cell lines and thus might be a promoter of tumor cell invasion [106]. Further evidence that irregular expression of BMPs is important in tumor progression came from investigation of Langenfeld et al. [20]. They showed that migration and invasion of lung cells are stimulated by BMP-2. In addition, BMP-2 induced vascular endothelial growth factor (VEGF) and thus might stimulate neo-vascularization [107]. This is supported by the finding that BMP-7 activated the VEGF promoter, whereas Noggin diminished VEGF promoter activity and protein expression [108]. BMPs are well known as chemoattractant for many cell types, e.g. monocytes, osteoblasts and mesenchymal progenitor cells [109–111]. Because almost all cells produce BMPs, alteration of BMP concentration is responsible for this effect and explains the movement of breast cancer cells towards a BMP-2 source. In case of tumors, this might not only affect the tumor cells. An elevated expression of BMP-2 or other members of the BMP family by the tumor cells might also be responsible for the recruitment of cells from the periphery of the tumor. A single application of BMPs led to a time-dependent response of the incubated cells. Does the observation that cells migrate due to changes of the BMP concentration hold true or is it the peak concentration which makes the effect and, as a consequence, does a continuously high level of BMPs alone elevate the migration rate of the cells? The growth behavior of primary ovary carcinoma cells is unaltered after short-term administration of BMP-4, but long-term treatment resulted in decreased cell density and an increased cell spreading and adherence [33]. Furthermore, A-549 lung cancer cells continuously treated with BMP-4 acquired a senescent phenotype and exhibited lower extracellular-signal-regulated kinase (ERK) activation and a reduced expression of VEGF and Bcl-2. These cells were less tumorigenic than untreated cells in xenograft studies [112]. Thus, short-term treatment with BMPs exerts different effects compared to continuous/long-term application.

BMPs and prognostic markers

Differentiation of tumor subtypes

Immunohistochemical analysis with a bovine BMP monoclonal antibody could differentiate osteosarcoma from fibrosarcoma and other non-osteogenic tumors, but also further classify osteosarcomas according to BMP content and distribution [113]. Radiolabelling of that antibody and subsequent injection into patients with osteosarcoma enabled visualization of the tumor areas of the osteosarcoma [114]. Monoclonal antibodies, specific for BMP-2 and BMP-4, were claimed to allow detection of malignant cells with primitive mesenchymal features like bone tumors (osteosarcomas, malignant histiocytomas) and some soft tissue tumors (liposarcomas, leiomyosarcomas, malignant fibrous histiocytomas) [31]. Monitoring of BMP-2 expression was claimed to be used as a powerful marker in differentiating pleomorphic adenoma from other salivary gland tumors [15]. BMP-7 expression in breast cancer is highly correlated with estrogen and progesterone receptor levels which are important markers for breast cancer prognosis and therapy [115]. BMPR-IA, BMPR-IB and BMPR-II are downregulated in poorly differentiated prostate cancer and thus might contribute to progression of prostate cancer [49].

BMP expression is upregulated in high-grade tumors

Elevated levels of BMPs might on the one hand suppress the initial steps of tumor formation, but on the other hand they might also be connected to the progression of tumors. During this process, more and more highly abnormal cells appear and large numbers of dividing cells tend to grow more rapidly. In addition, they have the tendency to spread to other organs more frequently and be less responsive to therapy. These tumors are designed as "high-grade" in contrast to tumors with fewer cell abnormalities which are named "low-grade". BMP-7 has been shown to be highly overexpressed in high-grade tumors [116] as well as BMP-2 and BMP-4, which achieved higher mRNA levels in gastric cancer cell lines derived from poorly differentiated (high-grade) tumors [17]. BMP-6 immunostaining was higher in high-grade prostate cancers than in low-grade tumors [36]. The same was described for esophageal tumors, where BMP-6 protein content was correlated with the grade of tumor cell differentiation and may add to indications of poor prognosis [41]. In contrast, studies in nephroblastoma showed a reduced expression of BMP-7 compared to normal kidney tissue [44]. Interestingly, this view is supported by a very recent report. Masuda et al. [45] demonstrated that BMP-7 mRNA levels are reduced during development of prostate cancer [45]. Other members of the BMP/Smad signalling system are also affected, like the BMPR-IB, as demonstrated in malignant gliomas [50]. The extracellular BMP antagonist Gremlin is downregulated in neuronal tumors which is in concordance with the hypothesis that downregulation of BMP antagonists supports tumor progression [57]. Smad7, an intracellular antagonist of BMP signalling, is overexpressed during skin carcinogenesis and the receptor Smads, Smad1–5, were downregulated which contributed to loss of growth inhibition, a prerequisite of tumor progression [52]. Smad7 overexpression accelerated tumor progression through inhibition of TGF-β superfamily signalling and upregulation of the EGF-like superfamily of growth factors [117]. This is supported by the observation that inactivation of Smad3 in knockout mice facilitated development of colorectal cancer [12]. The repressors of Smad signalling, Ski and SnoN, associate with receptor Smads and Smad4 and block BMP-dependent gene expression [118, 119]. During the progression of malignant melanomas the Ski mRNA and protein levels were highly elevated and the intracellular distribution of Ski changed [60]. Increased expression of BMPs, their receptors and intracellular signalling molecules in tumors was not only described in comparison to unaffected normal tissue, but also in comparison to benign lesions of various tissues, e.g. oral mucosa, salivary gland or lung [19, 20, 40].

Members of the BMP signalling network – putative predictors of prognosis?

Yoshikawa et al. [31] could show that BMP immunoreactivity is correlated with poor prognosis in osteosarcoma. In addition, expression of BMP-6 in tissue specimen from radical prostatectomy has been shown to correlate with increased recurrence rates and decreased survival [120]. On the other hand, Sulzbacher et al. [121] examined osteosarcomas immunohistochemically for BMP expression (BMP-2/4, BMP-3, BMP-4, BMP-5, BMP-6, BMP-7 and BMP-8) and could not find a significant correlation of any of the analyzed BMPs with clinical outcome [121]. In contrast, the intracellular distribution of Smad-interacting proteins like Ski and SnoN seems to be altered in tumor cells. Zhang et al. [59] reported that cytoplasmic SnoN levels were elevated in ductal breast carcinomas whereas high nuclear accumulation was observed in lobular breast carcinomas. Furthermore, reduced expression of SnoN correlated with a longer disease-free survival in estrogen receptor-positive patients.

BMPs in leukemia and lymphoma

During embryogenesis BMPs play an important role in the development of the hematopoietic system. Therefore, it is useful to extend BMP research on leukemias and lymphomas. As reported for solid tumors, BMP-2 causes cell-cycle arrest in the G_1 phase and subsequent apoptosis. Kawamura et al. [122] showed an upregulation of p21 and p27 and hypophosphorylation of the retinoblastoma protein as well as a downregulation of Bcl-XL. STAT3 was inactivated immediately after BMP-2 treatment. The BMP-2-dependent induction of apoptosis was not only observed in human myeloma cell lines, but also in primary samples from patients with multiple myeloma [122, 123]. Studies with a multiple myeloma cell line also showed that BMP-4 inhibited DNA synthesis in a dose-dependent manner with regard to interleukin-6 dependence of the cell lines. Interleukin (IL)-6-dependent multiple myeloma cell lines are sensitive for BMP-4 action, but not IL-6-independent cell lines. BMP-4 induced apoptosis in freshly isolated multiple myeloma cells in four out of 13 patients. Therefore, BMP-4 or analogues may be attractive therapeutic agents in multiple myeloma because of possible beneficial effects on both tumor burden and bone disease [124].

In bone marrow cells of patients suffering from acute promyeloblastic leukemia a strong expression of BMP-2, -4 and -7 and BMPR-IA, -IB and -II was observed at diagnosis. ATRA treatment and chemotherapy reduce these markers. This implicates a possible role of BMPs as markers of minimal residual disease in acute promyeloblastic leukemia [125]. In leukemias, another line of research was directed towards Smad5. The Smad5 gene is a tumor-suppressor candidate at 5q31.1 and is homozygously lost and not mutated in the retained allele in the human leukemia cell line HL60 [126]. Nevertheless, Smad5 does not have to be a common target of somatic inactivation in malignancy [127]. But the BMP-4/Smad5 signal transduction pathway may be impaired in anemia manifestations in myelodysplastic syndrome (MDS) and acute myelogenous leukemia (AML) patients with Smad5 haploinsufficiency [128]. SMAD5β, an isoform of Smad5 with a truncated MH2 domain and a unique C-terminal tail of 18 amino acids, which may be the functional equivalent of inactivating mutations, is expressed in undifferentiated CD34⁺ hematopoietic stem cells to a higher extent than in the terminally differentiated peripheral blood leukocytes. This might be a novel mechanism to protect malignant cells from the growth inhibitory and differentiation signals of BMPs [129].

Conclusion

BMPs act in an autocrine and paracrine fashion. For tumor cells themselves, these proteins are susceptibility factors, keeping the cells in an open state [22]. Tumor-derived BMPs induce an altered behavior in the surrounding cells, e.g. differentiation. In nearly all cases in which the expression and activity of molecules from the BMP signalling network has been described so far, more than one component is expressed differently when tumor and normal tissues are compared [18]. However, the analysis of primary tumor tissues is difficult. A major problem is to gain sufficient tumor cells beside other contaminating cells. Therefore, microdissection becomes an important technique and some data cited in this paper were obtained from material prepared with this technique. It remains an open question whether these results could be compared to data raised with "crude" samples from tumor tissues. The same holds true for the comparison of data obtained from frozen tissue samples or paraffin-embedded specimens. Furthermore, expression data based on the transcriptional (mRNA) or translational (protein) level might differ because of various independent regulatory processes. Unfortunately, even on the transcriptional level, data from standard RT-PCR or from real-time PCR approaches seem to be contradictory, indicating that the source from which RNAs were prepared must be very carefully considered, which may be different subtypes of the respective tumor.

Thus, careful evaluation and comparison of experimetal design, methods and results are necessary.

In the future, it needs to be evaluated whether the BMP signalling components contribute to the effects of anticancer drugs or might be useful as additional factors supporting tumor therapies. In this respect, rapamycin, a natural product, acts immunosuppressively and antiproliferatively by targeting mammalian target of rapamycin (mTOR), a phosphoinositide 3-kinase-related kinase [130]. It co-operated with TGF-β-induced growth inhibition, especially *via* inhibiting cyclin-dependent kinase 2 [131]. Rapamycin-induced BMP-4 gene expression showing that Smad signalling plays a role in the anticancer effects of rapamycin. Furthermore, combination with phosphoinositide 3-kinase inhibition improved the growth inhibitory action of rapamycin [132]. And in cancer after-care, BMPs may be a favorable tool for regeneration of skeletal loss due to cancer. Nussenbaum et al. [133] demonstrated that BMP-7 *ex vivo* gene therapy is capable of successfully regenerating bone in rat calvarial defects even after a therapeutic dose of radiation. They suggest that this approach may represent a new strategy for regenerating skeletal loss due to head and neck cancer.

In conclusion, the BMP signalling network with its various actors is a fascinating task for further investigations in cancer biology. It may provide important information: (i) for our understanding of the complex interaction and regulation of signalling pathways, (ii) on how molecules act as mediators of tumor-related processes, e.g. tumor invasion, tumor angiogenesis and metastasis formation, and (iii) on how members of the BMP signalling network can be exploited as targets for therapy or as putative therapeutic agents.

Acknowledgement
Our studies on the role of BMPs in tumor biology are supported by the Dr Rainald Stromeyer Foundation, the Deutsche Krebshilfe e.V and the Thüringer Ministerium für Wissenschaft, Forschung und Kultur (TMWFK). We appreciate the encouragement and support of our collegues and collaborators during the past decade.

References

1   Hogan BLM (1996) Bone morphogenetic proteins: Multifunctional regulators of vertebrate development. *Genes Dev* 10: 1580–1594

2   Clement JH, Fettes P, Knöchel S, Lef J, Knöchel W (1995) Bone morphogenetic protein 2 in the early development of *Xenopus laevis*. *Mech Dev* 52: 357–370

3   Wiles MV, Johansson BM (1997) Analysis of factors controlling primary germ layer formation and early hematopoiesis using embryonic stem cell *in vitro* differentiation. *Leukemia* 11 (Suppl 3): 454–456

4   Howe JR, Bair JL, Sayed MG, Anderson ME, Mitros FA, Petersen GM, Velculescu VE, Traverso G, Vogelstein B (2001) Germline mutations of the gene encoding bone morphogenetic protein receptor 1A in juvenile polyposis. *Nat Genet* 28: 184–187

5   Xie W, Mertens JC, Reiss DJ, Rimm DL, Camp RL, Haffty BG, Reiss M (2002) Alterations of Smad signaling in human breast carcinoma are associated with poor outcome: a tissue microarray study. *Cancer Res* 62: 497–505

6   Riggins GJ, Kinzler KW, Vogelstein B, Thiagalingam S (1997) Frequency of Smad gene mutations in human cancers. *Cancer Res* 57: 2578–2580

7   Hahn SA, Schutte M, Hoque AT, Moskaluk CA, da Costa LT, Rozenblum E, Weinstein CL, Fischer A, Yeo CJ, Hruban RH, Kern SE (1996) DPC4, a candidate tumor suppressor gene at human chromosome 18q21.1. *Science* 271: 350–353

8   Heldin CH, Miyazono K, ten Dijke P, 1997. TGF-beta signalling from cell membrane to nucleus through SMAD proteins. *Nature* 390: 465–471

9   Bouras M, Tabone E, Bertholon J, Sommer P, Bouvier R, Droz JP, Benahmed M (2000) A novel SMAD4 gene mutation in seminoma germ cell tumors. *Cancer Res* 60: 922–928

10 Gemma A, Hagiwara K, Vincent F, Ke Y, Hancock AR, Nagashima M, Bennett WP, Harris CC (1998) hSmad5 gene, a human hSmad family member: its full length cDNA, genomic structure, promoter region and mutation analysis in human tumors. *Oncogene* 16: 951–956

11 Nakao A, Roijer E, Imamura T, Souchelnytskyi S, Stenman G, Heldin CH, ten Dijke P (1997) Identification of Smad2, a human Mad-related protein in the transforming growth factor beta signaling pathway. *J Biol Chem* 272: 2896–2900

12 Bevan S, Woodford-Richens K, Rozen P, Eng C, Young J, Dunlop M, Neale K, Phillips R, Markie D, Rodriguez-Bigas M et al (1999) Screening SMAD1, SMAD2, SMAD3, and SMAD5 for germline mutations in juvenile polyposis syndrome. *Gut* 45: 406–408

13 Karoui M, Tresallet C, Julie C, Zimmermann U, Staroz F, Brams A, Muti C, Boulard C, Robreau AM, Puy H et al (2004) Loss of heterozygosity on 10q and mutational status of PTEN and BMPR1A in colorectal primary tumours and metastases. Germline mutations in BMPR1A/ALK3 cause a subset of cases of juvenile polyposis syndrome and of Cowden and Bannayan-Riley-Ruvalcaba syndromes. *Br J Cancer* 90: 1230–1234

14 Alexander JM, Bikkal HA, Zervas NT, Laws ER Jr, Klibanski A (1998) Genetic alterations of the transforming growth factor beta receptor genes in pancreatic and biliary adenocarcinomas. *J Clin Endocrinol Metab* 58: 5329–5332

15 Zhao M, Takata T, Ogawa I, Takekoshi T, Nikai H (1998) Immunohistochemical demonstration of bone morphogenetic protein-2 and type II collagen in pleomorphic adenoma of salivary glands. *J Oral Pathol Med* 27: 293–296

16 Kusafuka K, Yamaguchi A, Kayano T, Fujiwara M, Takemura T (1998) Expression of bone morphogenetic proteins in salivary pleomorphic adenomas. *Virchows Arch* 432: 247–253

17 Katoh M, Terada M (1996) Overexpression of bone morphogenic protein (BMP)-4 mRNA in gastric cancer cell lines of poorly differentiated type. *J Gastroenterol* 31: 137–139

18 Kleeff J, Maruyama H, Ishiwata T, Sawhney H, Friess H, Büchler MW, Korc M (1999) Bone morphogenetic protein 2 exerts diverse effects on cell growth *in vitro* and is expressed in human pancreatic cancer *in vivo*. *Gastroenterology* 116: 1202–1216

19 Jin Y, Lu HB, Liong E, Lau TY, Tipoe GL (2001) Transcriptional mRNA of BMP-2, 3, 4 and 5 in trigeminal nerve, benign and malignant peripheral nerve sheath tumors. *Histol Histopathol* 16: 1013–1019

20 Langenfeld EM, Calvano SE, Abou-Nukta F, Lowry SF, Amenta P, Langenfeld J (2003) The mature bone morphogenetic protein-2 is aberrantly expressed in non-small cell lung carcinomas and stimulates tumor growth of A549 cells. *Carcinogenesis* 24: 1445–1454

21 Arnold SF, Tims E, McGrath BE (1999) Identification of bone morphogenetic proteins and their receptors in human breast cancer cell lines: importance of BMP2. *Cytokine* 11: 1031–1037

22 Clement JH, Marr N, Meissner A, Schwalbe M, Sebald W, Kliche KO, Höffken K, Wölfl S (2000) Bone morphogenetic protein 2 (BMP-2) induces sequential changes of Id gene expression in the breast cancer cell line MCF-7. *J Cancer Res Clin Oncol* 126: 271–279

23 Bentley H, Hamdy FC, Hart KA, Seid JM, Williams JL, Johnstone D, Russell RG (1992) Expression of bone morphogenetic proteins in human prostatic adenocarcinoma and benign prostatic hyperplasia. *Br J Cancer* 66: 1159–1163

24 Harris SE, Harris MA, Mahy P, Wozney J, Feng JQ, Mundy GR (1994) Expression of bone morphogenetic protein messenger RNAs by normal rat and human prostate and prostate cancer cells. *Prostate* 24: 204–211

25 Kiyozuka Y, Miyazaki H, Yoshizawa K, Senzaki H, Yamamoto D, Inoue K, Bessho K, Okubo Y, Kusumoto K, Tsubura A (1999) An autopsy case of malignant mesothelioma with osseous and cartilaginous differentiation: bone morphogenetic protein-2 in mesothelial cells and its tumor. *Dig Dis Sci* 44: 1626–1631

26 Birzele J, Schmitz I, Müller KM (2003) Ossification in lung metastases of primary colorectal adenocarcinomas. *Pathologe* 24: 66–69

27 Kurokawa I, Kusumoto K, Bessho K, Okubo Y, Senzaki H, Tsubura A (2000) Immunohistochemical expression of bone morphogenetic protein-2 in pilomatricoma. *Br J Dermatol* 143: 754–758

28 Kypson AP, Morphew E, Jones R, Gottfried MR, Seigler HF (2003) Heterotopic ossification in rectal cancer: Rare finding with a novel proposed mechanism. *J Surg Oncol* 82: 132–136

29 Reinholz MM, Iturria SJ, Ingle JN, Roche PC (2002) Differential gene expression of TGF-beta family members and osteopontin in breast tumor tissue: analysis by real-time quantitative PCR. *Breast Cancer Res Treat* 74: 255–269

30 Hardwick JC, Van Den Brink GR, Bleuming SA, Ballester I, Van Den Brande JM, Keller JJ, Offerhaus GJ, Van Deventer SJ, Peppelenbosch MP (2004) Bone morphogenetic protein 2 is expressed by, and acts upon, mature epithelial cells in the colon. *Gastroenterology* 126: 111–121

31 Yoshikawa H, Rettig WJ, Takaoka K, Alderman E, Rup B, Rosen V, Wozney JM, Lane JM, Huvos AG, Garin-Chesa P (1994) Expression of bone morphogenetic proteins in

human osteosarcoma. Immunohistochemical detection with monoclonal antibody. *Cancer* 73: 85–91

32 Kim JS, Crooks H, Dracheva T, Nishanian TG, Singh B, Jen J, Waldman T (2002) Oncogenic beta-catenin is required for bone morphogenetic protein 4 expression in human cancer cells. *Cancer Res* 62: 2744–2748

33 Shepherd TG, Nachtigal MW (2003) Identification of a putative autocrine bone morphogenetic protein-signaling pathway in human ovarian surface epithelium and ovarian cancer cells. *Endocrinology* 144: 3306–3314

34 Giacomini D, Paez-Pereda M, Refojo D, Carbia Nagashima A, Chervin A, Goldberg V, Arzt E (2003) New mechanisms involved in the pathogenesis of pituitary adenomas. *Medicina (B Aires)* 63: 147–150

35 Imai N, Iwai A, Hatakeyama S, Matsuzaki K, Kitagawa Y, Kato S, Hokari R, Kawaguchi A, Nagao S, Miyahara T et al (2001) Expression of bone morphogenetic proteins in colon carcinoma with heterotopic ossification. *Pathol Int* 51: 643–648

36 Barnes J, Anthony CT, Wall N, Steiner MS (1995) Bone morphogenetic protein-6 expression in normal and malignant prostate. *World J Urol* 13: 337–343

37 Hamdy FC, Autzen P, Robinson MC, Horne CH, Neal DE, Robson CN (1997) Immunolocalization and messenger RNA expression of bone morphogenetic protein-6 in human benign and malignant prostatic tissue. *Cancer Res* 57: 4427–4431

38 Autzen P, Robson CN, Bjartell A, Malcolm AJ, Johnson MI, Neal DE, Hamdy FC (1998) Bone morphogenetic protein 6 in skeletal metastases from prostate cancer and other common human malignancies. *Br J Cancer* 78: 1219–1223

39 Tateyama S, Uchida K, Hidaka T, Hirao M, Yamaguchi R (2001) Expression of bone morphogenetic protein-6 (BMP-6) in myoepithelial cells in canine mammary gland tumors. *Vet Pathol* 38: 703–709

40 Heikinheimo KA, Laine MA, Ritvos OV, Voutilainen RJ, Hogan BL, Leivo IV, Heikinheimo AK (1999) Bone morphogenetic protein-6 is a marker of serous acinar cell differentiation in normal and neoplastic human salivary gland. *Cancer Res* 59: 5815–5821

41 Raida M, Sarbia M, Clement JH, Adam S, Gabbert HE, Höffken K (1999) Expression, regulation and clinical significance of bone morphogenetic protein 6 in esophageal squamous-cell carcinoma. *Int J Cancer* 83: 38–44

42 Takeda M, Otsuka F, Suzuki J, Kishida M, Ogura T, Tamiya T, Makino H (2003) Involvement of activin/BMP system in development of human pituitary gonadotropinomas and nonfunctioning adenomas. *Biochem Biophys Res Commun* 306: 812–818

43 Clement JH, Sänger J, Höffken K (1999) Expression of bone morphogenetic protein 6 in normal mammary tissue and breast cancer cell lines and its regulation by epidermal growth factor. *Int J Cancer* 80: 250–256

44 Higinbotham KG, Karavanova ID, Diwan BA, Perantoni AO (1998) Deficient expression of mRNA for the putative inductive factor bone morphogenetic protein-7 in chemically initiated rat nephroblastomas. *Mol Carcinog* 23: 53–61

45 Masuda H, Fukabori Y, Nakano K, Shimizu N, Yamanaka H (2004) Expression of bone morphogenetic protein-7 (BMP-7) in human prostate. *Prostate* 59: 101–106

46   Aldred MA, Ginn-Pease ME, Morrison CD, Popkie AP, Gimm O, Hoang-Vu C, Krause U, Dralle H, Jhiang SM, Plass C, Eng C (2003) Caveolin-1 and caveolin-2, together with three bone morphogenetic protein-related genes, may encode novel tumor suppressors down-regulated in sporadic follicular thyroid carcinogenesis. *Cancer Res* 63: 2864–2871

47   Thomas R, True LD, Lange PH, Vessella RL (2001) Placental bone morphogenetic protein (PLAB) gene expression in normal, pre-malignant and malignant human prostate: relation to tumor development and progression. *Int J Cancer* 93: 47–52

48   Kakehi Y, Segawa T, Wu XX, Kulkarni P, Dhir R, Getzenberg RH (2004) Down-regulation of macrophage inhibitory cytokine-1/prostate derived factor in benign prostatic hyperplasia. *Prostate* 59: 351–356

49   Kim IY, Lee DH, Ahn HJ, Tokunaga H, Song W, Devereaux LM, Jin D, Sampath TK, Morton RA (2000) Expression of bone morphogenetic protein receptors type-IA, -IB and -II correlates with tumor grade in human prostate cancer tissues. *Cancer Res* 60: 2840–2844

50   Yamada N, Kato M, ten Dijke P, Yamashita H, Sampath TK, Heldin CH, Miyazono K, Funa K (1996) Bone morphogenetic protein type IB receptor is progressively expressed in malignant glioma tumours. *Br J Cancer* 73: 624–629

51   Ide H, Yoshida T, Matsumoto N, Aoki K, Osada Y, Sugimura T, Terada M (1997) Growth regulation of human prostate cancer cells by bone morphogenetic protein-2. *Cancer Res* 57: 5022–5027

52   He W, Cao T, Smith DA, Myers TE, Wang XJ (2001) Smads mediate signaling of the TGFbeta superfamily in normal keratinocytes but are lost during skin chemical carcinogenesis. *Oncogene* 20: 471–483

53   Shim C, Zhang W, Rhee CH, Lee JH (1998) Profiling of differentially expressed genes in human primary cervical cancer by complementary DNA expression array. *Clin Cancer Res* 4: 3045–3050

54   Korchynskyi O, Landstrom M, Stoika R, Funa K, Heldin CH, ten Dijke P, Souchelnytskyi S (1999) Expression of Smad proteins in human colorectal cancer. *Int J Cancer* 82: 197–202

55   Das P, Maduzia LL, Padgett RW (1999) Genetic approaches to TGFbeta signaling pathways. *Cytokine Growth Factor Rev* 10: 179–186

56   Cerutti JM, Ebina KN, Matsuo SE, Martins L, Maciel RM, Kimura ET (2003) Expression of Smad4 and Smad7 in human thyroid follicular carcinoma cell lines. *J Endocrinol Invest* 26: 516–521

57   Topol LZ, Modi WS, Koochekpour S, Blair DG (2000) DRM/GREMLIN (CKTSF1B1) maps to human chromosome 15 and is highly expressed in adult and fetal brain. *Cytogenet Cell Genet* 89: 79–84

58   Sekiya T, Adachi S, Kohu K, Yamada T, Higuchi O, Furukawa Y, Nakamura Y, Nakamura T, Tashiro K, Kuhara S et al (2004) Identification of BMP and activin membrane-bound inhibitor (BAMBI), an inhibitor of transforming growth factor-beta signaling, as

a target of the beta-catenin pathway in colorectal tumor cells. *J Biol Chem* 279: 6840–6846

59   Zhang F, Lundin M, Ristimaki A, Heikkila P, Lundin J, Isola J, Joensuu H, Laiho M (2003) Ski-related novel protein N (SnoN), a negative controller of transforming growth factor-beta signaling, is a prognostic marker in estrogen receptor-positive breast carcinomas. *Cancer Res* 63: 5005–5010

60   Medrano EE (2003) Repression of TGF-beta signaling by the oncogenic protein SKI in human melanomas: consequences for proliferation, survival, and metastasis. *Oncogene* 22: 3123–3129

61   Yang LJ, Jin Y (1990) Immunohistochemical observations on bone morphogenetic protein in normal and abnormal conditions. *Clin Orthop* 1 (259): 249–256

62   Kübler N, Urist MR (1991) Isolation of bone morphogenetic protein from human osteosarcoma tissue. *Dtsch Z Mund Kiefer Gesichtschir* 15: 258–264

63   Kübler N, Urist MR (1993) Cell differentiation in response to partially purified osteosarcoma-derived bone morphogenetic protein *in vivo* and *in vitro*. *Clin Orthop* 1 (292): 321–328

64   Takaoka K, Yoshikawa H, Hasimoto J, Masuhara K, Miyamoto S, Suzuki S, Ono K, Matsui M, Oikawa S, Tsuruoka N et al (1993) Gene cloning and expression of a bone morphogenetic protein derived from a murine osteosarcoma. *Clin Orthop* 1 (294): 344–352

65   Takaoka K, Yoshikawa H, Hashimoto J, Miyamoto S, Masuhara K, Nakahara H, Matsui M, Ono K (1993) Purification and characterization of a bone-inducing protein from a murine osteosarcoma (Dunn type). *Clin Orthop* 1 (292): 329–336

66   Masuda H, Fukabori Y, Nakano K, Takezawa Y, Suzuki T, Yamanaka H (2003) Increased expression of bone morphogenetic protein-7 in bone metastatic prostate cancer. *Prostate* 54: 268–274

67   Lee Y, Schwarz E, Davies M, Jo M, Gates J, Wu J, Zhang X, Lieberman JR (2003) Differences in the cytokine profiles associated with prostate cancer cell induced osteoblastic and osteolytic lesions in bone. *J Orthop Res* 21: 62–72

68   Israel DI, Nove J, Kerns KM, Moutsatsos IK, Kaufman RJ (1992) Expression and characterization of bone morphogenetic protein-2 in Chinese hamster ovary cells. *Growth Factors* 7: 139–150

69   Lianjia Y, Yan J, Hitoshi N, Shinichiro S, Akihide K, Masahiko M (1993) An immunohistochemical study of bone morphogenetic protein in pleomorphic adenoma of the salivary gland. *Virchows Arch A Pathol Anat Histopathol* 422: 439–443

70   Hatakeyama S, Ohara-Nemoto Y, Kyakumoto S, Satoh M (1993) Expression of bone morphogenetic protein in human adenocarcinoma cell line. *Biochem Biophys Res Commun* 190: 695–701

71   Gitelman SE, Kobrin MS, Ye JQ, Lopez AR, Lee A, Derynck R (1994) Recombinant Vgr-1/BMP-6-expressing tumors induce fibrosis and endochondral bone formation *in vivo*. *J Cell Biol* 126: 1595–1609

72   Hatano H, Tokunaga K, Ogose A, Hotta T, Yamagiwa H, Hayami T, Endo N, Taka-

hashi HE (1998) Origin of bone-forming cells in human osteosarcomas transplanted into nude mice – which cells produce bone, human or mouse? *J Pathol* 185: 204–211

73 Maroulakou IG, Shibata MA, Anver M, Jorcyk CL, Liu M, Roche N, Roberts AB, Tsarfaty I, Reseau J, Ward J, Green JE (1999) Heterotopic endochondrial ossification with mixed tumor formation in C3(1)/Tag transgenic mice is associated with elevated TGF-beta1 and BMP-2 expression. *Oncogene* 18: 5435–5447

74 Hirata K, Tsukazaki T, Kadowaki A, Furukawa K, Shibata Y, Moriishi T, Okubo Y, Bessho K, Komori T, Mizuno A, Yamaguchi A (2003) Transplantation of skin fibroblasts expressing BMP-2 promotes bone repair more effectively than those expressing Runx2. *Bone* 32: 502–512

75 Tokunaga K, Ogose A, Endo N, Nomura S, Takahashi HE (1996) Human osteosarcoma (OST) induces mouse reactive bone formation in xenograft system. *Bone* 19: 447–454

76 Shimizu K, Yoshikawa H, Matsui M, Masuhara K, Takaoka K (1994) Periosteal and intratumorous bone formation in athymic nude mice by Chinese hamster ovary tumors expressing murine bone morphogenetic protein-4. *Clin Orthop* 1 (300): 274–280

77 Ogose A, Motoyama T, Hotta T, Watanabe H (1996) Expression of bone morphogenetic proteins in human osteogenic and epithelial tumor cells. *Pathol Int* 46: 9–14

78 Blessing M, Nanney LB, King LE, Hogan BL (1995) Chemical skin carcinogenesis is prevented in mice by the induced expression of a TGF-beta related transgene. *Teratog Carcinog Mutagen* 15: 11–21

79 Wach S, Schirmacher P, Protschka M, Blessing M (2001) Overexpression of bone morphogenetic protein-6 (BMP-6) in murine epidermis suppresses skin tumor formation by induction of apoptosis and downregulation of fos/jun family members. *Oncogene* 20: 7761–7769

80 Tamada H, Kitazawa R, Gohji K, Kitazawa S (2001) Epigenetic regulation of human bone morphogenetic protein 6 gene expression in prostate cancer. *J Bone Miner Res* 16: 487–496

81 Fan H, Oro AE, Scott MP, Khavari PA (1997) Induction of basal cell carcinoma features in transgenic human skin expressing Sonic Hedgehog. *Nat Med* 3: 788–792

82 Andela VB, Sheu TJ, Puzas EJ, Schwarz EM, O'Keefe RJ, Rosier RN (2002) Malignant reversion of a human osteosarcoma cell line, Saos-2, by inhibition of NFkappaB. *Biochem Biophys Res Commun* 297: 237–241

83 Midorikawa Y, Ishikawa S, Iwanari H, Imamura T, Sakamoto H, Miyazono K, Kodama T, Makuuchi M, Aburatani H (2003) Glypican-3, overexpressed in hepatocellular carcinoma, modulates FGF2 and BMP-7 signaling. *Int J Cancer* 103: 455–465

84 Hollnagel A, Oehlmann V, Heymer J, Rüther U, Nordheim A (1999) Id genes are direct targets of bone morphogenetic protein induction in embryonic stem cells. *J Biol Chem* 274: 19838–19845

85 Soda H, Raymond E, Sharma S, Lawrence R, Cerna C, Gomez L, Timony GA, Von Hoff DD, Izbicka E (1998) Antiproliferative effects of recombinant human bone morphogenetic protein-2 on human tumor colony-forming units. *Anticancer Drugs* 9: 327–331

86 Wen XZ, Miyake S, Akiyama Y, Yuasa Y (2004) BMP-2 modulates the proliferation and differentiation of normal and cancerous gastric cells. *Biochem Biophys Res Commun* 316: 100–106

87 Pouliot F, Blais A, Labrie C (2003) Overexpression of a dominant negative type II bone morphogenetic protein receptor inhibits the growth of human breast cancer cells. *Cancer Res* 63: 277–281

88 Dumont N, Arteaga CL (2003) A kinase-inactive type II TGFbeta receptor impairs BMP signaling in human breast cancer cells. *Biochem Biophys Res Commun* 301: 108–112

89 Ghosh-Choudhury N, Ghosh-Choudhury G, Celeste A, Ghosh PM, Moyer M, Abboud SL, Kreisberg J (2000) Bone morphogenetic protein-2 induces cyclin kinase inhibitor p21 and hypophosphorylation of retinoblastoma protein in estradiol-treated MCF-7 human breast cancer cells. *Biochim Biophys Acta* 1497: 186–196

90 Nakamura Y, Ozaki T, Koseki H, Nakagawara A, Sakiyama S (2003) Accumulation of p27 KIP1 is associated with BMP2-induced growth arrest and neuronal differentiation of human neuroblastoma-derived cell lines. *Biochem Biophys Res Commun* 307: 206–213

91 Franzen A, Heldin NE (2001) BMP-7-induced cell cycle arrest of anaplastic thyroid carcinoma cells *via* p21(CIP1) and p27(KIP1). *Biochem Biophys Res Commun* 285: 773–781

92 Brubaker KD, Corey E, Brown LG, Vessella RL (2004) Bone morphogenetic protein signaling in prostate cancer cell lines. *J Cell Biochem* 91: 151–160

93 Pouliot F, Labrie C (2002) Role of Smad1 and Smad4 proteins in the induction of p21WAF1,Cip1 during bone morphogenetic protein-induced growth arrest in human breast cancer cells. *J Endocrinol* 172: 187–198

94 Chen B, Athanasiou M, Gu Q, Blair DG (2002) Drm/Gremlin transcriptionally activates p21(Cip1) *via* a novel mechanism and inhibits neoplastic transformation. *Biochem Biophys Res Commun* 295: 1135–1141

95 Fu Y, O'Connor LM, Shepherd TG, Nachtigal MW (2003) The p38 MAPK inhibitor, PD169316, inhibits transforming growth factor beta-induced Smad signaling in human ovarian cancer cells. *Biochem Biophys Res Commun* 310: 391–397

96 Waite KA, Eng C (2003) BMP2 exposure results in decreased PTEN protein degradation and increased PTEN levels. *Hum Mol Genet* 12: 679–684

97 Paez-Pereda M, Giacomini D, Refojo D, Nagashima AC, Hopfner U, Grübler Y, Chervin A, Goldberg V, Goya R, Hentges ST et al (2003) Involvement of bone morphogenetic protein 4 (BMP-4) in pituitary prolactinoma pathogenesis through a Smad/estrogen receptor crosstalk. *Proc Natl Acad Sci USA* 100: 1034–1039

98 Ghosh-Choudhury N, Woodruff K, Qi W, Celeste A, Abboud SL, Ghosh-Choudhury G (2000) Bone morphogenetic protein-2 blocks MDA MB 231 human breast cancer cell proliferation by inhibiting cyclin-dependent kinase-mediated retinoblastoma protein phosphorylation. *Biochem Biophys Res Commun* 272: 705–711

99 Yamamoto T, Saatcioglu F, Matsuda T (2002) Cross-talk between bone morphogenic proteins and estrogen receptor signaling. *Endocrinology* 143: 2635–2642

100 Thomas R, Anderson WA, Raman V, Reddi AH (1998) Androgen-dependent gene expression of bone morphogenetic protein 7 in mouse prostate. *Prostate* 37: 236–245

101 Ong DB, Colley SM, Norman MR, Kitazawa S, Tobias JH (2004) Transcriptional regulation of a BMP-6 promoter by estrogen receptor alpha. *J Bone Miner Res* 19: 447–454

102 Glozak MA, Rogers MB (2001) Retinoic acid- and bone morphogenetic protein 4-induced apoptosis in P19 embryonal carcinoma cells requires p27. *Exp Cell Res* 268: 128–138

103 Hallahan AR, Pritchard JI, Chandraratna RA, Ellenbogen RG, Geyer JR, Overland RP, Strand AD, Tapscott SJ, Olson JM (2003) BMP-2 mediates retinoid-induced apoptosis in medulloblastoma cells through a paracrine effect. *Nat Med* 9: 1033–1038

104 Sumantran VN, Brederlau A, Funa K (2003) BMP-6 and retinoic acid synergistically differentiate the IMR-32 human neuroblastoma cells. *Anticancer Res* 23: 1297–1303

105 Cheng SL, Lecanda F, Davidson MK, Warlow PM, Zhang SF, Zhang L, Suzuki S, St John T, Civitelli R (1998) Human osteoblasts express a repertoire of cadherins, which are critical for BMP-2-induced osteogenic differentiation. *J Bone Miner Res* 13: 633–644

106 Comijn J, Berx G, Vermassen P, Verschueren K, van Grunsven L, Bruyneel E, Mareel M, Huylebroeck D, van Roy F (2001) The two-handed E box binding zinc finger protein SIP1 downregulates E-cadherin and induces invasion. *Mol Cell* 7: 1267–1278

107 Langenfeld EM, Langenfeld J (2004) Bone morphogenetic protein-2 stimulates angiogenesis in developing tumors. *Mol Cancer Res* 2: 141–149

108 Dai J, Kitagawa Y, Zhang J, Yao Z, Mizokami A, Cheng S, Nor J, McCauley LK, Taichman RS, Keller ET (2004) Vascular endothelial growth factor contributes to the prostate cancer-induced osteoblast differentiation mediated by bone morphogenetic protein. *Cancer Res* 64: 994–999

109 Cunningham NS, Paralkar V, Reddi AH (1992) Osteogenin and recombinant bone morphogenetic protein 2B are chemotactic for human monocytes and stimulate transforming growth factor beta1 mRNA expression. *Proc Natl Acad Sci USA* 89: 11740–11744

110 Lind M, Eriksen EF, Bünger C (1996) Bone morphogenetic protein-2 but not bone morphogenetic protein-4 and -6 stimulates chemotactic migration of human osteoblasts, human marrow osteoblasts, and U2-OS cells. *Bone* 18: 53–57

111 Willette RN, Gu JL, Lysko PG, Anderson KM, Minehart H, Yue T (1999) BMP-2 Gene expression and effects on human vascular smooth muscle cells. *J Vasc Res* 36: 120–125

112 Buckley S, Shi W, Driscoll B, Ferrario A, Anderson K, Warburton D (2004) BMP4 signaling induces senescence and modulates the oncogenic phenotype of A549 lung adenocarcinoma cells. *Am J Physiol Lung Cell Mol Physiol* 286: L81–L86

113 Jin Y (1991) A quantitative immunohistochemical analysis of bone morphogenetic protein (BMP) in osteosarcoma of jaw. *Zhonghua Kou Qiang Yi Xue Za Zhi* 26: 276–278, 316–317

114 Wang Z (1991) Radioimmunoimaging of osteosarcoma with BMP monoclonal antibodies. *Zhonghua Wai Ke Za Zhi* 29: 489–491

115 Schwalbe M, Sänger J, Eggers R, Naumann A, Schmidt A, Höffken K, Clement JH

(2003) Differential expression and regulation of bone morphogenetic protein 7 in breast cancer. *Int J Oncol* 23: 89–95

116 Weber KL, Bolander ME, Sarkar G (1998) Selective differential fingerprinting. A method for identifying differentially expressed genes in a family between two samples. *Mol Biotechnol* 10: 77–81

117 Liu X, Lee J, Cooley M, Bhogte E, Hartley S, Glick A (2003) Smad7 but not Smad6 cooperates with oncogenic ras to cause malignant conversion in a mouse model for squamous cell carcinoma. *Cancer Res* 63: 7760–7768

118 Wang W, Mariani FV, Harland RM, Luo K (2000) Ski represses bone morphogenic protein signaling in *Xenopus* and mammalian cells. *Proc Natl Acad Sci USA* 97: 14394–14399

119 Wu JW, Krawitz AR, Chai J, Li W, Zhang F, Luo K, Shi Y (2002) Structural mechanism of Smad4 recognition by the nuclear oncoprotein Ski: insights on Ski-mediated repression of TGF-beta signaling. *Cell* 111: 357–367

120 Thomas BG, Hamdy FC (2000) Bone morphogenetic protein-6: potential mediator of osteoblastic metastases in prostate cancer. *Prostate Cancer Prostatic Dis* 3: 283–285

121 Sulzbacher I, Birner P, Trieb K, Pichlbauer E, Lang S (2002) The expression of bone morphogenetic proteins in osteosarcoma and its relevance as a prognostic parameter. *J Clin Pathol* 55: 381–385

122 Kawamura C, Kizaki M, Yamato K, Uchida H, Fukuchi Y, Hattori Y, Koseki T, Nishihara T, Ikeda Y (2000) Bone morphogenetic protein-2 induces apoptosis in human myeloma cells with modulation of STAT3. *Blood* 96: 2005–2011

123 Kawamura C, Kizaki M, Ikeda Y (2002) Bone morphogenetic protein (BMP)-2 induces apoptosis in human myeloma cells. *Leuk Lymphoma* 43: 635–639

124 Hjertner O, Hjorth-Hansen H, Borset M, Seidel C, Waage A, Sundan A (2001) Bone morphogenetic protein-4 inhibits proliferation and induces apoptosis of multiple myeloma cells. *Blood* 97: 516–522

125 Grcevic D, Marusic A, Grahovac B, Jaksic B, Kusec R (2003) Expression of bone morphogenetic proteins in acute promyelocytic leukemia before and after combined all trans-retinoic acid and cytotoxic treatment. *Leuk Res* 27: 731–738

126 Zavadil J, Brezinova J, Svoboda P, Zemanova Z, Michalova K (1997) Smad5, a tumor suppressor candidate at 5q31.1, is hemizygously lost and not mutated in the retained allele in human leukemia cell line HL60. *Leukemia* 11: 1187–1192

127 Hejlik DP, Kottickal LV, Liang H, Fairman J, Davis T, Janecki T, Sexton D, Perry W 3rd, Tavtigian SV, Teng DH, Nagarajan L (1997) Localization of SMAD5 and its evaluation as a candidate myeloid tumor suppressor. *Cancer Res* 57: 3779–3783

128 Fuchs O, Simakova O, Klener P, Cmejlova J, Zivny J, Zavadil J, Stopka T (2002) Inhibition of Smad5 in human hematopoietic progenitors blocks erythroid differentiation induced by BMP4. *Blood Cells Mol Dis* 28: 221–233

129 Jiang Y, Liang H, Guo W, Kottickal LV, Nagarajan L (2000) Differential expression of a novel C-terminally truncated splice form of SMAD5 in hematopoietic stem cells and leukemia. *Blood* 95: 3945–3950

130 Hidalgo M, Rowinsky EK (2000) The rapamycin-sensitive signal transduction pathway as a target for cancer therapy. *Oncogene* 19: 6680–6686

131 Law BK, Chytil A, Dumont N, Hamilton EG, Waltner-Law ME, Aakre ME, Covington C, Moses HL (2002) Rapamycin potentiates transforming growth factor beta-induced growth arrest in nontransformed, oncogene-transformed, and human cancer cells. *Mol Cell Biol* 22: 8184–8198

132 van der Poel HG, Hanrahan C, Zhong H, Simons JW (2003) Rapamycin induces Smad activity in prostate cancer cell lines. *Urol Res* 30: 380–386

133 Nussenbaum B, Rutherford RB, Teknos TN, Dornfeld KJ, Krebsbach PH (2003) *Ex vivo* gene therapy for skeletal regeneration in cranial defects compromised by postoperative radiotherapy. *Hum Gene Ther* 14: 1107–1115

134 Fearon ER, Vogelstein B (1990) A genetic model for colorectal tumorigenesis. *Cell* 61: 759–767

135 Karan D, Chen SJ, Johansson SL, Singh AP, Paralkar VM, Lin MF, Batra SK (2003) Dysregulated expression of MIC-1/PDF in human prostate tumor cells. *Biochem Biophys Res Commun* 305: 598–604

136 Welsh JB, Sapinoso LM, Su AL, Kern SG, Wang-Rodriguez J, Moskaluk CA, Frierson HR Jr, Hampton GM (2001) Analysis of gene expression identifies candidate markers and pharmacological targets in prostate cancer. *Cancer Res* 61: 5974–5978

137 Welsh JB, Sapinoso LM, Kern SG, Brown DA, Liu T, Bauskin AR, Ward RL, Hawkins NJ, Quinn DI, Russell PJ et al (2003) Large-scale delineation of decreted protein biomarkers overexpressed in cancer tissue and serum. *Proc Natl Acad Sci USA* 100: 3410–3415

Bone healing: Bone morphogenetic proteins and beyond

Vishwas M. Paralkar[1], William A. Grasser[1], Keith A. Riccardi[1], David D. Thompson[1] and Slobodan Vukicevic[2]

[1]Pfizer Global Research and Development, Groton Laboratories, Groton, CT 06340, USA; [2]Laboratory of Mineralized Tissues, Department of Anatomy, Zagreb Medical School, University of Zagreb, Croatia

The skeleton has the unique ability to repair and heal itself following injury [1, 2]. This process is a cascade of synchronized events involving many systemic and local signaling molecules [3]. However, in approximately 10% of cases, fractured bones heal more slowly (malunion) or fail to heal (nonunion) requiring additional costly medical intervention to repair the fracture [4]. These malunions and nonunions cause significant patient morbidity, significantly limiting the quality of life and increasing healthcare costs. New therapies that could ensure rapid bone healing would lessen the need for further medical intervention and would greatly reduce the loss of independence and morbidity associated with immobilization. The discovery of bone morphogenetic proteins (BMPs) has increased our understanding of the cascade of events that take place during fracture healing. This Chapter will focus on the bone morphogenetic proteins (BMPs) as well as new therapies that could ensure rapid bone healing.

Bone morphogenetic proteins (BMPs) constitute a large subfamily of proteins within a group of structurally related proteins known as the transforming growth factor-β (TGF-β) superfamily [5]. In addition to the BMPs, this superfamily contains other morphogens and growth factors including the various TGF-βs, inhibins, activins, and Mullerian inhibiting substance. The BMPs are highly conserved, secreted molecules whose biologically active C-terminal peptide signals through cell surface receptors. The BMPs play a wide variety of roles in embryonic pattern formation, body plan establishment, organogenesis, and tissue maintenance and repair [1–3, 6] (see also the chapters by Simic et al. and Martinovic et al.). Activities including inhibition of cell growth, induction of cell differentiation, and activation cellular apoptosis have been demonstrated in species ranging from *Drosophila* and *Caenorhabditis elegans* through the mammals to humans. Animals or humans lacking or having mutations in various BMP family members exhibit a spectrum of phenotypes. These range from early embryonic death due to lack of mesodermal development to viable, but severely compromised animals with a variety of skeletal

Bone Morphogenetic Proteins: Regeneration of Bone and Beyond, edited by Slobodan Vukicevic and Kuber T. Sampath
© 2004 Birkhäuser Verlag Basel/Switzerland

defects, to human diseases such as fibrodysplasia ossificans progressiva and dentino-genesis imperfecta.

Bone has the ability to repair and renew itself throughout life, as demonstrated by fracture healing and bone remodeling. Fracture healing recapitulates events that occur during embryonic bone formation, and indicates that the cells and pattern-defining molecules which drive bone formation during embryogenesis are still present in adult bone. The ability to induce ectopic bone formation in an adult animal was demonstrated nearly 70 years ago. Huggins [7] discovered that urinary bladder epithelium, when transplanted into abdominal muscles in dogs, induces new cartilage and bone formation. This process, termed "epithelial osteogenesis", was the first indication of the existence of morphogens that can induce differentiation of cells along the chondrocytic and osteoblastic pathway. Urist [8] showed that such bone forming molecules named bone morphogenetic proteins by him, are also present in adult bone and their activity can be induced upon demineralization of bone matrix. While these early observations demonstrated the presence of molecules in both bone matrix and urinary bladder epithelium that are capable of inducing a cascade of events that ultimately culminates in endochondral ossification. Once Reddi and colleagues characterized and described the subcutaneous rat demineralized bone matrix (DBM) bone formation assay a hunt for BMPs in isolated protein fractions from non-collagenous bone proteins started. This led to the next step which was the purification and cloning of bone morphogenetic proteins. Cloning and expression of the BMPs by Wozney and co-workers [9] reveled the presence of multiple BMPs. Work by Reddi, Oppermann, Sampath and collaborators [5, 10, 11] resulted in sequency and/or cloning of additional BMPs such as osteogenin (BMP-3), BMP-7 (osteogenic protein-1 or OP-1) and BMP-8 (OP-2). Sequencing of these cDNAs established the BMPs as a sub-family within the TGF-β superfamily [5, 6]. Currently, approximately 40 different BMPs have been identified and consistent with the early work by Urist and Reddi many of these proteins have been shown to have the ability to induce ectopic bone formation [5]. Given the exuberant ability of the BMPs to induce bone formation there always had been a great deal of interest among scientists and clinicians in these proteins as therapeutic molecules for bone healing. Although multiple BMPs have been cloned and expressed clinically, the majority of the work has focused on only two BMPs – namely BMP-2 and BMP-7 (OP-1).

Several preclinical studies have demonstrated the capability of these proteins to induce and facilitate the process of bone repair [10, 12–14]. Preclinically both BMP-2 and OP-1 have been shown to have activity in diverse species such as rat, rabbit, sheep, dog and monkey [15–25]. These studies have been carried out in various models of bone defect such as critical defects in both the calvaria as well as long bones; fracture models; posterolateral lumbar spine fusion as well as in intervertibral disc space [18, 21–23, 26]. Given the robust activity of both BMP-2 and OP-1 in the preclinical bone healing models in various species both of them were advanced into the clinics to test the safety and efficacy of these proteins in humans.

The preclinical studies showed efficacy of BMP-2 and OP-1 in long bone fracture models (tibia, ulna etc.) as well as in spinal fusion and both the clinical programs stayed focused on these two orthopedic sites. These clinical studies confirmed that the efficacy shown in preclinical models translate into human efficacy. Both BMP-2 and OP-1 were shown to have a clear ability to induce growth and repair in a clinical setting (see the chapter by Friedlaender). Based upon a long clinical development program that spanned almost a decade, BMP-2 (InductOs, [27]) was approved in Europe for "the treatment of acute tibia fractures as an adjunct to standard care using open fracture reduction and intermedullary nail fixation". Similarly, BMP-2 along with an implant cage device (INFUSE™) was approved for "lumbar interbody spinal fusion". The subjects treated with the BMP-2 product (INFUSE™) showed a statistically higher rate of fusion in comparison to patients undergoing standard of care with autogenous iliac crest bone graft (see the chapter by McKay). The development program of OP-1 although focused on the same anatomical site (tibia) involved a different patient population. Although OP-1 in a collagen carrier has efficacy in healing of human fibular critical sized defects, the patient population looked at in the regulatory studies in the US and Europe involved patients with tibial nonunions. These data resulted in OP-1 being approved in Europe for the "treatment of non-union of tibia of at least 9 month duration, secondary to trauma, in skeletally mature patients in case where previous treatment with autograft has failed or use of autograft is unfeasible" [28]. In the US, OP-1 has been approved for the treatment of tibial non-unions for humanitarian purposes only. This approval was based on radiographic and clinical end-points. Since OP-1 treatment is equivalent to and therefore avoids the need for autograft, the major advantage of this therapy is a lack of donor site morbidity.

Throughout the course of quality control testing BMPs were implanted ectopically into rats in combination with type I collagen, which was used as a carrier to deliver BMPs [29]. Consistent with early studies, the majority of the preclinical animal work has been done with collagen type I as a carrier for either BMP-2 or OP-1 [30]. As a result, even clinically BMP-2 and OP-1 have been used only in combination with collagen type I. This has significantly limited the applications of the BMP product to open reductions. Subsequently, more recent preclinical work has focused on identifying other carriers to deliver BMPs. The focus of these efforts is to overcome some of the shortcoming of the clinical BMP product. Recent published work has demonstrated the ability of injectable calcium phosphate paste to deliver BMP-2 [31] and preclinical work has demonstrated the efficacy of BMP-2 in a single percutaneous injection in the rat fracture model [32]. Whether these, or some other depot formulations, succeed in delivering BMPs and if such dosing paradigms are efficacious in larger animal models or in humans remains to be determined. The success in this area would depend on whether the carrier is needed to only retain the BMPs at the surgical site or if they are also required to serve as scaffolds for cell attachment and subsequent differentiation.

The trailblazing work on BMPs in the area of bone healing has clearly under-scored the importance and the medical need for agents that would enhance bone healing not only at the tibia and the spine but also at other sites such as the hip, the radius etc. Optimizing delivery with optimal injectable carriers, ease of use, and costs of treatment are just some of the areas that could lead to improvement in agents that would follow BMPs to the marketplace. This has been the focus of our research in the bone healing area. We were interested in developing a small mole-cule that can be used in an open or closed surgical setting, for the enhancement of bone healing irrespective of the anatomical site. In short, we felt that the clinical data with BMP-2 and OP-1 indicated a need for a pharmaceutical approach to bone healing. These issues prompted us to identify additional mechanisms and pathways involved in bone formation which could be modulated with a non-peptidyl small molecule. PGE_2 has been shown to have multiple biological effects in many tissues including bone. PGE_2 causes significant increases in bone mass and bone strength when administered systemically or locally to the skeleton. However, due to a severe side effect profile that includes diarrhea, lethargy, and flushing, PGE_2 is an unac-ceptable therapeutic option for bone healing. PGE_2 binds to and affects its pharma-cological activity via four different G-protein coupled cell surface receptor subtypes – EP1, EP2, EP3, and EP4. These four receptors are responsible for mediating the tissue-specific actions of PGE_2. Of these four receptors, three are involved in mod-ulation of cyclic adenosine 5'-monophosphate (cAMP) levels. Activation of the EP2 and EP4 receptor subtypes results in elevation of intracellular cAMP levels, where-as activation of the EP3 receptor results in a reduction of intracellular cAMP levels. The fourth receptor, EP1, is involved in regulating intracellular calcium levels. In bone, PGE_2 has an anabolic action, which has been linked, to an elevated level of cAMP, thereby implicating the EP2 and/or the EP4 receptor subtypes in bone for-mation. We initiated a discovery effort to identify EP2 and EP4 receptor selective agonists.

Early work with receptor selective agonists led to the discovery that the EP2 receptor was sufficient for the local bone anabolic activity of PGE_2. This work resulted in the discovery of CP-533,536 [12] a highly selective and potent function-al EP2 receptor agonist, which is able to mimic the effects of PGE_2 upon local administration to bone. An important prerequisite for an EP2 receptor agonist is high affinity for the EP2 receptor. The binding affinity of CP-533,536 to the EP2 receptor was assessed using radiolabeled PGE_2 [33]. As shown in Table 1, CP-533,536 bound the EP2 receptor with high affinity. CP-533,536 was approximate-ly 250× more selective for the human EP2 receptor subtype compared with the other EP receptor subtypes, EP1, EP3, or EP4. Similarly, CP-533,536 was selective for EP2 binding when measured against other prostanoid receptors including the prostaglandin D2 receptor (DP), prostaglandin $F_{2\alpha}$ receptor (FP), prostacyclin receptor (IP), and thromboxane receptor (TP) in the prostanoid receptor family (Tab. 1). CP-533,536 is a functional EP2 receptor agonist and induces cAMP pro-

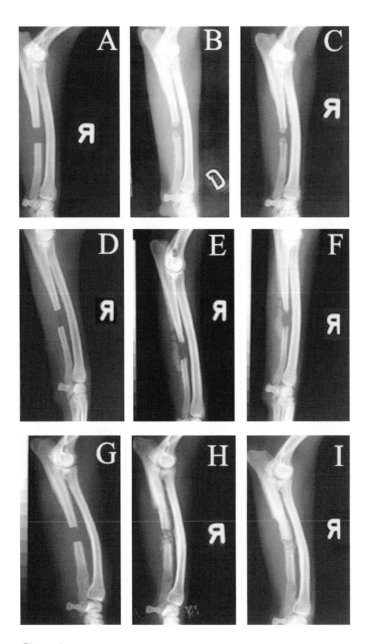

Figure 1

Canine critical defect model treated with 3 (A–C) 7 (D–F) and 14 (G–I) daily injections of CP-533,536 (100 mg/injection). The injections were initiated on the 4th day following surgery. The X-rays show time course of healing on 2 weeks, (A, D and G) 6 weeks (B, E and H) and 8 weeks (C, F and I) following surgery.

Table 1 - CP-533,536 binding activity for EP and prostanoid receptors

Receptors	EP1	EP2	EP3	EP4	DP	FP	IP	TP
IC_{50} (nM)	> 2800	50	2800	> 3200	820	> 3200	> 3200	> 3200

CP-533,536 was selective for EP$_2$ binding when measured against other prostanoid receptors including those for prostaglandin E$_2$ (EP1, EP3, EP4), prostaglandin D$_2$ (DP), prostaglandin F$_{2\alpha}$(FP), prostacyclin (IP) and thromboxane (TP).

duction in EP2 transfected cells with an EC_{50} of 5 nM [12]. The ability of CP-533,536 to stimulate local bone formation and enhance bone healing in vivo was tested in multiple animal models. These models included application of CP-533,536 locally to the bone marrow and to the periosteal surface, tibial fracture healing in rats and dogs, and critical bone defects in dogs.

A rat bone marrow injection model was used to investigate the ability of CP-533,536 upon local administration to stimulate bone formation *in vivo*. Sprague-Dawley male rats at 6 weeks of age were used in this assay. Using this model, a single dose of CP-533,536 was directly injected into the marrow cavity of the proximal tibial metaphysis. Seven days later, the injected proximal tibia was harvested. The injected site was evaluated by the following methods: peripheral quantitative computerized tomography (pQCT), a noninvasive radiological method of evaluating bone mass [34], histomorphometric analyses of new bone formation [35], and bone strength testing (indentation test of injection site [36]). It was found that CP-

Figure 2
Canine critical defect model treated with 1.0 ml of PLGH matrix shows no healing/rebridging sequence at (A) 2 weeks, (B) 12 weeks and (C) 24 weeks following surgery as demonstrated by X-rays and (D) histology at 24 weeks. Critical size defects treated with 10 mg of CP-533,536 and dissolved in 1.0 ml of matrix showed a healing/rebridging sequence at (E) 2 weeks, (F) 12 weeks and (G) 24 weeks following surgery. The histological slide shows full rebridgement and remodeling of the newly formed bone at 24 weeks following surgery (H). Intense remodeling of newly formed cortical bone consisting of osteons (I, arrowhead) and active vascular channels with hematopoietic marrow, rows of osteoblasts and newly deposited osteoid (O) on the surface of mineralized lamellar bone (J, arrow). (K, L) Intense remodeling of newly formed trabeculae with newly deposited osteoid (O), active vascular channels with hematopoietic marrow and rows of osteoblasts on the surface of mineralized lamellar bone (arrows).

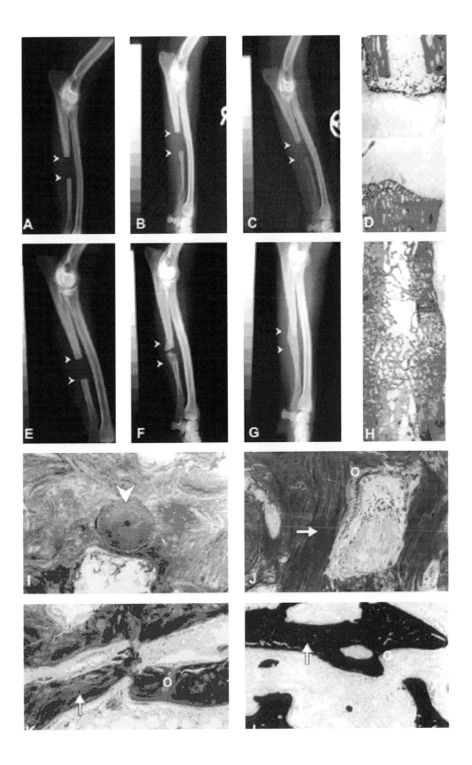

533,536 could increase bone formation upon a single injection into the marrow cavity. Similarly CP-533,536 showed positive activity on bone formation in rat when injected onto the periosteum of femur and also in the femoral fracture model.

These findings led us to test the ability of CP-533,536 to heal critical size bone defects and osteotomies in a canine model. Initially, we used the canine ulnar critical defect model and surgically implanted a collagen carrier (Helistat) in the defect site. This was followed by daily injections of CP-533,536 [100 mg/ml dissolved in calcium magnesium-free phosphate-buffered saline (PBS)] into the collagen sponges for 2, 6 or 8 weeks. This experiment yielded a rebridgement success rate of 20%, 30% and 60%, respectively (Fig. 1). When injected intravenously as an aqueous solution, CP-533,536 had a half-life of approximately 1.7 h in dogs. These results suggested that sustained exposure of CP-533,536 would lead to improved efficacy in bone healing. In order to obtain this sustained exposure, CP-533,536 was formulated with the ATRIGEL® Delivery System consisting of poly(D,L-lactide-co-glycolide) (PLGH) dissolved in N-methyl-2-pyrrolidone (NMP) [12]. After placement in the body, the NMP diffuses out of the implant, encapsulating the drug in a PLGH matrix. Administration of this CP-533,536 formulation at the site of tibial osteotomy or ulnar critical defect resulted in an extended release of the compound as measured by plasma drug levels over a period of about 7 days [12].

This application of formulated CP-533,536 to the canine ulnar critical size defect showed a decrease in the total dose of CP-533,536 needed to heal as well as markedly accelerate the rate of fracture healing. None of control dogs treated with PLGH matrix alone exhibited healing of the ulnar critical size defects (Fig. 2A–D). Animals treated with a single injection of 10 mg of CP-533,536 dissolved in the matrix, showed a healing success rate of 75% at 24 weeks following surgery (Fig. 2E–H). While no healing was observed at 2 weeks (Fig. 2E), in a majority of treated animals bone formation was evident as early as 4 weeks (data not shown). At 12 weeks postsurgery, 50% of animals showed pronounced new bone formation within the defect (Fig. 2F); while at 24 weeks, in 75% of the animals bone regeneration was completed as revealed by x-rays and histology (Fig. 2G, H). The newly formed bone at the rebridgement site was remodeled with cortices and the medullar cavity fully restored (Fig. 2H). Histological examination at 24 weeks revealed that the newly formed cortical bone was undergoing remodeling consisting of osteons (Fig. 2I) with osteoblasts and active vascular channels containing hematopoietic marrow (Fig. 2J, K, L).

In another set of studies, we tested the ability of CP-533,536 to accelerate fracture healing in the canine tibial osteotomy model. Animals with tibial osteotomy treated with CP-533,536 in the PLGH matrix showed rebridgement of defects within 8 weeks following surgery (Fig. 3B). The defects were filled with new bone of uniform radiodensity, which approximated the density of normal bone. Control animals, untreated or treated with 0.5 ml of matrix alone, failed to rebridge the defect within 8 weeks following surgery (Fig. 3A).

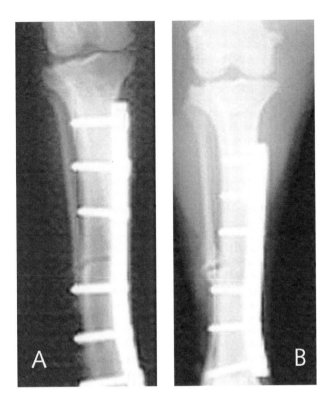

Figure 3
Canine tibial osteotomy model treated with 0.5 ml of PLGH matrix alone shows no healing/rebridging sequence at (A) 8 weeks following surgery as demonstrated by X-rays. Defects treated with 5 mg of CP-533,536 dissolved in 0.5 ml of matrix showed healing/rebridging at (B) 8 weeks following surgery.

There is clearly a high unmet medical need for fracture healing therapy given the current limitations of therapeutic procedures such as autographs and allographs. BMPs have been successfully used in reconstruction of skeletal defects but due to high costs and potential problems with protein therapy we sought to discover a new bone healing therapy that resulted with a novel prostaglandin, E_2, against which successfully healed critical sized defects in the canine ulna and dramatically accelerated healing in a tibial osteotomy model. Thus, the potent bone anabolic capacity of CP-533,536 offers a promising new therapeutic alternative for the enhancement of bone healing and treatment of bone defects and fractures in patients. Given the rapid pace of science in the fracture healing area, the next decade holds promise for the development of a truly novel pharmaceutical for the treatment of fracture healing that would enhance the quality of life of patients throughout the world.

References

1   Lieberman JR, Daluiski A, Einhorn TA (2002) The role of growth factors in the repair of bone. Biology and clinical applications. *J Bone Joint Surgery* 84–A: 1032–1044

2   Bostrom MPG, Yang X, Koutras I (2000) Biologics in bone healing. *Curr Opin Orthrop* 11: 403–412

3   Barnes GL, Kosrenuik PL, Gerstenfeld LC, Einhorn TL (1999) Growth factor regulation of fracture repair. *J Bone Miner Res* 14: 1805–1815

4   Einhorn TA (1995) Enhancement of fracture-healing. *J Bone Joint Surgery* 77–A: 940–976

5   Paralkar VM, Grasser WA, Baumann AP, Castleberry TA, Owen TA, Vukicevic SA (2002) Prostate-derived factor and growth and differentiation factor-8: newly discovered members of the TGF-beta superfamily. In: S Vukicevic, KT Sampath (eds): *Bone morphogenetic proteins: From laboratory to clinical practice*. Birkhäuser Verlag, Basel, Switzerland, 19–30

6   Paralkar VM, Owen TA (2002) Bone morphogenetic proteins. In: *Wiley Encyclopedia of molecular medicine*. John Wiley & Sons, Indianapolis, Indiana, 384–388

7   Huggins C (1931) The formation of bone under the influence of epithelium of the urinary tract. *Arch Surg* 22: 395–397

8   Urist MR (1965) Bone: formation by autoinduction. *Science* 150: 893–899

9   Wozney J, Rosen V, Celeste AJ, Mitsock LM, Kriz RW, Hewick RM, Wang EA (1988) Novel regulators of bone formation: molecular clones and activities. *Science* 242: 1528–1534

10 Sampath TK, Maliakal JC, Hauschka PV, Jones WK, Sasak H, Tucker RF, White KH, Coughlin JE, Tucker MM, Pang RHL et al (1992) Recombinant human osteogenic protein-1 (hOP-1) induces new bone formation *in vivo* with a specific activity comparable with natural bovine osteogenic protein and stimulates osteoblast proliferation and differentiation *in vitro*. *J Biol Chem* 267: 20352–20362

11 Reddi AH (1997) Bone Morphogenetic Proteins: an unconventional approach to isolation of first mammalian morphogens. *Cytokines and Growth Factors* 8: 11–20

12 Paralkar VM, Borovecki F, Ke HZ, Cameron KA, Lefker BA, Grasser WA, Owen TA, Li M, DaSilva-Jardine P, Zhou M et al (2003) An EP2 receptor-selective prostaglandin E2 agonist induces bone healing. *PNAS* 100: 6736–6740

13 Reddi AH (2001) Bone morphogenetic proteins: from basic science to clinical applications. *J Bone Joint Surgery* 83–A: S1–S6

14 Gerhart TN, Kirker-Head CA, Kriz MJ et al (1993) Healing segmental femoral defects in sheep using recombinant human bone morphogenetic protein-2. *Clin Orthop* 293: 317–326. [Context Link] 293: 317–326

15 Seeherman H, Li R, Blake C et al (2002) A single injection of rhBMP-2/calcium phosphate paste given one week after surgery accelerates osteotomy healing by 50% in a nonhuman primate fibula osteotomy model. In: *Transactions of the 48th Annual Meeting of the Orthopaedic Research Society*. Dallas, Texas, 237

16   Seeherman H, Aiolova M, Bouxsein M et al (2001) A single injection of rhBMP-2/cal-cium phosphate paste accelerates healing in a nonhuman primate osteotomy model. In: *Transactions of the 47th Annual Meeting of the Orthopaedic Research Society*. San Francisco, California, 251

17   Radomsky ML, Aufdemorte TB, Swain LD et al (1999) Novel formulation of fibroblast growth factor-2 in a hyaluronan gel accelerates fracture healing in nonhuman primates. *J Orthop Res* 17: 607–614

18   Popich LS, Salkeld SL, Rueger DC et al (1997) Critical and noncritical size defect heal-ing with osteogenic protein-1. In: *Transactions of the 43rd Annual Meeting of the Orthopaedic Research Society*. San Francisco, California, 600

19   Blokhuis TJ, den Boer FC, Bramer JAM, Jenner JMGT, Bakker FC, Patka P, Haarman HJTM (2001) Biomechanical and histological aspects of fracture healing stimulated with osteogenic protein-1. *Biomaterials* 22: 725–730

20   Bouxsein ML, Turek TJ, Blake CA et al (1999) rhBMP-2 accelerates healing in a rabbit ulnar osteotomy model. In: *Transactions of the 46th Annual Meeting of the Orthopaedic Research Society*. Anaheim, California, 138

21   Cook SD, Baffes GC, Wolfe MW, Sampath TK, Rueger DC (1994) Recombinant human bone morphogenetic protein-7 induces healing in a canine long-bone segmental defect model. *Clin Orthop* 301: 302–312

22   Cook SD, Wolfe MW, Salkeld SL et al (1995) Effect of recombinant osteogenic protein-1 on healing of segmental defects in nonhuman primates. *J Bone Joint Surg Am* 77: 734–750

23   Sciadini MF, Johnson KD (2000) Evaluation of recombinant human bone morpho-genetic protein-2 as a bone-graft substitute in a canine segmental defect model. *J Orthop Res* 18: 289–302

24   Cook SD, Dalton JE, Tan EH, Whitecloud III TS, Rueger DC (1994) *In vivo* evaluation of recombinant human osteogenic protein (rhOP-1) implants as a bone graft substitute for spinal fusions. *Spine* 19: 1655–1663

25   Ripamonti U, Van den Heever B, Sampath TK, Tucker MM, Rueger DC, Reddi AH (1996) Complete regeneration of bone in the baboon by recombinant human osteogenic protein-1 (hOP-1, bone morphogenetic protein-7). *Growth Factors* 13: 273–289

26   Margolin MD, Cogan AG, Taylor M, Buck D, McAllister TN, Toth C, McAllister BS (1998) Maxillary sinus augmentation in the non-human primate: a comparative radio-graphic and histologic study between recombinant human osteogenic protein-1 and natural bone mineral. *J Periodontol* 69: 911–919

27   European Agency for the Evaluation of Medicinal Products, CfPMP, European Public Assessment Report (EPAR) (2002) Inductos. Genetics Institute of Europe BV

28   European Agency for the Evaluation of Medicinal Products, CfPMP, European Public Assessment Report (EPAR) (2001) Osigraft. Howmedica International S de RL (Ireland)

29   Sampath TK, Reddi AH (1983) Homology of bone-inductive proteins from human, monkey, bovine and rat extracellular matrix. *PNAS* 80: 6591–6595

30 Seeherman H, Wozney J, Li R (2002) Bone morphogenetic protein delivery systems. *Spine* 27: S16–S23

31 Riedel GE (1999) Clinical evaluation of rhBMP-2/ACS in orthopedic trauma: A progress report. *Orthopedics* 22: 663–665

32 Einhorn TA, Majeska RJ, Mohaideen A, Kagel EM, Bouxsein ML, Turek TJ, Wozney JM (2003) A Single percutaneous injection of recombinant human bone morphogenetic protein-2 accelerates fracture repair. *J Bone Joint Surgery* 85-A: 1425–1435

33 Regan JW, Bailey TJ, Pepperl DJ, Pierce KL, Bogardus AM, Donello JE, Fairbairn CE, Kedzie KM, Woodward DF, Gil DW (1994) Cloning of a novel human prostaglandin receptor with characteristics of the pharmacologically defined EP2 subtype. *Mol Pharmacology* 46: 213–220

34 Ke HZ, Qi H, Chidsey-Frink KL, Crawford DT, Thompson DD (2001) Lasofoxifene (CP-336,156) Protects against the age-related changes in bone mass, bone strength and total serum cholesterol in intact aged male rats. *J Bone Miner Res* 16: 765–773

35 Jee WSS, Ma YF (1997) The *in vivo* anabolic actions of prostaglandins in bone. *Bone* 21: 297–304

36 Ke HZ, Chen HK, Simmons HA, Qi H et al (1998) Prostaglandin E_2 increases bone strength in intact rats and in ovariectomized rats with established osteopenia. *Bone* 23: 249–255

Index

The PIR-Series
Progress in Inflammation Research

Homepage: http://www.birkhauser.ch

Up-to-date information on the latest developments in the pathology, mechanisms and therapy of inflammatory disease are provided in this monograph series. Areas covered include vascular responses, skin inflammation, pain, neuroinflammation, arthritis cartilage and bone, airways inflammation and asthma, allergy, cytokines and inflammatory mediators, cell signalling, and recent advances in drug therapy. Each volume is edited by acknowledged experts providing succinct overviews on specific topics intended to inform and explain. The series is of interest to academic and industrial biomedical researchers, drug development personnel and rheumatologists, allergists, pathologists, dermatologists and other clinicians requiring regular scientific updates.

Available volumes:
T Cells in Arthritis, P. Miossec, W. van den Berg, G. Firestein (Editors), 1998
Chemokines and Skin, E. Kownatzki, J. Norgauer (Editors), 1998
Medicinal Fatty Acids, J. Kremer (Editor), 1998
Inducible Enzymes in the Inflammatory Response,
    D.A. Willoughby, A. Tomlinson (Editors), 1999
Cytokines in Severe Sepsis and Septic Shock, H. Redl, G. Schlag (Editors), 1999
Fatty Acids and Inflammatory Skin Diseases, J.-M. Schröder (Editor), 1999
Immunomodulatory Agents from Plants, H. Wagner (Editor), 1999
Cytokines and Pain, L. Watkins, S. Maier (Editors), 1999
In Vivo *Models of Inflammation*, D. Morgan, L. Marshall (Editors), 1999
Pain and Neurogenic Inflammation, S.D. Brain, P. Moore (Editors), 1999
Anti-Inflammatory Drugs in Asthma, A.P. Sampson, M.K. Church (Editors), 1999
Novel Inhibitors of Leukotrienes, G. Folco, B. Samuelsson, R.C. Murphy (Editors), 1999
Vascular Adhesion Molecules and Inflammation, J.D. Pearson (Editor), 1999
Metalloproteinases as Targets for Anti-Inflammatory Drugs,
    K.M.K. Bottomley, D. Bradshaw, J.S. Nixon (Editors), 1999
Free Radicals and Inflammation, P.G. Winyard, D.R. Blake, C.H. Evans (Editors), 1999
Gene Therapy in Inflammatory Diseases, C.H. Evans, P. Robbins (Editors), 2000
New Cytokines as Potential Drugs, S. K. Narula, R. Coffmann (Editors), 2000
High Throughput Screening for Novel Anti-inflammatories, M. Kahn (Editor), 2000
Immunology and Drug Therapy of Atopic Skin Diseases,
    C.A.F. Bruijnzeel-Komen, E.F. Knol (Editors), 2000
Novel Cytokine Inhibitors, G.A. Higgs, B. Henderson (Editors), 2000
Inflammatory Processes. Molecular Mechanisms and Therapeutic Opportunities,
    L.G. Letts, D.W. Morgan (Editors), 2000